PC-based Instrumentation and Control

PC-based Instrumentation and Control

Second edition

MIKE TOOLEY

Newnes
An imprint of Butterworth-Heinemann
Linacre House, Jordan Hill, Oxford OX2 8DP
225 Wildwood Avenue, Woburn, MA 01801-2041
A division of Reed Educational and Professional Publishing Ltd

A member of the Reed Elsevier plc group

OXFORD BOSTON JOHANNESBURG
MELBOURNE NEW DELHI SINGAPORE

First published 1991
Second edition 1995
Reprinted 1996, 1997, 1998

British Library Cataloguing in Publication Data
Tooley, Michael H.
PC-based Instrumentation and Control. –
2Rev.ed
I. Title
629.895416

ISBN 0 7506 2093 5

Library of Congress Cataloguing in Publication Data
Tooley, Michael H.
PC-based Instrumentation and Control / Mike Tooley. – 2nd ed.
p. cm.
Bibliography: p.
Includes bibliographical references and index
ISBN 0 7506 2093 5
1. Automatic control – Data processing. – 2. Microprocessors.
I. Title
TJ223.M53T65 1994
629.8'9–dc20 94–25119
CIP

Typeset by Vision Typeseting, Manchester
Printed and bound in Great Britain by MPG Books Ltd, Bodmin, Cornwall

Contents

Preface		vii
1	The IBM PC and compatibles	1
2	PC expansion systems	60
3	The operating system	104
4	Programming	149
5	Assembly language programming	164
6	BASIC programming	181
7	C programming	200
8	The IEEE-488 bus	217
9	Interfacing	232
10	Software packages	277
11	Applications	289
12	Reliability and fault-finding	315
13	System configuration	332
Appendix A	*Glossary of terms*	346
Appendix B	*SI units*	354
Appendix C	*Multiples and sub-multiples*	356
Appendix D	*Decimal, binary, hexadecimal and ASCII table*	356

Appendix E Bibliography 363

Appendix F List of suppliers 368

Index 378

Preface

Ask any production engineer or control or instrumentation specialist to define his objectives and his reply will probably include increasing efficiency without compromising on quality or reliability. Ask him what his most pressing problems are and lack of suitably trained personnel will almost certainly be high on the list. Happily, both of these perenniel problems can be solved with the aid of a PC (or PC-compatible) acting as an intelligent controller. All that is required is sufficient peripheral hardware and the necessary software to provide an interface with the production/test environment.

As an example, consider the procedure used for testing and calibrating an item of electronic equipment. Traditional methods involve the use of a number of items of stand-alone test equipment (each with its own peculiarities and set-up requirements). A number of adjustments may then be required and each will require judgement and expertise on the part of the calibration technician or test engineer. The process is thus not only time consuming but also demands the attention of experienced personnel. Furthermore, in today's calibration laboratory and production test environment, the need is for a cluster of test equipment rather than for a number of stand-alone instruments. Such an arrangement is an ideal candidate for computer control.

The computer (an ordinary PC or PC-compatible) control each item of external instrumentation and automates the test and calibration procedure, increasing throughput, consistency, and reliability, freeing the test engineer for higher level tasks. A PC-based arrangement thus provides a flexible and highly cost-effective alternative to traditional methods. Furthermore, systems can be easily configured to cope with the changing requirements of the user.

In general, PC-based instrumentation and control systems offer the following advantages:

- Flexible and adaptable (the system can be easily extended or reconfigured for a different application).
- The technology of the PC is well known and understood and most companies already have such equipment installed in a variety of locations.
- Low-cost (PC-based systems can be put together at a fraction of the cost associated with dedicated controllers).
- Availability of an extensive range of PC-compatible expansion cards from an increasingly wide range of suppliers.
- Ability to interface with standard bus systems (including the immensely popular IEEE-488 General Purpose Instrument Bus).
- Support for a variety of popular network and asynchronous data communications standards (allowing PC-based systems to become fully integrated within larger manufacturing and process control systems).
- Internationally accepted standard (the PC has set the standard for entry-level microcomputer systems throughout the world).

Typical applications for PC-based instrumentation and control systems include:

- Data acquisition and data logging.
- Automatic component and QA acceptance testing.
- Signal monitoring.
- Production monitoring and control.
- Environmental control.
- Security and alarm systems.
- Control of test and calibration clusters.
- Process control systems.
- Factory automation systems.
- Automated monitoring and performance measurement.
- Small-scale production management systems.

Aims

The book aims to provide readers with sufficient information to be able to select the necessary hardware and software to implement a wide range of practical PC-based instrumentation and control systems. Wherever possible the book contains examples of practical configurations and working circuits (all of which have been rigorously tested). Representative software is also included in a variety of languages including 8086 assembly language, BASIC and C. Furthermore, a number of popular software packages for control, instrumentation and data analysis have been described in some detail.

Information has been included so that circuits and software routines can be readily modified and extended by readers to meet their own particular needs. Overall, the aim has been that of providing the reader with sufficient information so that he or she can solve a wide variety of control and instrumentation problems in the shortest possible time and without recourse to any other texts.

Readership

This book is aimed primarily at the professional control and instrumentation specialist. It does not assume any previous knowledge of microprocessors or microcomputer systems and thus should appeal to a wide audience (including mechanical and production engineers looking for new solutions to control and instrumentation problems).

Chapter 1 provides an introduction to microcomputer systems and the IBM PC compatible equipment. The Intel range of microprocessors is introduced as are the variety of VLSI support devices found in the generic PC.

Chapter 2 describes various expansion systems which can be used to extend the I/O capability of the PC. These systems include the Industry Standard Architecture (8- and 16-bit PC expansion bus), Micro Channel Architecture (found in PS/2 equipment), and the popular IEEE-1000 (STE) backplane bus system (for which PC-compatible processor cards are widely available). Representative expansion cards and STE bus processor cards are discussed in some detail.

Chapter 3 is devoted to the ever-popular MS-DOS/PC-DOS operating system. Each of the most popular MS-DOS commands is described and details are provided which should assist readers in creating batch files (which can be important in unattended systems which must be capable of initializing themselves and automanning a control program in the event of power failure). The chapter concludes with a detailed example of the use of the MS-DOS debugger, DEBUG.

Programming techniques are introduced in Chapter 4. This chapter is intended for those who may be developing programs for their own specialized applications and for whom no 'off-the-shelf' software is available. The virtues of modular and structured programming are stressed and various control structures are discussed in some detail. Some useful pointers are included for those who need to select a language for control and instrumentation applications.

Chapter 5 deals with assembly language programming. The 8086 instruction set is briefly explained and several representative assembly language routines (written using the popular Microsoft Macro Assembler) are included.

The BASIC programming language is introduced in Chapter 6 whilst Chapter 7 is devoted to C programming. These chapters aim to provide readers with a brief introduction to both languages and numerous examples are included taken from applications within the general field of control and instrumentation. Here again, the example routines were written using the highly recommended Microsoft programming tools (QuickBASIC and QuickC respectively).

The ever-popular IEEE-488 instrument bus is introduced in Chapter 8. A representative PC adapter card is described which allows a PC to be used as an IEEE-488 bus controller. Typical software routines for bus control are also provided.

Chapter 9 deals with the general principles of interfacing analogue and digital signals to PC expansion bus modules, Analogue-to-digital and digital-to-analogue conversion. A variety of sensors, transducers and practical interface circuits have been included.

Several commercial software packages are now available to deal with specific data acquisition and instrumentation requirements. Chapter 10 provides details of a number of these packages and has been designed to assist the newcomer in the selection of a package which will satisfy his or her needs.

The general procedure for selection and specification of system hardware and software is described in Chapter 11. Several PC-based system applications are described in detail. Finally, Chapter 12 deals with reliability and fault tolerance. Basic quality procedures are described together with some basic guidelines for fault-location.

Since the configuration of a system can be crucial in determining its overall performance, Chapter 13 describes various techniques for optimizing a system by means of changes to its configuration files. This chapter also shows you how to install and configure device drivers, how to set up a disk cache or a RAM drive, and how to make optimum use of extended and/or expanded memory.

A glossary is included in Appendix A while Appendices B and C deal with fundamental SI units, multiples and sub-multiples. A binary, hexadecimal and ASCII conversion table appears in Appendix D. A bibliography is provided in Appendix E and a list of equipment, component software suppliers appears in Appendix F.

This book is the end result of several thousand hours of research and development and I should like to extend my thanks and gratitude to all those, too numerous to mention, who have helped and assisted in its production. May it now be of benefit to many!

Mike Tooley

1 The IBM PC and compatibles

Ever since IBM entered the personal computer scene, it was clear that its 'PC' (first announced in 1981) would gain an immense following. In a specification that now seems totally inadequate, the original PC had an 8088 processor, 64 to 256 kilobyte of system board RAM (expandable to 640 kilobyte with 384 kilobyte fitted in expansion slots). It supported two 360 kilobyte floppy disk drives, an 80 column × 25 line display, and 16 colours with an IBM colour graphics adapter.

The original PC was quickly followed by the PC-XT. This machine, an improved PC, with a single 5¼ inch 360 kilobyte floppy disk drive and a 10 megabyte hard disk, was introduced in 1983. In 1984, the PC-XT was followed by a yet further enhanced machine, the PC-AT (where XT and AT stood for eXtended and Advanced Technology, respectively). The PC-AT used an 80286 microprocessor and catered for a 5¼ inch 1.2 megabyte floppy drive together with a 20 megabyte hard disk.

While IBM were blazing a trail, many other manufacturers were close behind. The standards set by IBM attracted much interest from other manufacturers, notable among whom were Compaq and Olivetti. These companies were not merely content to produce machines with an identical specification but went on to make further significant improvements. Other manufacturers were happy to 'clone' the PC; indeed, one could be excused for thinking that the highest accolade that could be offered by the computer press was that a machine was 'IBM compatible'.

This chapter sets out to introduce the PC and provide an insight into the architecture and operation of a 'generic PC'. It should, perhaps, be stated that the term 'PC' now applies to such a wide range of equipment that it is difficult to pin down the essential ingredients of such a machine. However, at the risk of oversimplifying matters, a 'PC' need only satisfy two essential criteria:

- Be based upon an Intel 16-, 32- or 64-bit processor (such as a 386, 486, Pentium) or a compatible device.
- Be able to support the Microsoft MS-DOS (or compatible) operating system.

Other factors, such as available memory size, disk capacity and display technology, remain secondary. Typical specifications for various types of PC are listed in Table 1.1 but, for the benefit of the newcomer, we

Table 1.1 Typical PC specifications

Standard	*Processor*	*RAM*	Cache (kbyte)	*Floppy disk*	*Hard disk* (Mbyte)	*Graphics*	*Parallel port(s)*	*Serial port(s)*	*Clock speed* (MHz)	*Bus*
PC	8088	256 kbyte	Nil	1 or 2, 360 kbyte	None	Text or CGA	1 or 2	1 or 2	8	ISA (8-bit)
XT	8088 or 80286	640 kbyte	Nil	1 or 2, 5.25 in., 360 kbyte	10	Text and CGA	1 or 2	1 or 2	8 or 10	ISA (8-bit)
AT	80286	1 Mbyte	Nil	1 or 2, 5.25 in., 1.2 Mbyte	20	Text, CGA or EGA	1 or 2	1 or 2	12 or 16	ISA (16-bit)
386SX based	80386SX	1–8 Mbyte	64	1 or 2, 3.5 in., 1.44 Mbyte; or 5.25 in., 1.2 Mbyte	80	Text, VGA or SVGA	1 or 2	1 or 2	16 or 20	ISA (16-bit)
386DX based	80386DX	1–16 Mbyte	128	1 or 2, 3.5 in., 1.44 Mbyte; or 5.25 in., 1.2 Mbyte	120	Text, VGA or SVGA	1 or 2	1 or 2	25 or 33	ISA (16-bit)
486SX based	80486SX	4–16 Mbyte	256	1 or 2 3.5 in., 1.44 Mbyte; or 5.25 in., 1.2 Mbyte	230	Text, VGA or SVGA	1 or 2	1 or 2	25 or 33	ISA/VL
486DX based	80486DX	4–64 Mbyte	256	1 or 2, 3.5 in., 1.44 Mbyte	340	Text, VGA or SVGA	1 or 2	1 or 2	33, 50 or 66	ISA/VL/ EISA
Pentium based	P24T	8–64 Mbyte	512	1 or 2, 3.5 in., 1.44 Mbyte	528	Text, VGA or SVGA	1 or 2	1 or 2	66	PCI
PS/2	80286 or 80386	1–16 Mbyte	Nil	1, 3.5 in., 720 kbyte, or 1.44 Mbyte	44, 70 or 117	Text, EGA or VGA	1 or 2	1 or 2	8, 10, 16 or 20	MCA
PS/1	80286 or 80386	1–16 Mbyte	Nil	1, 3.5 in., 1.44 Mbyte	85 or 130	Text, VGA or SVGA	1 or 2	1 or 2	8, 10, 16 or 20	MCA

shall begin by briefly describing the basic elements of a microcomputer system before considering PC architecture in more detail.

Microcomputer systems

The principal elements within a microcomputer system consist of a central processing unit (CPU), read/write memory (RAM), read-only memory (ROM), together with one (or more) input/output (I/O) devices. These elements are connected together by a bus system along which data, address, and control signals are passed, as shown in Figure 1.1.

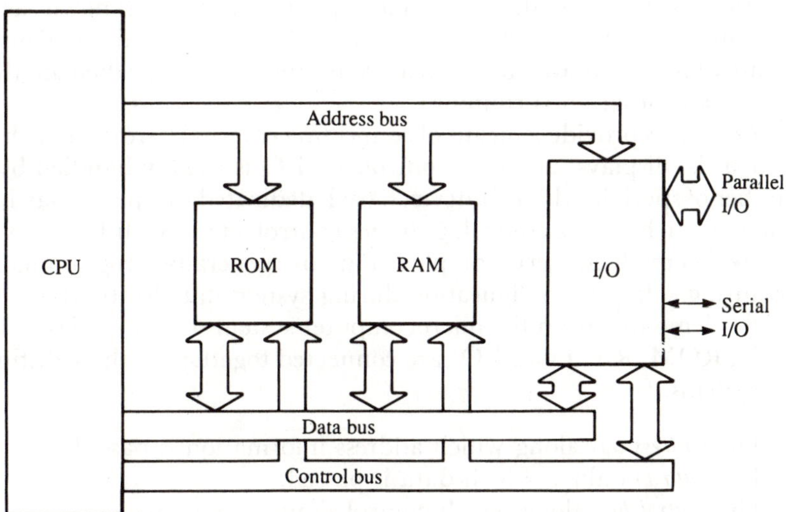

Figure 1.1 *Elements of a basic microcomputer system*

The CPU is the microprocessor itself (e.g. a 286, 386 or 486), whilst the read/write and read-only memory is implemented using a number of semiconductor memory devices (RAM and ROM, respectively). The semiconductor ROM provides non-volatile storage for part of the operating system code (the code remains intact when the power supply is disconnected), whereas the semiconductor RAM provides storage for the remainder of the operating system code, applications programs and transient data. It is important to note that this memory is volatile and any program or data stored within it will be lost when the power supply is disconnected.

The operating system is a collection of programs and software utilities that provide an environment in which applications software can easily interact with system hardware. The operating system also provides the user with a means of carrying out general housekeeping tasks such as disk formatting, disk copying, etc. In order to provide a means of interaction with the user (via keyboard entered commands and on-screen prompts and messages), the operating system incorporates a shell program (e.g. the COMMAND.COM program provided within MS-DOS).

Part of the semiconductor RAM is reserved for operating system use and for storage of a graphic/text display (as appropriate). In order to optimize the use of the available memory, most modern operating systems employ memory management techniques which allocate memory to transient programs and then release the memory when the program is terminated. A special type of program (known as a 'terminate and stay resident' program) can, however, remain resident in memory for immediate execution at some later stage (e.g. when another application program is running).

I/O devices provide a means of connecting external hardware such as keyboards, displays, and disk controllers. I/O is usually handled by a number of specialized VLSI devices, each dedicated to a particular I/O function (such as disk control, graphics control, etc.). Such I/O devices are, in themselves, very complex and are generally programmable (requiring software configuration during system initialization).

The elements within the microcomputer system shown in Figure 1.1 (CPU, ROM, RAM and I/O) are connected together by three distinct bus systems:

1 The *address bus* along which address information is passed.
2 The *data bus* along which data is passed.
3 The *control bus* along which control signals are passed.

Signals

Signals present on the bus lines are digital and have only two states, logic 1 (or *high*) and logic 0 (or *low*). Addresses and data values are coded in binary format with the most significant bit (MSB) present on the uppermost address or data line and the least significant bit (LSB) on the lowermost address or data line (labelled A0 and D0, respectively).

The bus lines (whether they be address, data, or control) are common to all four elements of the system. Data is passed via the data bus line in parallel groups of either 8 or 16 bits. An 8-bit group of data is commonly known as a *byte* whereas a 16-bit group is usually referred to as a *word*.

As an example, assume that the state of the eight data bus lines in a system at a particular instant of time is as shown below:

	(MSB)							(LSB)
Data bus line:	D7	D6	D5	D4	D3	D2	D1	D0
Value:	2^7 (=128)	2^6 (=64)	2^5 (=32)	2^4 (=16)	2^3 (=8)	2^2 (=4)	2^1 (=2)	2^0 (=1)
Logic state:	1	0	1	0	0	1	1	1
Hex. equiv.:		A				7		

The binary value (MSB first, LSB last) is 10100111 and its decimal value (found by adding together the decimal equivalents wherever a '1' is present in the corresponding bit position) is 167.

It is often more convenient to express values in hexadecimal (base 16) format (see Appendix D). The value of the byte (found by grouping the binary digits into two four-bit nibbles and then converting each to its corresponding hexadecimal character) is A7 (variously shown as A7H, &HA7, or $A7_{16}$ in order to indicate that the base is 16).

The data bus invariably comprises eight (or 16) separate lines labelled D0 to D7 (or D0 to D16), the address bus in the PC, PC-XT, and PC-AT (and compatibles), whilst the address bus comprises 20 lines labelled A0 to A19. Modern systems (i.e. those based on 80486DX processors) provide for extended 32-bit data and address busses.

The system shown in Figure 1.1 can be expanded by making the three bus systems accessible to a number of expansion modules, as shown in Figure 1.2. These modules (which invariably take the form of plug-in printed circuit cards) can take the form of additional memory (expansion memory), I/O, or they may provide additional functionality

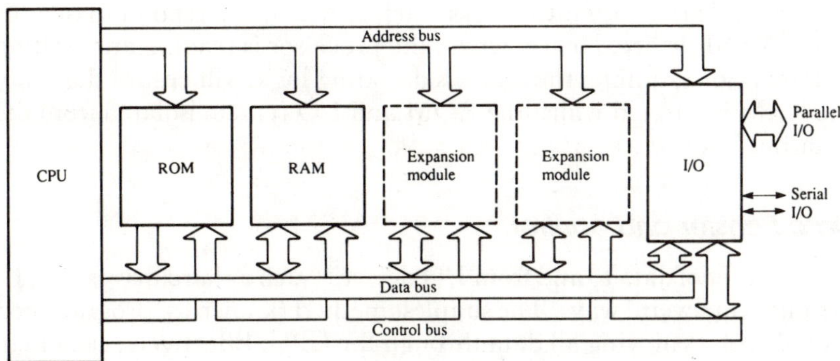

Figure 1.2 *Microcomputer system with bus expansion capability*

associated with graphics or disk control. Expansion cards are often referred to as 'option cards' or 'adapter cards' and they provide a means of configuring a basic microcomputer system for a particular application.

Microprocessor operation

The majority of operations performed by a microprocessor involve the movement of data. Indeed, the program code (a set of instructions stored in ROM or RAM) must itself be fetched from memory prior to execution. The microprocessor thus performs a continuous sequence of instruction fetch and execute cycles. The act of fetching an instruction code (or operand or data value) from memory involves a read operation whilst the act of moving data from the microprocessor to a memory location involves a write operation.

Microprocessors determine the source of data (when it is being read) and the destination of data (when it is being written) by placing a unique address on the address bus. The address at which the data is to be placed (during a write operation) or from which it is to be fetched (during a read operation) can either constitute part of the memory of the system (in which case it may be within ROM or RAM) or it can be considered to be associated with input/output (I/O).

Since the data bus is connected to a number of VLSI devices, an essential requirement of such chips (e.g. ROM or RAM) is that their data outputs should be capable of being isolated from the bus whenever necessary. These VLSI devices are fitted with select or enable inputs which are driven by address decoding logic (not shown in Figures 1.1 and 1.2). This logic ensures that ROM, RAM and I/O devices never simultaneously attempt to place data on the bus!

The inputs of the address decoding logic are derived from one, or more, of the address bus lines. The address decoder effectively divides the available memory into blocks, each of which correspond to one (or more) VLSI device. Hence, where the processor is reading and writing to RAM, for example, the address decoding logic will ensure that only the RAM is selected whilst the ROM and I/O remain isolated from the data bus.

Data transfer and control

The transfer of data to and from I/O devices (such as hard drives) can be arranged in several ways. The simplest method (known as 'programmed I/O') involves moving all data through the CPU. Effectively, each item of data is first read into a CPU register and then written to its destination. This form of data transfer is straightforward but slow, particularly where a large volume of data has to be transferred. The method is also somewhat inflexible as the transfer of data has to be included specifically within the main program flow.

An alternative method allows data to be transferred 'on demand' in response to an 'interrupt request'. Essentially, an interrupt request (IRQ) is a signal that is sent to the CPU when a peripheral device requires attention (this topic is described in greater detail later in this chapter). The advantage of this method is that CPU intervention is only required when data is actually ready to be transferred or is ready to be accepted (the CPU can thus be left to perform more useful tasks until data transfer is necessary).

The final method, 'direct memory access' (DMA), provides a means of transferring data between I/O and memory devices *without* the need for direct CPU intervention. Direct memory access provides a means of achieving the highest possible data transfer rates and it is instrumental in minimizing the time taken to transfer data to and from the hard disk or any other mass storage device. Additional 'DMA request' (DRQ) and 'DMA acknowledge' (DACK) signals are necessary so that the CPU is made aware that other devices require access to the bus. Furthermore, as with IRQ signals, several different DMA channels must be provided in order to cater for the needs of several devices that may be present within a system. This topic is dealt with in greater detail later in this chapter.

Parallel versus serial I/O

Most microcomputer systems have provision for both parallel (e.g. a parallel printer) and serial (e.g. RS-232) I/O. Parallel I/O involves transferring data one byte at a time between the microcomputer and peripheral along multiple wires (usually eight plus a common ground connection). Serial I/O, on the other hand, involves transferring one bit after another along a pair of lines (one of which is usually a ground connection).

In order to transmit a byte (or group of bytes) the serial method of I/O must comprise a sequence or stream of bits. The stream of bits will continue until all of the bytes concerned have been transmitted and additional bits may be added to the stream in order to facilitate decoding and provide a means of error detection.

Since data present on a microprocessor data bus exists in parallel form, it should be apparent that a means of parallel-to-serial and serial-to-parallel conversion will be required in order to implement a serial data link between microcomputers and peripherals.

Serial data may be transferred in either synchronous or asynchronous mode. In the former case, all transfers are carried out in accordance with a common clock signal (the clock must be available at both ends of the transmission path). Asynchronous operation involves transmission of data in packets; each packet containing the necessary information required to decode the data which it contains. Clearly this technique is more complex but it has the considerable advantage that a separate clock signal is not required.

As with parallel I/O, signals from serial I/O devices are invariably TTL compatible. It should be noted that, in general, such signals are unsuitable for anything other than the shortest of transmission paths (e.g. between a keyboard and a computer system enclosure). Serial data transmission over any appreciable distance requires additional line drivers to provide buffering and level shifting between the serial I/O device and the physical medium. In addition, line receivers are required to condition and modify the incoming signal to TTL levels.

The CPU

The CPU is crucial in determining the performance of a PC and the 80*x*86 family (see Tables 1.2 and 1.3) has been consistently upgraded. The lastest members of the family offer vastly improved performance when compared with their predecessors. Despite this, a core of common features has been retained in order to preserve compatability, and hence all current CPU devices provide a superset of the basic 8086 registers.

The 8086 and 8088

The original member of the 80*x*86 family, the 8086, was designed with modular internal architecture. This approach to microprocessor design has allowed Intel to produce a similar microprocessor with identical internal architecture but employing an 8-bit external bus. This device, the 8088, shares the same 16-bit internal architecture as its 16-bit bus counterpart. Both devices are packaged in 40-pin DIL encapsulations, the pin connections for which are shown in Figure 1.3. The CPU signal lines are described in Table 1.4.

8086/8088 architecture

The 8086/8088 can be divided internally into two functional blocks comprising an Execution Unit (EU) and a Bus Interface Unit (BIU), as shown in Figure 1.4. The EU is responsible for decoding and executing instructions, whilst the BIU prefetches instructions from memory and places them in an instruction queue where they await decoding and execution by the EU.

The EU comprises a general and special purpose register block,

Table 1.2 Processors used in PC equipment

CPU type	*Manu-facturer*	*Speed*	*Data bus*	*Notes*
8088	Intel	5, 8, 10	16-bit	Used in the IBM PC and IBM PC-XT (now obsolete)
8086	Intel	5, 8, 10	8-bit	16-bit data bus version of the 8086 (now obsolete)
80286	Intel	6, 8, 12	16-bit	First used in the IBM PC-AT. Supports only 1 Mbyte of directly addressable RAM. No internal coprocessor. No internal cache
80386SX (i386SX)	Intel	16, 20, 25, 33	32-bit internal, 16-bit external	Cut down and lower cost version of the 386DX
Am386SX	AMD	33	32-bit internal, 16-bit external	Faster than Intel's 386SX
80386DX (i386DX)	Intel	16, 20, 25, 33	32-bit	Intel's first 32-bit CPU launched in 1985
Am386DX	AMD	40	32-bit	Faster than Intel's 386DX
486SLC	IBM	20	32-bit internal, 16-bit external	Clock-doubled chip. No internal maths coprocessor
486SLC2	IBM	50	32-bit internal, 16-bit external	Clock-doubled chip. No internal maths coprocessor
486DLC2	IBM	33/66	32-bit internal	Clock-doubled chip. No internal maths coprocessor
Cx486S	Cyrix	40	32-bit	Replacement for Intel's 486SX; no internal maths coprocessor and only 2 kbyte of internal cache
Am486DX	AMD	33, 40	32-bit	Similar to Intel's i486DX but faster
Am486DX2	AMD	25/50, 33/66	32-bit	Similar to Intel's i486DX but faster. Clock doubled

i486SX2	Intel	50	32-bit	Can offer about 20% faster performance than a 486DX-33 but without the internal maths coprocessor
i486DX	Intel	25, 33, 50	32-bit	Has an internal 8 kbyte cache and internal maths co-processor
Cx486S	Cyrix	33, 50	32-bit	Similar to 486DX
i486SX	Intel	16, 20, 25, 33	32-bit	Cut-down version of the 486DX. No internal maths coprocessor
Cx486DRx2	Cyrix	16/32, 20/40, 25/50	32-bit	Replaces an existing 386DX
Cx486SRx2	Cyrix	16/32, 20/40, 25/50	16-bit	Fits over an existing 386SX
OverDrive	Intel	20/40, 25/50, 33/66	16/32-bit	Intel's upgrade processor available in two versions (SX and DX) to plug into the maths coprocessor socket
i486DX2	Intel	25/50, 33/66	32-bit	Internal clock doubling processor; 33 MHz is doubled to 66 MHz, and so on. (Also known as an 'Intel OverDrive' chip)
i486DX4	Intel	75, 100	32-bit	Internal clock tripled processor; 25 MHz is tripled to 75 MHz, and so on
Pentium	Intel	60, 66, 90	64-bit	Intel's first 64-bit processor
PowerPC	IBM	60, 66, 80	64-bit	Powerful RISC processor. Requires emulation software to run DOS, Windows and associated applications programs

Table 1.3 Intel's 80*x*86 family

	CPU type						
	8086	*8088*	*80186*	*80286*	*80386 (i386)*	*80486 (i486)*	*Pentium*
Data bus width (bits)	16	8	16	16	32	32	32
Internal data bus width (bits)	16	16	16	16	32	32	32
Typical clock rate (MHz)	5, 8	5, 8	6, 8	12, 16, 20	16, 20, 25, 33	25, 33, 40, 50	25, 33, 50, 66
Linear memory addressing range (bytes)	1M	1M	1M	1M	16M	4G	4G
I/O addressing range (bytes)	64K	64K	64K	64K	64K	64K	64K
Internal FPU	No	No	No	No	No	Yes	Yes
Internal data cache	No	No	No	No	No	Yes	Yes
Protected mode	No	No	No	Yes	Yes	Yes	Yes

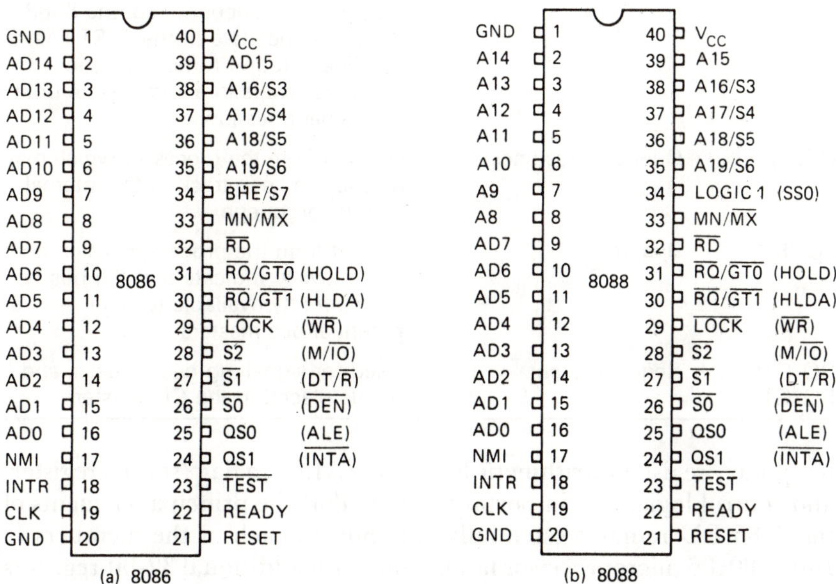

Figure 1.3 *8086 and 8088 pin connections: (a) 8086; (b) 8088*

Table 1.4 8086/8088 signals

Signal	*Function*	*Notes*
AD0–AD7 (8088)	Address/data lines	Multiplexed 8-bit address/data bus
AD0–AD15 (8086)	Address/data lines	Multiplexed 16-bit address/data bus
A8–A19 (8088)	Address lines	Address bus
A16–A19 (8086)	Address lines	Address bus
S0–S7	Status lines	S0–S2 are only available in Maximum Mode and are connected to the 8288 bus controller (U6). S3–S7 all share pins with other signals
INTR	Interrupt line	Level-triggered, active high interrupt request input
NMI	Non-maskable interrupt line	Positive edge-triggered non-maskable interrupt input
RESET	Reset line	Active high reset input
READY	Ready line	Active high ready input
TEST	Test	Input used to provide synchronization with external processors. When a WAIT instruction is encountered, the 8088 examines the state of the TEST line. If this line is found to be high, the processor waits in an 'idle' state until the signal goes low
QS0, QS1	Queue status lines	Outputs from the processor which may be used to keep track of the internal instruction queue
LOCK	Bus lock	Output from the processor which is taken low to indicate that the bus is not currently available to other potential bus masters
RQ/GT0–RQ/GT1	Request/Grant	Used for signalling bus requests and grants placed in the CL register

temporary registers, arithmetic logic unit (ALU), a flag (status) register, and control logic. It is important to note that the principal elements of the 8086 EU remains essentially common to each of the members of Intel's 80*x*86 microprocessor family, but with additional 32-bit registers in the case of the 386, 486 and Pentium.

The BIU architecture varies according to the size of the external data

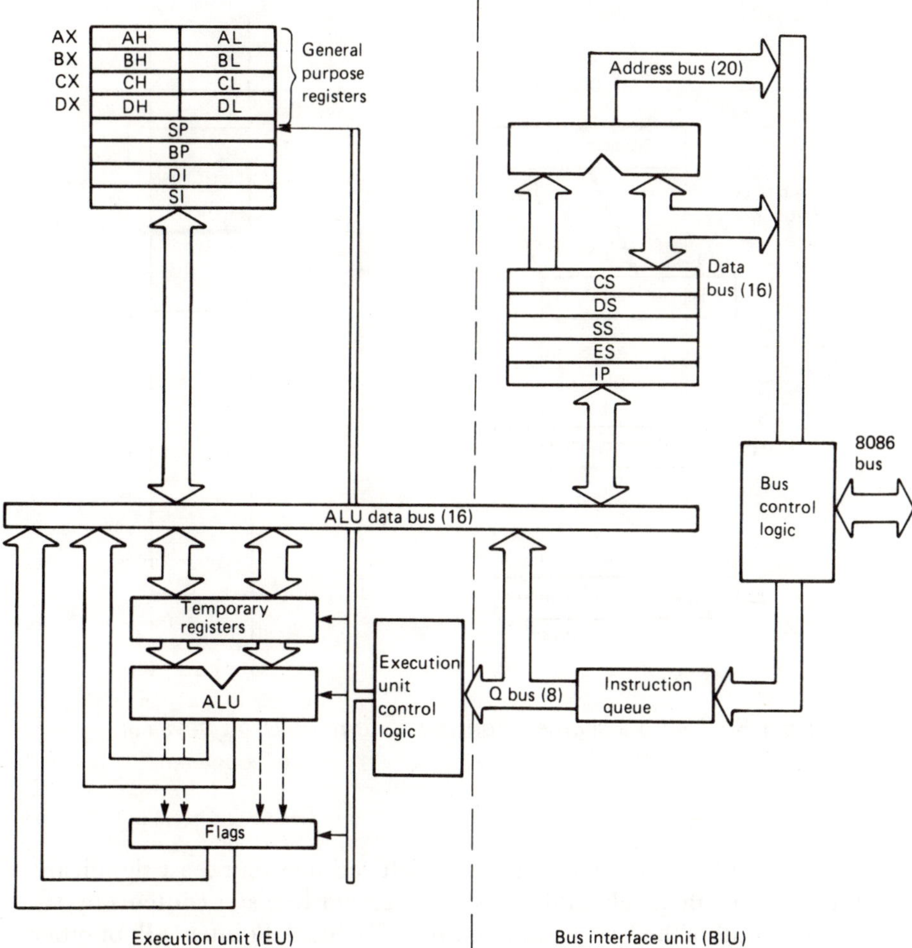

Figure 1.4 *Internal architecture of the 8086*

bus. The BIU comprises four segment registers and an instruction pointer, temporary storage for instructions held in the instruction queue, and bus control logic.

Addressing

The 8086 has 20 address lines and thus provides for a physical 1 megabyte memory address range (memory address locations 00000 to FFFFF hex.). The I/O address range is 64 kilobytes (I/O address locations 0000 to FFFF hex.).

The actual 20-bit physical memory address is formed by shifting the

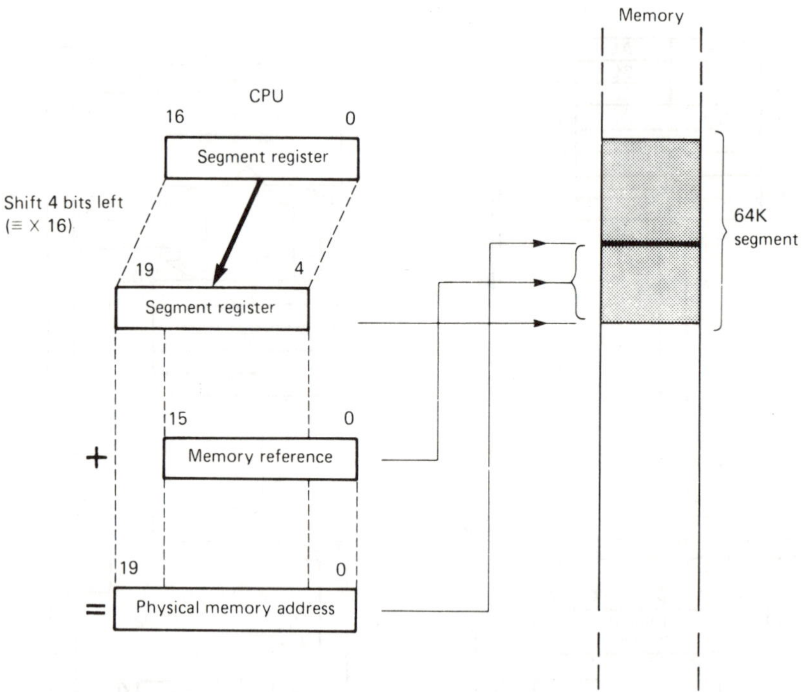

Figure 1.5 *Using a segment register to form a 20-bit physical address*

segment address four zero bits to the left (adding four least significant bits), which effectively multiplies the Segment Register contents by 16. The contents of the Instruction Pointer (IP), Stack Pointer (SP) or other 16-bit memory reference is then added to the result. This process is illustrated in Figure 1.5.

As an example of the process of forming a physical address reference, Table 1.5 shows the state of the 8086 registers after the RESET signal is applied. The instruction referenced (i.e. the first instruction to be executed after the RESET signal is applied) will be found by combining the Instruction Pointer (offset address) with the Code Segment register (paragraph address). The location of the instruction referenced is FFFF0 (i.e. F0000 + FFF0). Note that the PC's ROM physically occupies addresses F0000 to FFFFF and that, following a power-on or hardware reset, execution commences from address FFFF0 with a jump to the initial program loader.

The NEC V20 and V30 processors are pin-compatible replacements for the Intel 8088 and 8086, respectively. These chips are enhanced

Table 1.5 Contents of the 8086 registers after a reset

Register	*Content* (hex.)
Flag	0002
Instruction pointer	FFF0
Code segment	F000
Data segment	0000
Extra segment	0000
Stack segment	0000

versions of their Intel counterparts and they offer an increase in processing speed for certain operations.

The 80286

Intel's 80286 CPU was first employed in the PC-AT and PS/2 Models 50 and 60. The 80286 offers a 16 megabyte physical addressing range but incorporates memory management capabilities that can map up to a gigabyte of virtual memory. Depending upon the application, the 80286 is up to six times faster than the standard 5 MHz 8086 while providing upward software compatability with the 8086 and 8088 processors.

The 80286 has 15 16-bit registers, of which 14 are identical to those of the 8086. The additional Machine Status Word (MSW) register controls the operating mode of the processor and also records when a task switch takes place.

The bit functions within the MSW are summarized in Table 1.6. The MSW is initialized with a value of FFF0H upon reset, the remainder of the 80286 registers being initialized as shown in Table 1.5. The 80286 is packaged in a 68-pin JEDEC type-A plastic leadless chip carrier (PLCC), as shown in Figure 1.6.

The 80386

The 80386 (or 386) was designed as a *full* 32-bit device capable of manipulating data 32 bits at a time and communicating with the outside world through a 32-bit address bus. The 80386 offers a 'virtual 8086' mode of operation in which memory can be divided into 1-megabyte chunks with a different program allocated to each partition.

The 80386 is available in two basic versions. The 80386SX operates internally as a 32-bit device but presents itself to the outside world through only 16 data lines. This has made the CPU extremely popular for use in low-cost systems which could still boast the processing power of a 386 (despite the obvious limitation imposed by the reduced number of

Table 1.6 Bit functions in the 80286 machine status word

Bit	*Name*	*Function*
0	Protected mode (PE)	Enables protected mode and can only be cleared by asserting the RESET signal
1	Monitor processor (MP)	Allows WAIT instructions to cause a 'processor extension not present' exception (Exception 7)
2	Emulate processor (EP)	Causes a 'processor extension not present' exception (Exception 7) on ESC instructions to allow *emulation* of a processor extension
3	Task switched (TS)	Indicates that the next instruction using a processor extension will cause Exception 7 (allowing software to test whether the current processor extension context belongs to the current task)

Table 1.7 Power supply requirements and packages for CPU chips

	CPU type					
	8086	*8088*	*80186*	*80286*	*80386 (i386)*	*80486 (i486)*
Supply current (mA)	340–360	340–360	415–550	415–550	370–550	750–900
Typical supply power (W)	1.75	1.75	2.5	2.5	2.5	5
Packages	DIP	DIP	PLCC, PGA	PLCC PGA	PGA	PGA
Pins	40	40	68	68	132	168

data lines, the 'SX' version of the 80386 runs at approximately 80% of the speed of its fully fledged counterpart).

80386 architecture

The 80386 comprises a Bus Interface Unit (BIU), a Code Prefetch Unit, an Instruction Decode Unit, an Execution Unit (EU), a Segmentation Unit and a Paging Unit. The Code Prefetch Unit performs the program 'lookahead' function.

When the BIU is not performing bus cycles in the execution of an instruction, the Code Prefetch Unit uses the BIU to fetch sequentially the instruction stream. The prefetched instructions are stored in a 16-byte 'code queue' where they await processing by the Instruction Decode Unit.

The prefetch queue is fed to the Instruction Decode Unit which translates the instructions into microcode. These microcoded instruc-

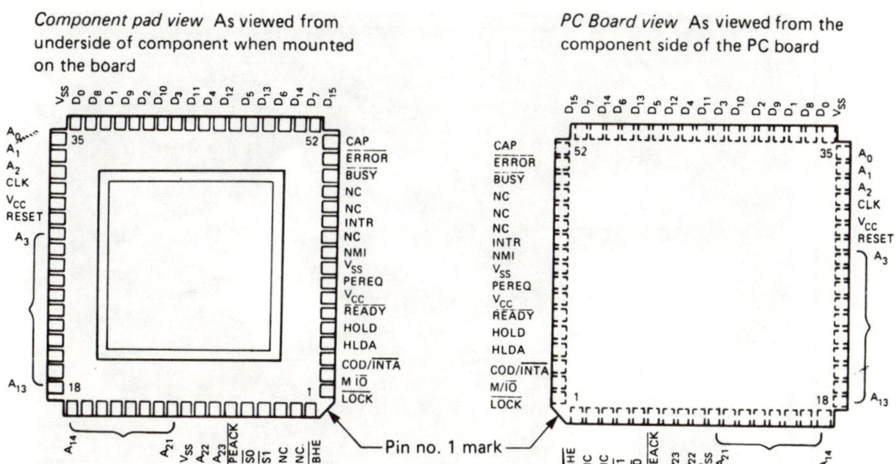

Figure 1.6 *80286 pin connections*

tions are then stored in a three-deep instruction queue on a first-in first-out (FIFO) basis. This queue of instructions awaits acceptance by the EU. Immediate data and opcode offsets are also taken from the prefetch queue.

The 80486

The 80486 CPU is not merely an upgraded 80386 processor; its redesigned architecture offers significantly faster processing speeds when running at the *same* clock speed as its predecessor. Enhancements include a built-in maths coprocessor, internal cache memory and cache memory control. The internal cache is responsible for a significant increase in processing speed. As a result, a 486 operating at 25 MHz can achieve a faster processing speed than a 386 operating at 33 MHz.

The 486 CPU (Plate 1.1) uses a large number of additional signals associated with parity checking (PCHK) and cache operation (AHOLD, FLUSH, etc.). The cache comprises a set of four 2-kilobyte blocks (128 × 16 bytes) of high-speed internal memory. Each 16-byte line of memory has a matching 21-bit 'tag'. This tag comprises a 17-bit linear address together with four protection bits. The cache control block contains 128 sets of seven bits. Three of the bits are used to implement the 'least recently used' (LRU) system for replacement and the remaining four bits are used to indicate valid data.

Interrupt handling

Interrupt service routines are subprograms stored away from the main body of code that are available for execution whenever the relevant

Plate 1.1 *Intel 486DX CPU chip in a PGA connector*

interrupt occurs. However, since interrupts may occur at virtually any point in the execution of a main program, the response must be automatic; the processor must suspend its current task and save the return address so that the program can be resumed at the point at which it was left. Note that the programmer must assume responsibility for preserving the state of any registers which may have their contents altered during execution of the interrupt service routine.

The Intel processor family uses a table of 256 4-byte pointers stored in the bottom 1 kilobyte of memory (addresses 0000H to 03FFH). Each of the locations in the Interrupt Pointer Table can be loaded with a pointer to a different interrupt service routine. Each pointer contains 2 bytes for loading into the Instruction Pointer (IP). This allows the programmer to place his or her interrupt service routines in any appropriate place within the 1 megabyte physical address space. Further details can be found in Chapter 5.

PC architecture

The generic PC, whether a 'desktop' or 'tower' system, comprises three units: System Unit, Keyboard and Display. The System Unit itself comprises three items: System Board, Power Supply and Floppy/Hard Disk Drives.

The original IBM PC System Board employed approximately 100 IC devices including an 8088 CPU, an 8259A Interrupt Controller, an optional 8087 Maths Coprocessor, an 8288 Bus Controller, an 8284A Clock Generator, an 8253 Timer/Counter, an 8237A DMA Controller, and an 8255A Parallel Interface together with a host of discrete logic (including bus buffers, latches and transceivers). Figure 1.7 shows the simplified bus architecture of the system.

Much of this architecture was carried forward to the PC-XT and the PC-AT. This latter machine employed an 80286 CPU, 80287 Maths Coprocessor, two 8237A DMA Controllers, 8254-2 Programmable Timer, 8284A Clock Generator, two 8259A Interrupt Controllers, and a 74LS612N Memory Mapper. In order to significantly reduce manufacturing costs as well as to save on space and increase reliability, more recent AT-compatible microcomputers are based on a significantly smaller number of devices (many of which may be surface mounted types). This trend has been continued with today's powerful 386- and 486-based systems. However, the functions provided by the highly integrated chip sets are usually a superset of those provided by the much larger number of devices found in their predecessors.

IBM PS/2

IBM's second generation of personal computers, Personal System/2, is based on four machines: Models 30, 50, 60 and 80. The smallest of these (Model 30) is based on an 8086 CPU and has dual 720/1.4 megabyte 3.5-inch floppy disk drives and 640 kilobyte of RAM. Models 50 and 60 use the 80286 CPU with 1 megabyte of RAM (expandable) and fixed disk drives of 20 (Model 50), 44 or 70 megabyte (Model 60).

IBM's top-of-the-range machine, the Model 80, employs an 80386 CPU and a 1.4 megabyte 3.5-inch floppy disk drive, Three different fixed disk/RAM configurations are available for the Model 80 and these are based on 44 megabyte fixed disk/1 megabyte RAM, 70 megabyte fixed disk/2 megabyte RAM, and 115 megabyte fixed disk/2 megabyte RAM. Models 30 and 50 are designed for conventional desk-top operation whilst Models 60 and 80 both feature 'vertical' floor-standing systems units with front-access panels for the fixed disk tray. The PS/2 family has an impressive expansion capability and the following devices are supported by Models 50, 60 and 80:

- Monochrome display (8503)
- Colour display (8512)
- Colour display (8513) (medium resolution)

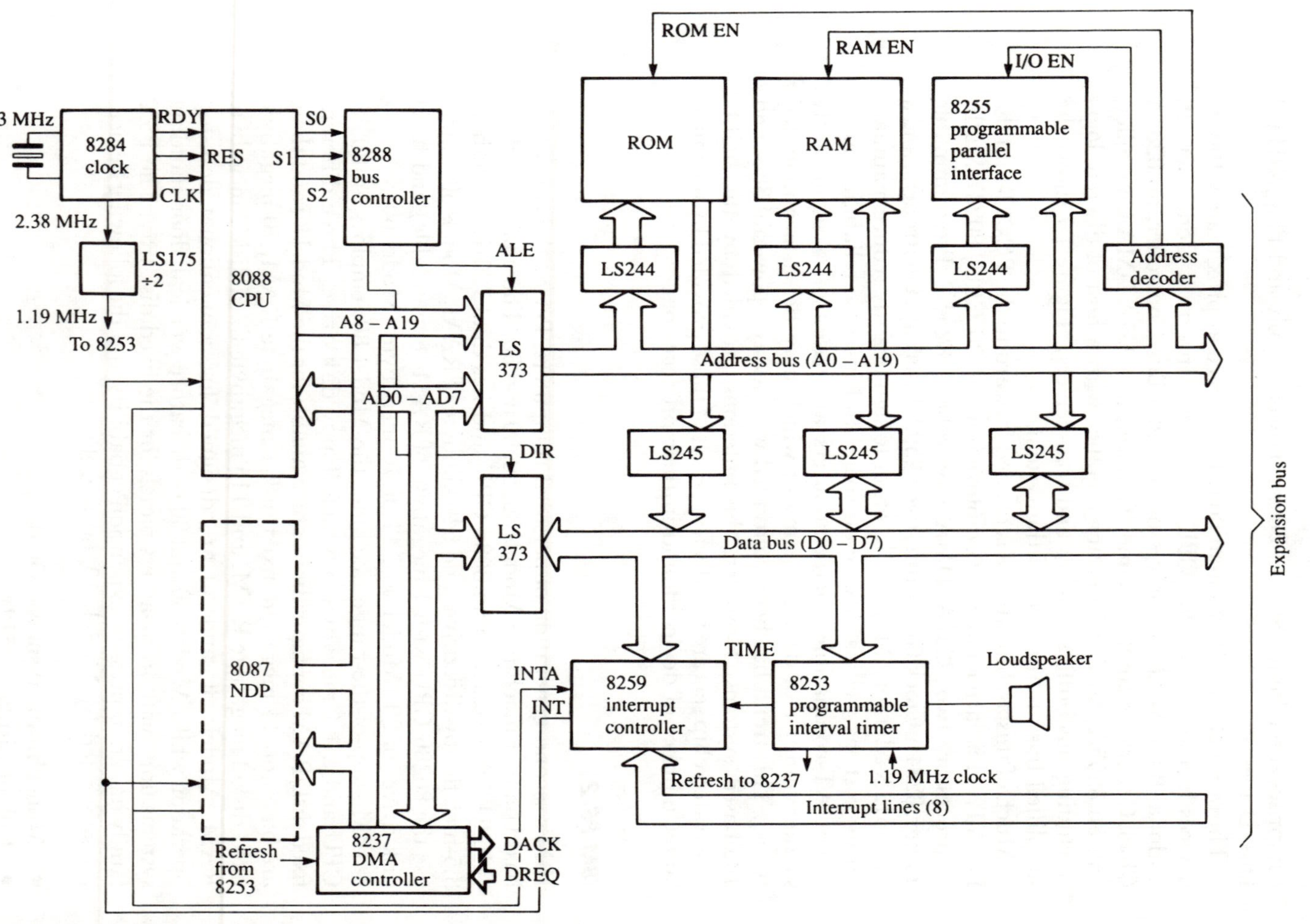

Figure 1.7 *Simplified architecture of the original PC*

- Colour display (8514) (high resolution)
- Memory expansion kits/cards
- Second fixed disk (44/70/115 megabyte)
- External 5.25-inch floppy disk drive
- Tape streamer (6157)
- Optical disk (internal or external 3363)
- Dual asynchronous communications adapter
- Internal modem (300/1200 baud)
- IBM PC network (LAN) adapter
- IBM token-ring (LAN) adapter
- Multi-protocol adapter (asynchronous/BSC/SDLC/HDLC)
- System 36/38 workstation emulator adapter

The 'generic' PC

The system architecture of a generic 8088-based PC is shown in Figure 1.7. There is more to this diagram than mere historical interest as all modern PCs can trace their origins to this particular arrangement. It is, therefore, worth spending a few moments developing an understanding of the configuration.

The 'CPU bus' (comprising lines A8 to A19 and AD0 to AD7 on the left-hand side of Figure 1.7) is separated from the 'system bus' which links the support devices and expansion cards. The eight least significant address and all eight of the data bus lines share a common set of eight CPU pins. These lines are labelled AD0 to AD7. The term used to describe this form of bus (where data and address information take turns to be present on a shared set of bus lines) is 'multiplexing'. This saves pins on the CPU package and it allowed Intel to make use of standard 40-pin packages for the 8088 and 8086 processors.

The system address bus (available on each of the expansion connectors) comprises 20 address lines (A0 to A19). The system data bus comprises eight lines (D0 to D7). Address and data information is alternately latched onto the appropriate set of bus lines by means of the four 74LS373 8-bit data latches. The control signals, ALE (address latch enable) and DIR (direction) derived from the 8288 bus controller are used to activate the two pairs of data latches.

The CPU bus is extended to the 8087 numeric data processor (maths coprocessor). This device is physically located in close proximity to the CPU in order to simplify the PCB layout.

The original PC required a CPU clock signal of 4.773 MHz from a dedicated Intel clock generator chip. The basic timing element for this device is a quartz crystal which oscillates at a fundamental frequency of 14.318 MHz. This frequency is internally divided by three in order to produce the CPU clock. The CPU clock frequency is also further divided by two internally and again by two externally in order to produce a clock signal for the 8253 programmable interrupt timer. This device provides three important timing signals used by the system. One (known appropriately as TIME) controls the 8259 programmable interrupt controller, another (known as REFRESH) provides a timing input for the 8237 DMA controller, whilst the third is used (in conjunction with some extra logic) to produce an audible signal at the loudspeaker.

74LS244 8-bit bus drivers and 74LS245 8-bit bus transceivers link each of the major support devices with the 'system address bus' and 'system data bus', respectively. Address decoding logic (with input signals derived from the system address bus) generates the chip enable lines which activate the respective ROM, RAM and I/O chip select lines.

The basic system board incorporates a CPU, provides a connector for the addition of a maths coprocessor, incorporates bus and DMA control, and provides the system clock and timing signals. The system board also houses the BIOS ROM, main system RAM, and offers some limited parallel I/O. It does not, however, provide a number of other essential facilities including a video interface, disk and serial I/O. These important functions must normally be provided by means of adapter cards (note that some systems which offer only limited expansion may have some or all of these facilities integrated into the system board).

Adapter cards are connected to the expansion bus by means of a number of expansion slots. The adapter cards are physically placed so that any external connections required are available at the rear (or side) of the unit. Connections to internal subsystems (such as hard and floppy disk drives) are usually made using lengths of ribbon cables and PCB connectors.

Typical system board layout

Figures 1.8 and 1.9 show typical system board layouts for 386- and 486-based PCs. This general layout started with the original PC and has been carried forward with improvements, enchancements, and a reduced chip count into a wide range of modern PC compatibles.

The system board RAM uses eight single in-line memory modules (SIMM). This memory is arranged in two banks (Bank 0 and Bank 1) and the system can be configured for 256X × 9, 9 × 1 megabyte and 9 × 4 megabyte devices. Hence up to 32 megabyte of system board memory can be fitted.

The system board also supports static cache RAM in two banks. U22 to U25 form bank 1 whilst U31 to U34 form bank 0. This cache memory can be configured for 64, 128 or 256 kilobyte and greatly improves the CPU performance.

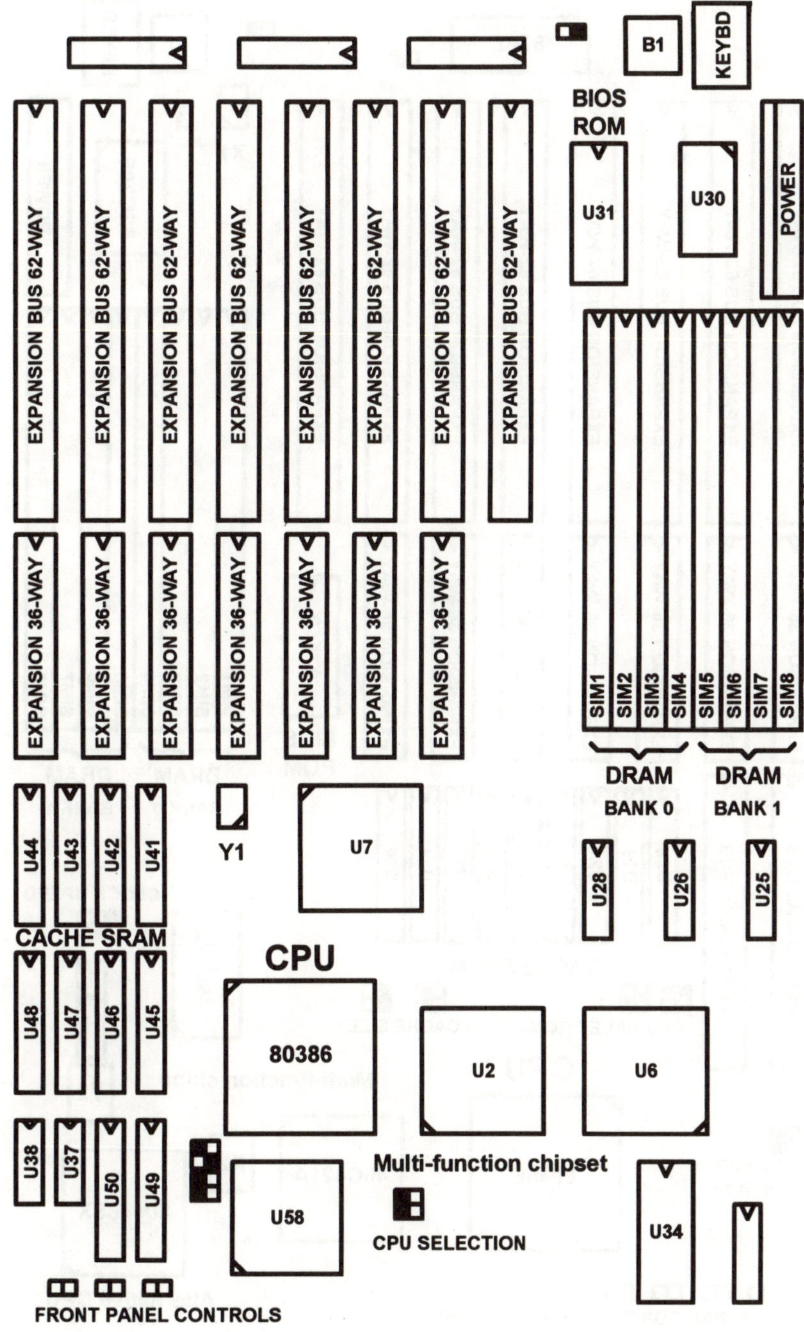

Figure 1.8 *Typical system board layout for a 386-based PC*

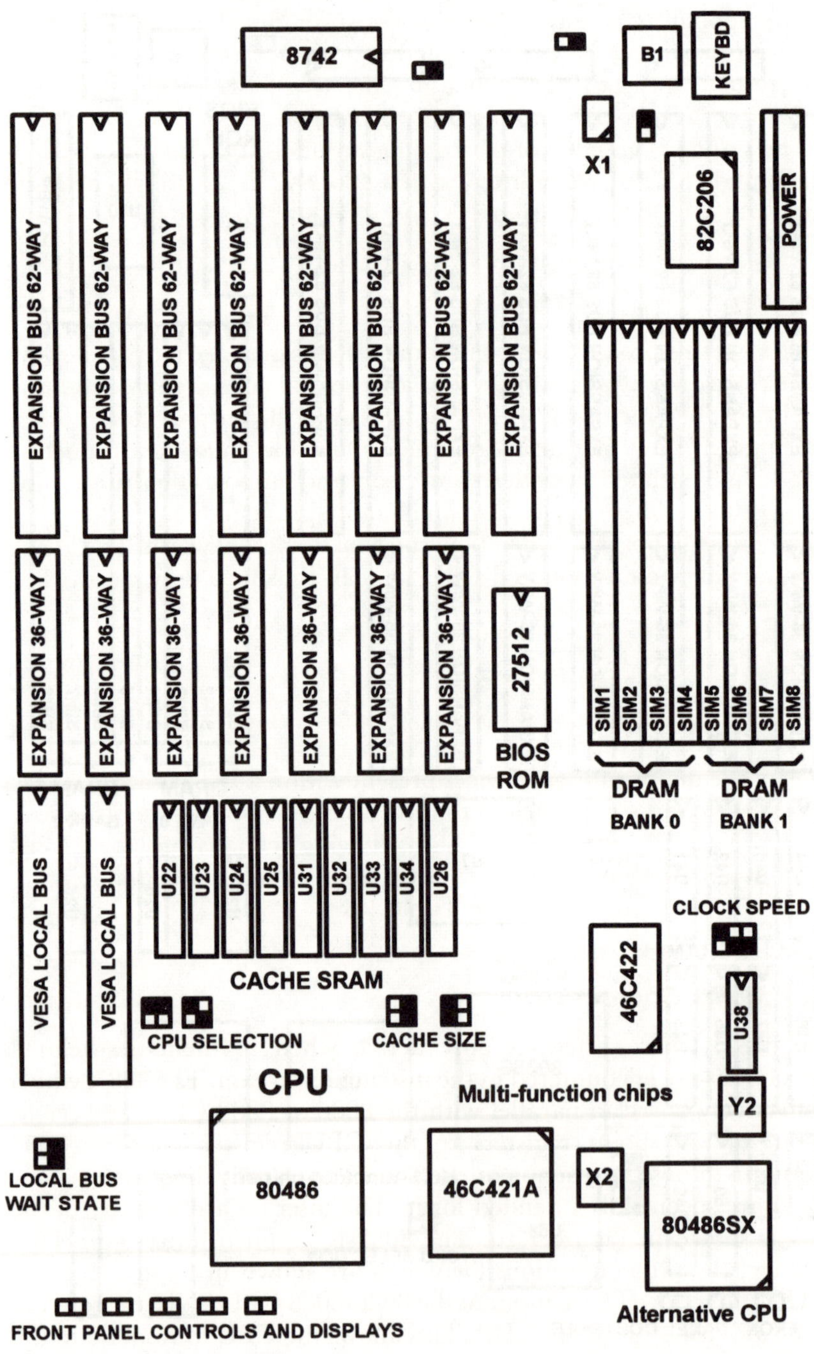

Figure 1.9 *Typical system board layout for a 486-based PC*

The 486 chip set (two VLSI devices) provides the functions of the major support devices (bus control, direct memory access control, etc). The system may be expanded using the two 32-bit VESA bus slots and/or the five 16-bit and one 8-bit expansion slots (ISA).

The system is capable of performing around 20 million instructions per second (MIPS) when fitted with a 486DX2 CPU using a 50 MHz clock and a 64 kilobyte cache.

Cooling

All PC systems produce heat, but some systems produce more heat than others. Adequate ventilation is thus an essential consideration and fans are invariably included within system units to ensure that there is adequate air flow. Furthermore, internal air flow must be arranged so that it is unrestricted as modern processors and support chips run at high temperatures. These devices are much more prone to failure when they run excessively hot than when they run cool or merely warm. The fitting of an auxiliary chip-mounted CPU fan can thus be a very worthwhile investment.

PC support devices

Each of the major support devices present within a PC has a key role to play in off-loading a number of routine tasks that would otherwise have to be performed by the CPU. The following sections provide a brief introduction to each generic device together with internal architecture and pin connecting details (see Table 1.8).

8087 Maths Coprocessor

The 8087, where fitted, is only active when mathematics related instructions are encountered in the instruction stream. The 8087, which is effectively wired in parallel with the 8086 or 8088 CPU, adds eight 80-bit floating point registers to the CPU register set. The 8087 maintains its own instruction queue and executes only those instructions which are specifically intended for it. The internal architecture of the 8087 is shown in Figure 1.10. The 8087 is supplied in a 40-pin DIL package, the pin connections for which are shown in Figure 1.11.

The active-low TEST input of the 8086/8088 CPU is driven from the BUSY output of the 8087 NDP. This allows the CPU to respond to the WAIT instruction (inserted by the assembler/compiler) which occurs before each coprocessor instruction. An FWAIT instruction follows

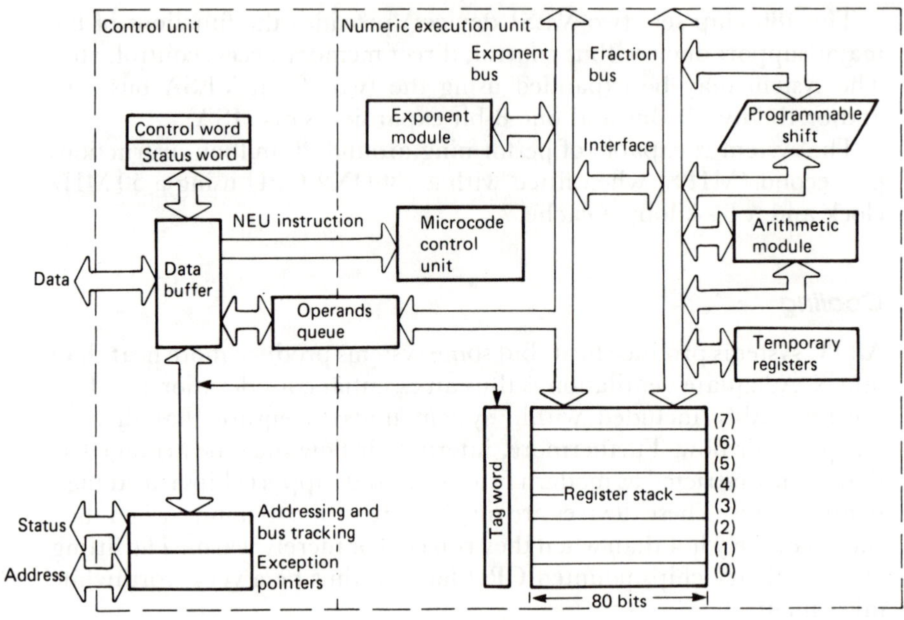

Figure 1.10 *Internal architecture of the 8087*

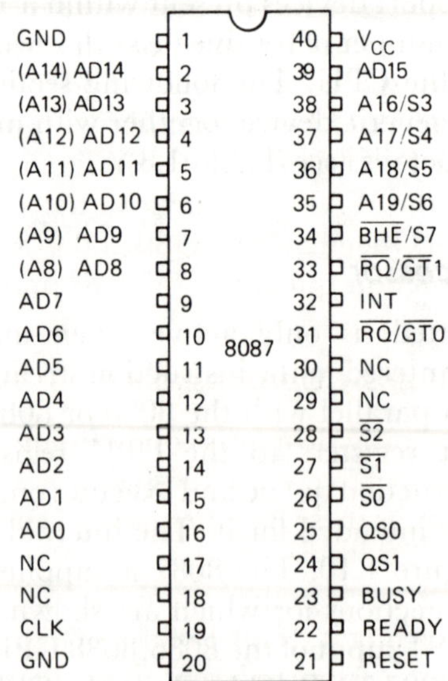

Figure 1.11 *8087 pin connections*

Table 1.8 Support chips used with early 80x86 processors

	CPU type				
	8086	*8088*	*80186*	*80286*	*80386 (386)*
Clock generator	8284A	8284A	On-chip	82284	82384
Bus controller	8288	8288	On-chip	82288	82288
Integrated support chips				82230/ 82231, 82335	82230/ 82231, 82335
Interrupt controller	8259A	8259A	On-chip	8259A	8259A
DMA controller	8089/ 82258	8089/ 8237/ 82258	On-chip/ 82258	8089/ 82258	8237/ 82258
Timer/counter	8253/ 8254	8253/ 8254	On-chip	8253/ 8254	8253/ 8254
Maths coprocessor	8087	8087	8087	80287	80287/ 80387
Chip select/ wait state logic	TTL	TTL	On-chip	TTL	TTL

each coprocessor instruction which deposits data in memory for immediate use by the CPU. The instruction is then translated to the requisite 8087 operation (with the preceding WAIT) and the FWAIT instruction is translated as a CPU WAIT instruction.

During coprocessor execution, the BUSY line is taken high and the CPU (responding to the WAIT instruction) halts its activity until the line goes low. The two Queue Status (QS0 and QS1) signals are used to synchronize the instruction queues of the two processing devices.

80287 and 80387 Maths Coprocessors

80287 and 80387 chips provide maths coprocessing facilities within AT- and 386-based PCs, respectively. In 486-based systems there is no need for a maths coprocessor as these facilities have been incorporated within the CPU.

You can easily find out whether a coprocessor is present within a system without dismantling it. If bit 1 of the byte stored at address 0410 hex. is set, a maths coprocessor is present. If bit 1 is reset, no coprocessor is fitted. You can check the bit in question using DEBUG or by using the following fragment of QuickBASIC code:

```
DEF SEG = 0
byte=PEEK(&H410)
IF byte AND 2 THEN
  PRINT "Coprocessor fitted"
ELSE
  PRINT "No coprocessor fitted"
END IF
```

8237A Direct Memory Access Controller

The 8237A Direct Memory Access Controller (DMAC) can provide service for up to four independent DMA channels, each with separate registers for Mode Control, Current Address, Base Address, Current Word Count and Base Word Count (see Figure 1.12). The DMAC is designed to improve system performance by allowing external devices to directly transfer information to and from the system memory. The 8237A offers a variety of programmable control features to enhance data throughput and allow dynamic reconfiguration under software control.

The 8237A provides four basic modes of transfer: block, demand, single word, and cascade. These modes may be programmed as required; however, channels may be autoinitialized to their original condition following an end of process (EOP) signal.

The 8237A is designed for use with an external octal address latch such as the 74LS373. A system's DMA capability may be extended by cascading further 8237A DMAC chips, and this feature is exploited in the PC-AT which has two such devices.

The least significant four address lines of the 8237A are bi-directional: when functioning as inputs, they are used to select one of the DMA controller's 16 internal registers. When functioning as outputs, however, a 16-bit address is formed by taking the eight address lines (A0 to A7) to form the least significant address byte whilst the most significant address byte (A8 to A15) is multiplexed onto the data bus lines (D0 to D7). The requisite address latch enable signal (ADSTB) is available from pin 8. The upper four address bits (A16 to A19) are typically supplied by a 74LS670 4 × 4 register file. The requisite bits are placed in this device (effectively a static RAM) by the processor before the DMA transfer is completed.

DMA channel 0 (highest priority) is used in conjunction with the 8253 programmable interval timer (PIT) in order to provide a memory refresh facility for the PC's dynamic RAM. DMA channels 1 to 3 are connected to the expansion slots for use by option cards.

The refresh process involves channel 1 of the PIT producing a negative going pulse with a period of approximately 15 μs. This pulse sets a bistable which, in turn, generates a DMA request at the channel-0 input of the DMAC (pin-19). The processor is then forced into a wait state and the

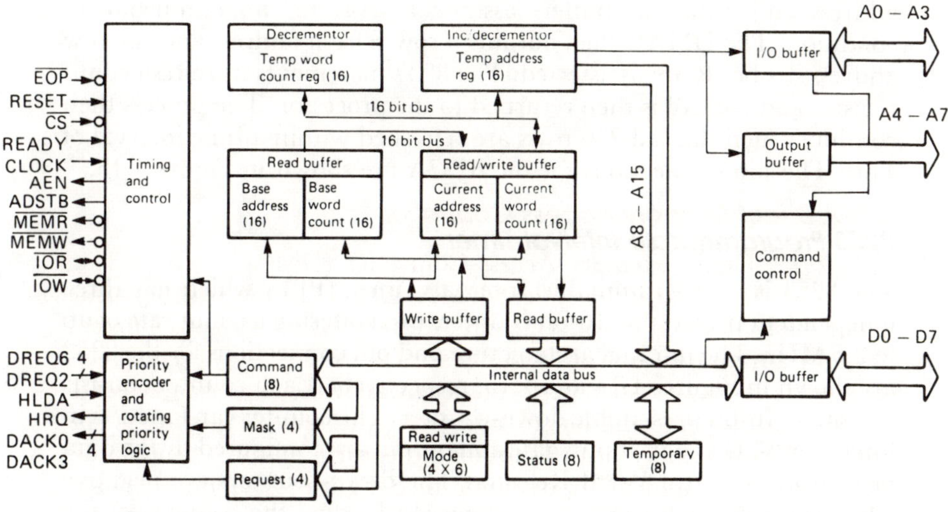

Figure 1.12 *Internal architecture of the 8237A*

IOR	1	8237A	40	A7
IOW	2		39	A6
MEMR	3		38	A5
MEMW	4		37	A4
LOGIC 1	5		36	EOP
READY	6		35	A3
HLDA	7		34	A2
ADSTB	8		33	A1
AEN	9		32	A0
HRQ	10		31	V_{CC} (+5 V)
CS	11		30	DB0
CLK	12		29	DB1
RESET	13		28	DB2
DACK2	14		27	DB3
DACK3	15		26	DB4
DREQ3	16		25	DACK0
DREQ2	17		24	DACK1
DREQ1	18		23	DB5
DREQ0	19		22	DB6
(GND) V_{SS}	20		21	DB7

Figure 1.13 *8237A pin connections*

address and data bus buffers assume a tri-state (high impedance) condition. The DMAC then outputs a row refresh address and the row address strobe (RAS) is asserted. The 8237 increments its refresh count register and control is then returned to the processor. The process then continues such that all 256 rows are refreshed within a time interval of 4 ms. The pin connections for the 8237A are shown in Figure 1.13.

8253 Programmable Interval Timer

The 8253 is a Programmable Interval Timer (PIT) which has three independent presettable 16-bit counters each offering a count rate of up to 2.6 MHz. The internal architecture and pin connections for the 8253 are shown in Figures 1.14 and 1.15, respectively. Each counter consists of a single 16-bit presettable down-counter. The counter can function in binary or BCD and its input, gate and output are configured by the data held in the Control Word Register. The down-counters are negative edge triggered such that, on a falling clock edge, the contents of the respective counter are decremented.

The three counters are fully independent and each can have a separate mode configuration and counting operation, binary or BCD. The contents of each 16-bit count register can be loaded or read using simple software referencing; the relevant port addresses are shown in Table 1.9. The truth table for the chip's active-low chip select (CS), read (RD), write (WR) and address lines (A1 and A0) is shown in Table 1.10.

8255A Programmable Peripheral Interface

The 8255A Programmable Peripheral Interface (PPI) is a general-purpose I/O device which provides no less than 24 I/O lines arranged as three 8-bit I/O ports. The internal architecture and pin connections of the 8255A are shown in Figures 1.16 and 1.17, respectively. The Read/Write and Control Logic block manages all internal and external data transfers. The port addresses used by the 8255A are given in Table 1.9.

The functional configuration of each of the 8255's three I/O ports is fully programmable. Each of the control groups accepts commands from the Read/Write Control Logic, receives Control Words via the internal data bus and issues the requisite commands to each of the ports. At this point, it is important to note that the 24 I/O lines are, for control purposes, divided into two logical groups (A and B). Group A comprises the entire eight lines of Port A together with the four upper (most significant) lines of Port B. Group B, on the other hand, takes in all eight lines from Port B together with the four lower (least significant) lines of Port C. The upshot of all this is simply that Port C can be split into two in

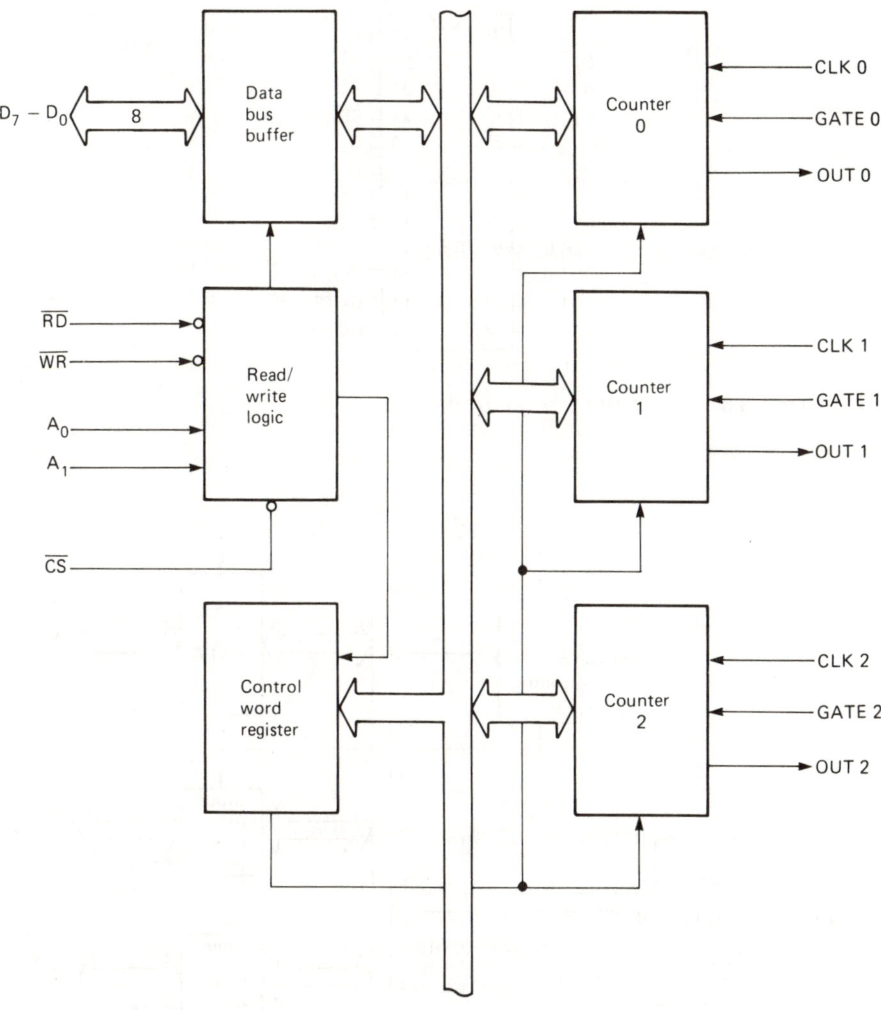

Figure 1.14 *Internal architecture of the 8253*

order to allow its lines to be used for status and control (handshaking) when data is transferred to or from Ports A or B.

8259A Programmable Interrupt Controller

The 8259A Programmable Interrupt Controller (PIC) was designed specifically for use in real-time interrupt driven microcomputer systems.

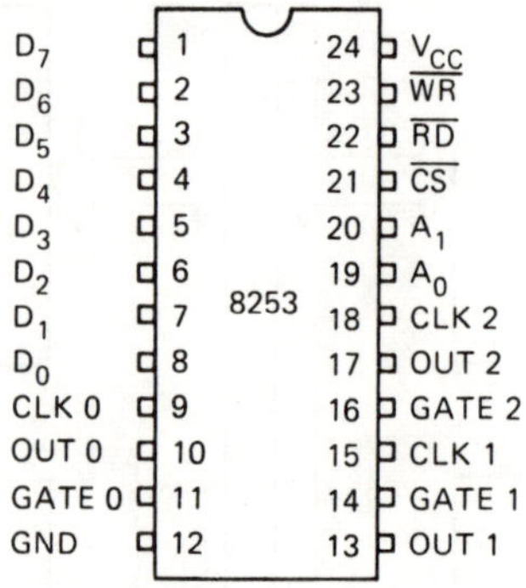

1	D_7	24	V_{CC}
2	D_6	23	$\overline{WR}$
3	D_5	22	$\overline{RD}$
4	D_4	21	$\overline{CS}$
5	D_3	20	A_1
6	D_2	19	A_0
7	D_1	18	CLK 2
8	D_0	17	OUT 2
9	CLK 0	16	GATE 2
10	OUT 0	15	CLK 1
11	GATE 0	14	GATE 1
12	GND	13	OUT 1

8253

Figure 1.15 *8253 pin connections*

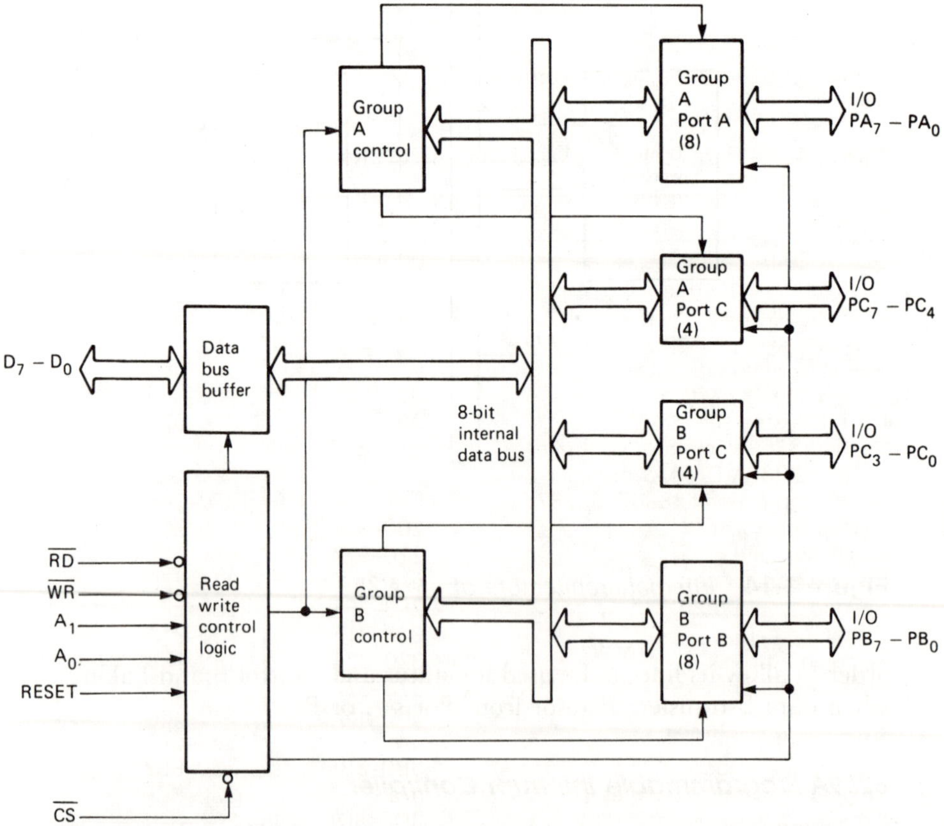

$D_7 - D_0$
Data bus buffer
Read write control logic
$\overline{RD}$
$\overline{WR}$
A_1
A_0
RESET
$\overline{CS}$
Group A control
Group B control
8-bit internal data bus
Group A Port A (8)
Group A Port C (4)
Group B Port C (4)
Group B Port B (8)
I/O $PA_7 - PA_0$
I/O $PC_7 - PC_4$
I/O $PC_3 - PC_0$
I/O $PB_7 - PB_0$

Figure 1.16 *Internal architecture of the 8255A*

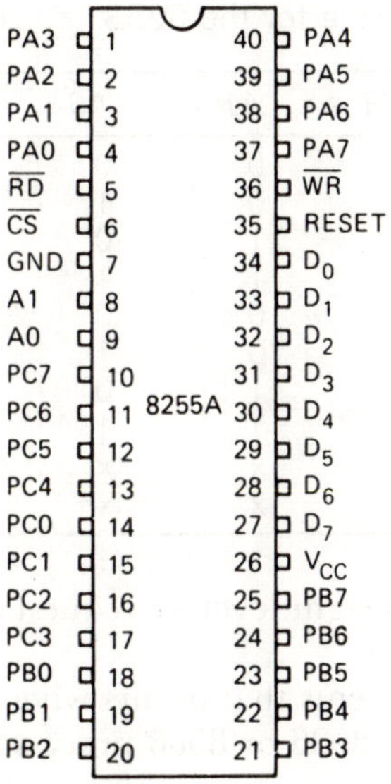

Figure 1.17 *8255A pin connections*

Table 1.9 Port addresses used in the PC family

Device	*PC-XT*	*PC-AT, etc.*
8237A DMA controller	000–00F	000–01F
8259A interrupt controller	020–021	020–03F
8253/8254 timer	040–043	040–05F
8255 parallel interface	060–063	n/a
8042 keyboard controller	n/a	060–06F
DMA page register	080–083	080–09F
NMI mask register	0A0–0A7	070–07F
Second 8259A interrupt controller	n/a	0A0–0BF
Second 8237A DMA controller	n/a	0C0–0DF
Maths coprocessor (8087, 80287)	n/a	0F0–0FF
Games controller	200–20F	200–207
Expansion unit	210–217	n/a
Second parallel port	n/a	278–27F
Second serial port	2F8–2FF	2F8–2FF
Prototype card	300–31F	300–31F
Fixed (hard) disk	320–32F	1F0–1F8
First parallel printer	378–37F	378–37F
SDLC adapter	380–38F	380–38F
BSC adapter	n/a	3A0–3AF
Monochrome adapter	3B0–3BF	3B0–3BF
Enhanced graphics adapter	n/a	3C0–3CF
Colour graphics adapter	3D0–3DF	3D0–3DF
Floppy disk controller	3F0–3F7	3F0–3F7
First serial port	3F8–3FF	3F8–3FF

Table 1.10 Truth table for the 8253

CS	*RD*	*WR*	*A1*	*A0*	*Function*
0	1	0	0	0	Load counter 0
0	1	0	0	1	Load counter 1
0	1	0	1	0	Load counter 2
0	1	0	1	1	Write mode word
0	0	1	0	0	Read counter 0
0	0	1	0	1	Read counter 1
0	0	1	1	0	Read counter 2
0	0	1	1	1	No-operation (tri-state)
1	x	x	x	x	Disable tri-state
0	1	1	x	x	No-operation (tri-state)

The device manages eight levels of request and can be expanded using further 8259A devices.

The sequence of events that occurs when an 8259A device is used in conjunction with an 8086 or 8088 processor is as follows:

1 One or more of the interrupt request lines (IR0–IR7) are asserted (note that these lines are active-high) by the interrupting device(s).
2 The corresponding bits in the IRR register become set.
3 The 8259A evaluates the requests on the following basis:
   (a) If more than one request is currently present, determine which of the requests has the highest priority.
   (b) Ascertain whether the successful request has a higher priority than the level currently being serviced.
   (c) If the condition in (b) is satisfied, issue an interrupt to the processor by asserting the active-high INT line.
4 The processor acknowledges the interrupt signal and responds by asserting the interrupt acknowledge by pulsing the interrupt acknowledge (INTA) line.
5 Upon receiving the INTA pulse from the processor, the highest priority ISR bit is set and the corresponding IRR bit is reset.
6 The processor then initiates a second interrupt acknowledge (INTA) pulse. During this second period for which the INTA line is taken low, the 8259 outputs a pointer on the data bus to be read by the processor.

The internal architecture and pin connections for the 8259A are shown in Figures 1.18 and 1.19 respectively.

8284A Clock Generator

The 8284A is a single-chip clock generator/driver designed specifically for use by the 8086 family of devices. The chip contains a crystal

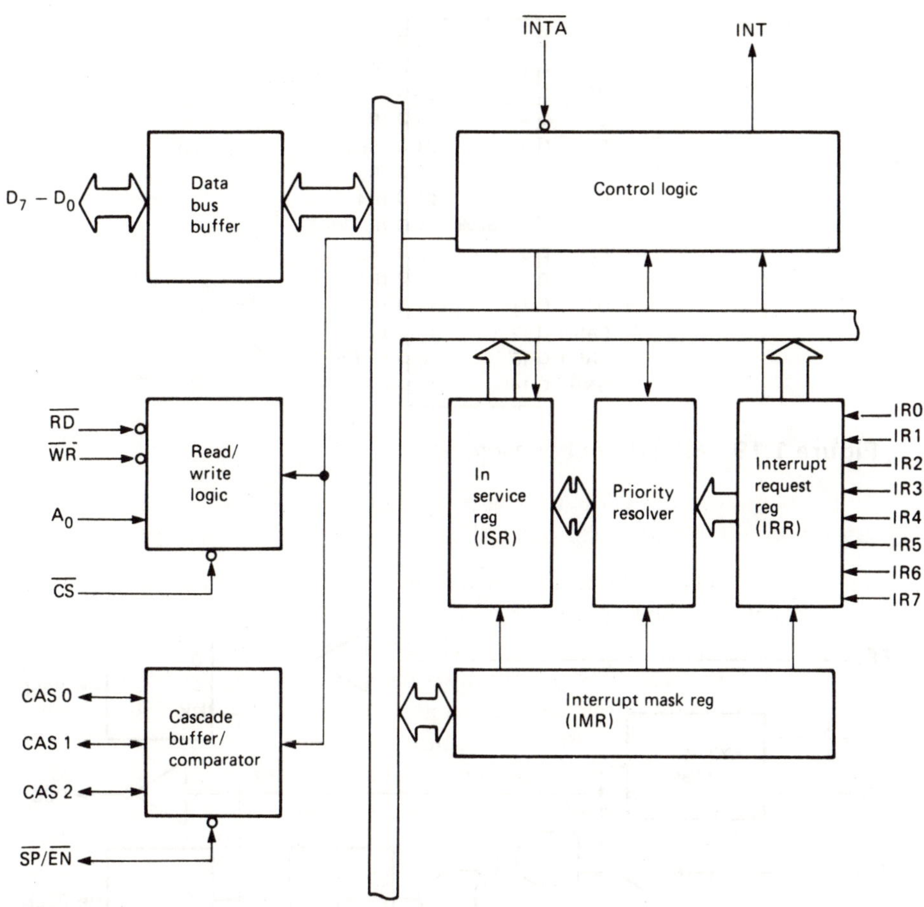

Figure 1.18 *Internal architecture of the 8259A*

oscillator, ÷3 counter, ready and reset logic as shown in Figure 1.20. The pin connections are shown in Figure 1.21. On the original PC, the quartz crystal is a series mode fundamental device which operates at a frequency of 14.312818 MHz. The output of the ÷3 counter takes the form of a 33% duty cycle square wave at precisely one-third of the fundamental frequency (i.e. 4.77 MHz). This signal is then applied to the processor's clock (CLK) input. The clock generator also produces a signal at 2.38 MHz which is externally divided to provide a 5.193 MHz 50% duty cycle clock signal for the 8253 Programmable Interval Timer (PIT), as shown in Figure 1.22.

$\overline{CS}$	1		28	V_{CC}
$\overline{WR}$	2		27	A_0
$\overline{RD}$	3		26	$\overline{INTA}$
D_7	4		25	IR7
D_6	5		24	IR6
D_5	6		23	IR5
D_4	7		22	IR4
D_3	8	8259A	21	IR3
D_2	9		20	IR2
D_1	10		19	IR1
D_0	11		18	IR0
CAS 0	12		17	INT
CAS 1	13		16	$\overline{SP}/\overline{EN}$
GND	14		15	CAS 2

Figure 1.19 *8259A pin connections*

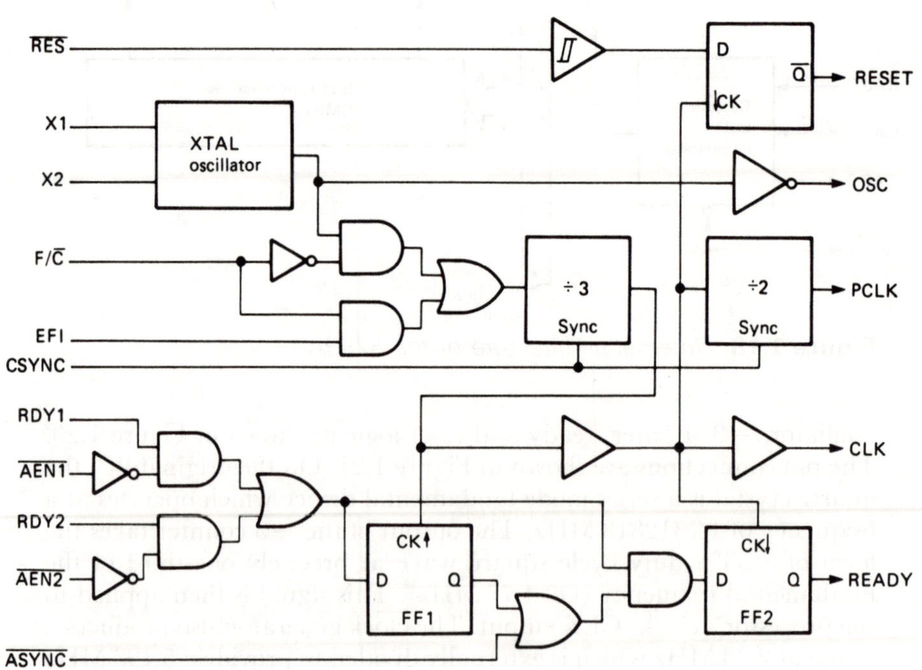

Figure 1.20 *Internal arrangement of the 8284A*

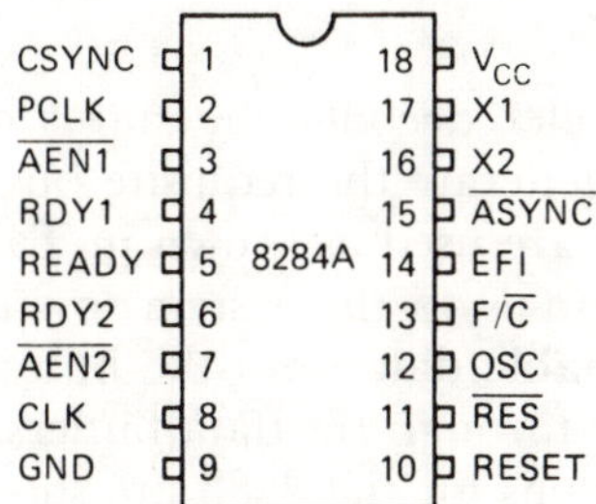

Figure 1.21 *8284A pin connections*

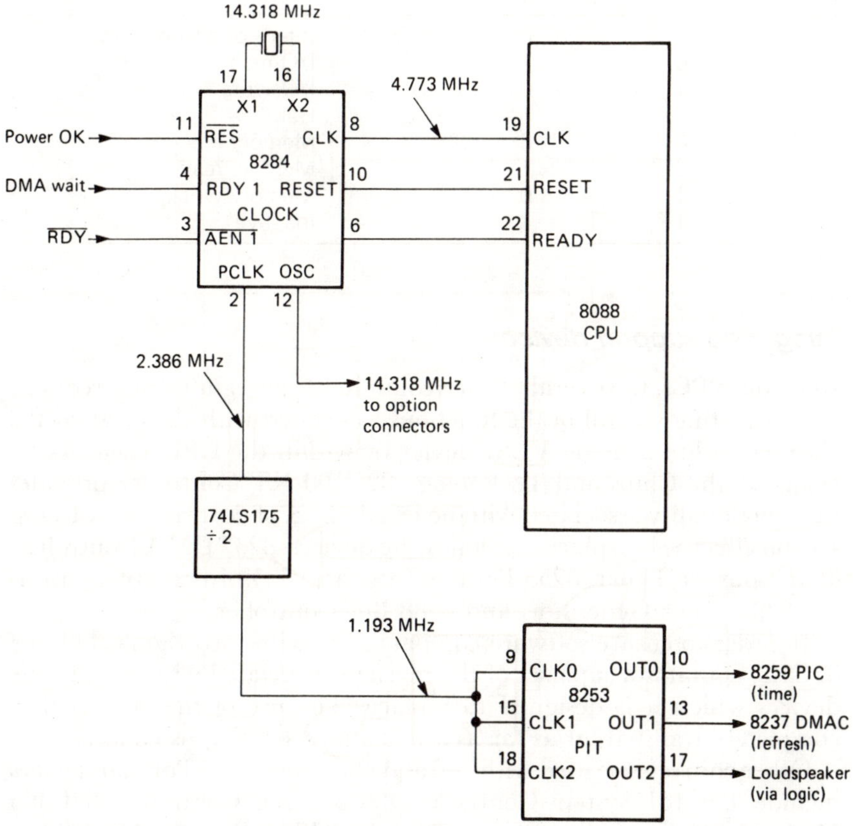

Figure 1.22 *Clock signals in the PC*

8288 Bus Controller

The 8288 bus controller decodes the status outputs from the CPU (S0–S1) in order to generate the requisite bus command and control signals. These signals are used as shown in Table 1.11.

The 8288 issues signals to the system to strobe addresses into the address latches, to enable data onto the buses, and to determine the direction of data flow through the data buffers. The internal architecture and pin connections for the 8288 are shown in Figures 1.23 and 1.24, respectively.

Table 1.11 8288 bus controller status inputs

CPU status line			
S2	*S1*	*S0*	*Condition*
0	0	0	Interrupt acknowledge
0	0	1	I/O read
0	1	0	I/O write
0	1	1	Halt
1	0	0	Memory read
1	0	1	Memory read
1	1	0	Memory write
1	1	1	Inactive

Integrated support devices

In modern PCs, the overall device count has been significantly reduced by integrating several of the functions associated with the original PC chip set within a single VLSI device or within the CPU itself. As an example, the Chips and Technology 82C100 XT Controller provides the functionality associated with no less than six of the original XT chip set and effectively replaces the following devices: 8237 DMA Controller, 8253 Counter/Timer, 8255 Parallel Interface, 8259 Interrupt Controller, 8284 Clock Generator, and 8288 Bus Controller.

In order to ensure software compatability with the original PC, the 82C100 contains a superset of the registers associated with each of the devices which it is designed to replace. The use of the chip is thus completely transparent as far as applications software is concerned.

Other chip sets are used with 386 and 486 computers. Popular devices include: 82C461 System Controller, 82C362 Bus Controller, 82C365 Cache Controller, 85C310 Cache/Memory Controller, 85C320 AT-bus Controller, and 85C330 Data Bus Buffer.

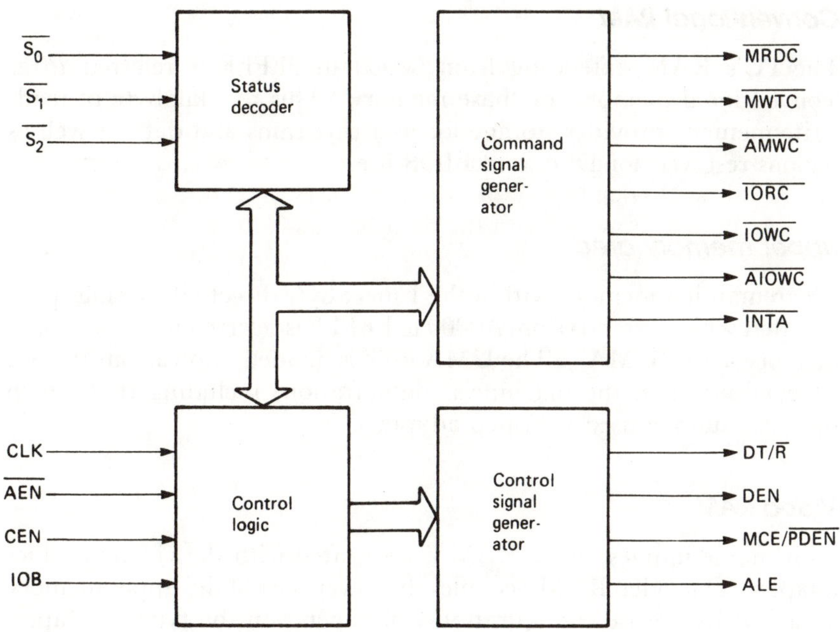

Figure 1.23 *Internal architecture of the 8288*

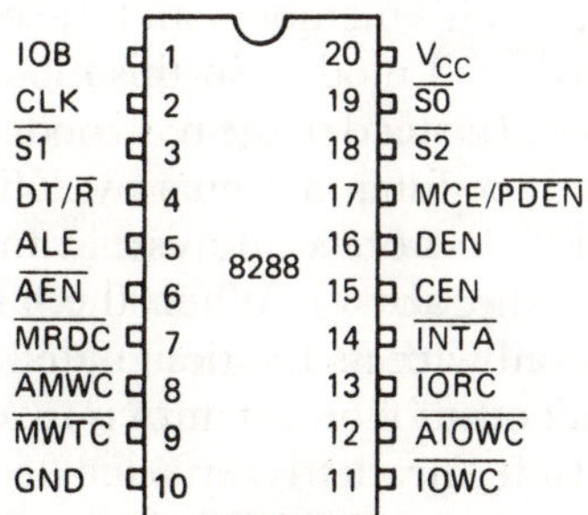

Figure 1.24 *8288 pin connections*

PC memory

Memory terminology

The following terminology is commonly used to describe the various types of memory present within a PC or PC-compatible system.

Conventional RAM

The PC's RAM extending from 00000 to 9FFFF is referred to as 'conventional memory' or 'base memory'. This 640 kilobyte of read/write memory provides storage for user programs and data as well as regions reserved for DOS and BIOS use.

Upper memory area

The remaining memory within the 1 megabyte direct addressing space (i.e. that which extends from A0000 to FFFFF is referred to as the 'upper memory area' (UMA)). The UMA itself is divided into various regions (depending upon the machine's configuration) including that which provides storage used by video adapters.

Video RAM

As its name implies, video RAM is associated with the video/graphics adapter. The video RAM occupies the lower part of the upper memory area and its precise configuration will depend on the type of adapter fitted.

Extended memory

Memory beyond the basic 1 megabyte direct addressing space is referred to as 'extended memory'. This memory can be accessed by a 286 or later CPU operating in 'protected mode'. In this mode, the CPU is able to generate 24-bit addresses (instead of the real mode's 20-bit addresses) by multiplying the segment register contents by 256 (instead of 16). This scheme allows the CPU to access addresses ranging from 000000 to FFFFFF (a total of 16 megabyte). When the CPU runs in protected mode, a program can only access the designated region of memory. A processor exception will occur if an attempt is made to write to a region of memory that is outside the currently allocated block.

Expanded memory

Expanded memory was originally developed for machines based on 8088 and 8086 CPUs which could not take advantage of the protected mode provide by the 286 and later processors. Expanded memory is accessed through a 64 kilobyte 'page frame' located within the upper memory area. This page frame acts as a window into a much larger area of memory.

Expanded memory systems are based on a standard developed by three manufacturers (Lotus, Intel and Microsoft). This 'LIM standard' is also known as the Expanded Memory Standard (EMS). In order to make use of EMS, a special expanded memory driver is required. EMS has largely been superseded by the advanced memory management facilities provided by the 386 and 486 CPU chips.

CMOS RAM

The PC-AT and later machine's CMOS memory is 64 byte of battery-backed memory contained within the real-time clock chip (a Motorola MC146818). Sixteen byte of this memory are used to retain the real-time clock settings (date and time information), whilst the remainder contains important information on the configuration of the system. When the CMOS battery fails or when power is inadvertently removed from the real-time clock chip, all data becomes invalid and the set-up program has to be used to restore the settings of the system. This can be a real problem *unless* you know what the settings should be!

The organization of the CMOS memory is shown in Table 1.12 (note that locations marked 'reserved' may have different functions in some non-IBM systems).

BIOS ROM

The BIOS ROM is programmed during manufacture. The programming data is supplied to the semiconductor manufacturer by the BIOS originator. This process is cost-effective for large-scale production. However, programming of the ROM is irreversible; once programmed, devices cannot be erased in preparation for fresh programming. Hence, the only way of upgrading the BIOS is to remove and discard the existing chips and replace them with new ones. This procedure is fraught with problems, not least of which is compatability of the BIOS upgrade with existing DOS software.

The BIOS ROM invariably occupies the last 64 or 128 kilobytes of memory (from F0000 to FFFFF or E0000 to FFFFF, respectively). It is normally based on two chips: one for the odd addresses and one for the even addresses.

BIOS variations

Several manufacturers (e.g. Compaq) have produced ROM BIOS code for use in their own equipment. This code must, of course, be compatible with IBM's BIOS code. Several other companies (e.g. American Megatrends (AMI), Award Software, and Phoenix Software) have developed generic versions of the BIOS code which have been incorporated into numerous clones and compatibles.

Table 1.12 CMOS memory organization

Offset (hex.)	*Contents*
00	Seconds
01	Seconds alarm
02	Minutes
03	Minutes alarm
04	Hours
05	Hours alarm
06	Day of the week
07	Day of the month
08	Month
09	Year
0A	Status register A
0B	Status register B
0C	Status register C
0D	Status register D
0E	Diagnostic status byte
0F	Shutdown status byte
10	Floppy disk type (drives A and B)
11	Reserved
12	Fixed disk type (drives 0 and 1)
13	Reserved
14	Equipment byte
15	Base memory (low byte)
16	Base memory (high byte)
17	Extended memory (low byte)
18	Extended memory (high byte)
19	Hard disk 0 extended type
1A	Hard disk 1 extended type
1B–2D	Reserved
2E–2F	Check-sum for bytes 10 to 1F
30	Actual extended memory (low byte)
31	Actual extended memory (high byte)
32	Date century byte (in BCD format)
33–3F	Reserved

There are, of course, minor differences between these BIOS versions. Notably these exist within the power-on self-test (POST), the set-up routines, and the range of hard disk types supported.

System board RAM

The PC system board's read/write memory provides storage for the DOS and BIOS as well as transient user programs and data. In addition, read/write memory is also used to store data which is displayed on the screen. Depending upon the nature of the applications software, this memory may be either character mapped (text) or bit-mapped (graphics). The former technique involves dividing the screen into a number of character-sized cells.

In text mode, each displayed character cell requires one byte of memory for storage of the character plus a further byte for 'attributes'. Hence a screen having 80 columns × 25 lines (corresponding to the IBM's basic text mode) will require 2000 byte: 1000 for storage of the characters and a further 1000 for the attributes.

Bit-mapped graphic displays require a very much larger amount of storage. Each display pixel is mapped to a particular bit in memory and the bit may be a 1 or a 0, depending upon whether it is to be light or dark. Where a colour display is to be produced, several colour planes must be implemented, and consequently an even larger amount of memory is required. A VGA graphics adapter, for example, may be fitted with up to 1 megabyte of video RAM.

In modern machines, DIL packaged DRAM devices have been replaced by memory modules. These units are small cards which usually contain surface-mounted DRAM chips that simply plug into the system board.

Single in-line memory modules (SIMMs) are available in various sizes including 256 kilobyte 1, 4, 8 and 16 megabyte. The most common type of SIMM (which was first used in the IBM XT-286 machine) has nine DRAM devices and it locates with a 30-pin header. Four of these 256 kilobyte modules are required to populate a full 1 megabyte memory map.

The integrity of stored data is checked by providing an additional 'parity bit' for each address location. This bit is either set or reset according to whether the number of '1's present within the byte are even or odd (i.e. 'even parity' and 'odd parity'). Parity bits are automatically written to memory during a memory write cycle and read from memory during a memory read cycle. A non-maskable interrupt (NMI) is generated if a parity error is detected, and thus users are notified if RAM faults develop during normal system operation.

PC memory allocation

The allocation of memory space within a PC can be usefully illustrated by means of a 'memory map'. An 8086 microprocessor can address any one of 1 048 576 different memory locations with its 20 address lines. It thus has a memory which ranges from 00000 (the lowest address) to FFFFF (the highest address). We can illustrate the use of memory using a diagram known as a 'memory map'. Figure 1.25 shows a representative memory map for a PC.

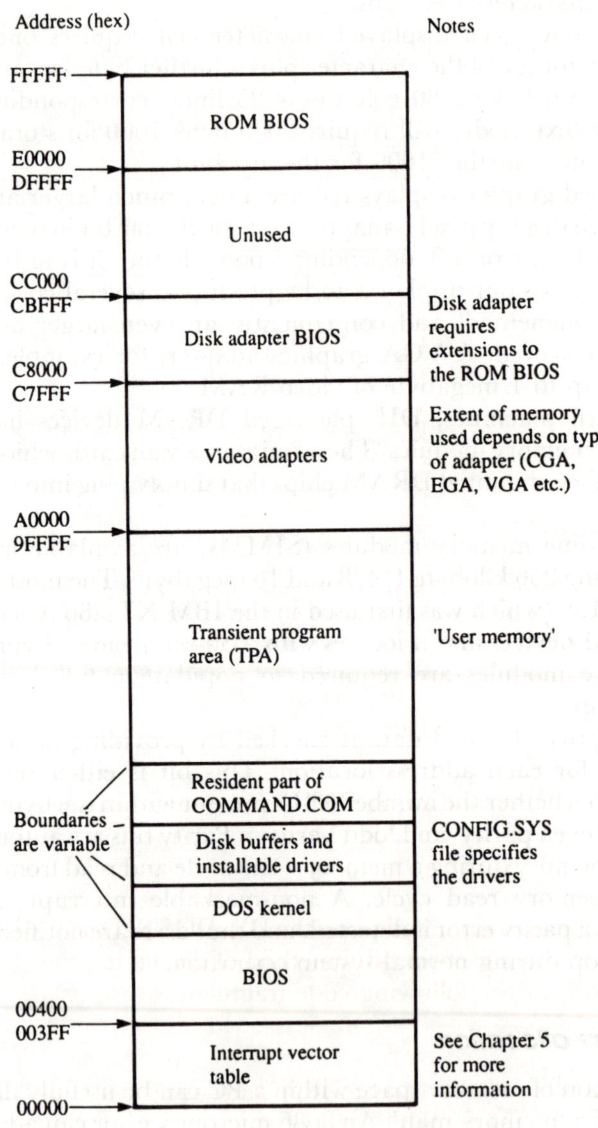
Address (hex)
Notes
FFFFF
ROM BIOS
E0000
DFFFF
Unused
CC000
CBFFF
Disk adapter BIOS
Disk adapter requires extensions to the ROM BIOS
C8000
C7FFF
Video adapters
Extent of memory used depends on type of adapter (CGA, EGA, VGA etc.)
A0000
9FFFF
Transient program area (TPA)
'User memory'
Resident part of COMMAND.COM
Boundaries are variable
Disk buffers and installable drivers
CONFIG.SYS file specifies the drivers
DOS kernel
BIOS
00400
003FF
Interrupt vector table
See Chapter 5 for more information
00000

Figure 1.25 *Representative PC memory map*

Useful PC memory locations

A number of memory locations can be useful in determining the current state of a PC or PC-compatible microcomputer. You can display the contents of these memory locations (summarized in Table 1.13) using the MS-DOS DEBUG utility or using a short routine written in QuickBASIC.

Table 1.13 Useful RAM locations

Address (hex.)	*No. of bytes*	*Function*
0410	2	Installed equipment list
0413	2	Usable base memory
0417	2	Keyboard status
043E	1	Disk calibration (see Chapter 10)
043F	1	Disk drive motor status (see Chapter 10)
0440	1	Drive motor count (see Chapter 10)
0441	2	Disk status (see Chapter 10)
0442	2	Disk controller status (see Chapter 10)
0449	1	Current video mode (see Chapter 13)
044A	2	Current screen column width (see Chapter 13)
046C	4	Master clock count (incremented by 1 on each clock 'tick')
0472	2	Set to 1234 hex. during a keyboard re-boot (this requires <CTRL–ALT–DEL> keys)
0500	1	Screen print byte (00 indicates normal ready status, 01 indicates that a screen print is in operation, FF indicates that an error has occurred during the screen print operation)

As an example, the following DEBUG command can be used to display the contents of 10 byte of RAM starting at memory location 0410:

```
D0:0410L0A
```

(NB: the equivalent command using the DR-DOS SID utility is D0:410, 419.) A rather more user-friendly method of displaying the contents of RAM is shown in the following code fragment:

```
DEF SEG = 0
CLS
INPUT "Start address (in hex)"; address$
address$ = "&H" + address$
address = VAL(address$)
INPUT "Number of bytes to display"; number
PRINT
PRINT "Address", "Byte"
PRINT "(hex)", "(hex)"
PRINT
FOR i% = 0 TO number - 1
```

```
    v = PEEK(address + i%)
    PRINT HEX$(address + i%), HEX$(v)
NEXT i%
PRINT
END
```

This QuickBASIC program prompts the user for a start address (expressed in hexadecimal) and the number of bytes to display. Figure 1.26 shows a typical example of running the program on a modern 486-based PC. The program has been used to display the contents of 10 bytes of RAM from address 0410 onwards.

```
Start address (in hex.)?        410
Number of bytes to display?      10

Address          Byte
(hex.)           (hex.)

410              63
411              44
412              BF
413              80
414              2
415              0
416              18
417              20
418              0
419              0
```

Figure 1.26 *Output produced by the memory dump program*

```
Address  (hex.):        0411            0410
Contents (hex.):       4     2         2     D
         (binary):   0100  0010      0010  1101
                     |        |      |        |
Bit position:        15       8      7        0
```

Figure 1.27 *Deciphering the equipment list*

The installed equipment list

The machine's Installed Equipment List (see Table 1.14) can tell you what hardware devices are currently installed in your system. The equipment list is held in the word (2 byte) starting address 0410. Figure 1.27 and Table 1.15 show you how to decipher that Equipment List word.

Amount of base memory available

The amount of usable base memory can be determined from the 2 byte starting at address 0413. The extent of memory is found by simply

Table 1.14 Equipment list word at address 0410

Bit number	*Meaning*		
0	Set if disk drives are present		
1	Unused		
2 and 3	System Board RAM size:		
	Bit 3	*Bit 2*	*RAM size* (kilobyte)
	0	0	16
	0	1	32
	1	1	64/256
	(NB: on modern systems this coding does not apply)		
4 and 5	Initial video mode:		
	Bit 5	*Bit 4*	*Mode*
	0	1	40 column colour
	1	0	80 column colour
	1	1	80 column monochrome
6 and 7	Number of disk drives plus 1:		
	Bit 7	*Bit 6*	*No. of drives*
	0	0	1
	0	1	2
	1	0	3
	1	1	4
8	Reset if DMA chip installed (standard)		
9 to 11	Number of serial ports installed		
12	Set if an IBM Games Adapter is installed		
13	Set if a serial printer is installed		
14 and 15	Number of printers installed		

adding the binary weighted values of each set bit position. Figure 1.28 and Table 1.16 show how this works.

BIOS ROM release date

The BIOS ROM release date and machine identification can be found by examining the area of read-only memory extending between absolute locations FFFF5 and FFFFC. The ROM release information (not found in all compatibles) is presented in American date format

Table 1.15 Example equipment list

Bit position	*Status*	*Comment*
0	1 = set	Disk drives are present
1	1 = set	This bit is not used
2	0 = reset	These bits have no meaning with systems having greater than 256 kilobyte RAM
3	0 = reset	
4	0 = reset	Initial video mode is 80 column colour
5	1 = set	
6	1 = set	Two disk drives installed
7	0 = reset	
8	0 = reset	DMA controller fitted (standard)
9	0 = reset	Two serial ports installed
10	1 = set	
11	0 = reset	
12	0 = reset	No IBM Games Adapter installed
13	0 = reset	No serial printer installed
14	1 = set	One printer attached
15	0 = reset	

Table 1.16 Determining the base memory

Bit position	*Status*	*Value** (kilobyte)
9	Set	512
8	Reset	256
7	Set	128
6	Reset	64
5	Reset	32
4	Reset	16
3	Reset	8
2	Reset	4
1	Reset	2
0	Reset	1

*Adding together the values associated with each of the set bits gives (512 + 128) = 640 kilobyte.

```
Address  (hex.):        0414            0413
Contents (hex.):       0    2          8    0
         (binary):  0000  0010      1000  0000
                     |       |       |       |
Bit position:       15       8       7       0
```

Figure 1.28 *Determining the usable base memory*

Table 1.17 IBM major ROM release dates

ROM date	*PC version*
04/24/81	Original PC
10/19/81	Revised PC
08/16/82	Original XT
10/27/82	PC upgrade to XT BIOS level
11/08/82	PC–XT
06/01/83	Original PC Junior
01/10/84	Original AT
06/10/85	Revised PC–AT
09/13/85	PC Convertible
11/15/85	Revised PC–AT
01/10/86	Revised PC–XT
04/10/86	XT 286
06/26/86	PS/2 Model 25
09/02/86	PS/2 Model 30
12/12/86	Revised PS/2 Model 30
02/13/87	PS/2 Models 50 and 60
12/05/87	Revised PS/2 Model 30
03/30/87	PS/2 Model 80 (16 MHz)
10/07/87	PS/2 Model 80 (20 MHz)
01/29/88	PS/2 Model 70
12/01/89	PS/1
11/21/89	PS/2 Model 80 (25 MHz)

using ASCII characters. Various ROM release dates for various IBM models are shown in Table 1.17.

Machine identification byte

The type of machine (whether PC-XT, AT, etc.) is encoded in the form of an identification (ID) byte which is stored at address FFFFE. Table 1.18 gives the ID bytes for each member of the PC family (non-IBM machines may have ID bytes which differ from those listed).

The ROM release date and machine ID byte can be displayed by using the MS-DOS DEBUG utility or by using the simple QuickBASIC program shown below:

Table 1.18 ID bytes for various IBM machines

ID byte (hex.)	*Machine*
F8	PS/2 Models 35, 40, 65, 70, 80 and 90 (386 and 486 CPU)
F9	PC Convertible
FA	PS/2 Models 25 and 30 (8086 CPU)
FB	PC-XT (revised versions, post-1986)
FC	AT, PS/2 Models 50 and 60 (286 CPU)
FD	PC Junior
FE	XT and Portable PC
FF	Original PC

```
DEF SEG = &HFFF0
CLS
PRINT "ROM address", "Byte", "ASCII"
PRINT "(hex)", "(hex)"
PRINT
FOR i% = &HF0 TO &HFF
  v = PEEK(i%)
  PRINT HEX$(i%), HEX$(v), ;
  IF v >31 AND v <128 THEN
    PRINT CHR$(v)
  ELSE
    PRINT " "
  END IF
NEXT i%
PRINT
END
```

An example of the output produced by this program is shown in Table 1.19 (the machine in question has an ID byte of FC and a ROM release date of 06/06/92).

Power supplies

A fully populated 386 system motherboard (including a 80387 coprocessor) requires approximately 5 A and 2 A from the +5 V and +12 V rails, respectively. A VGA graphics adapter and two standard floppy drives will demand an additional 2.4 A and 1.5 A from the +5V and +12 V rails, respectively. A cooling fan will require a further 0.3 A or so from the +12 V rail. The total load on the system power supply is thus 7.4 A from the +5 V rail and 4.1 A from the +12 V rail. With a standard XT power supply, reserves of only 7.6 A from the +5 V supply and a mere 400 mA from the +12 V supply will be available.

Modern 486 and Pentium based systems use significantly higher rated power supplies. However, problems may still arise when systems are upgraded or expanded as this may cause excessive loading on the power

Table 1.19 Output produced by the BIOS ROM dump program

ROM address (hex.)	*Byte* (hex.)	*ASCII*	*Comment*
F0	EA		
F1	5B	[	
F2	E0		
F3	0		
F4	F0		
F5	30	0	ROM release date 06/06/92
F6	36	6	
F7	2F	/	
F8	30	0	
F9	36	6	
FA	2F	/	
FB	39	9	
FC	32	2	
FD	0		
FE	FC		Machine ID byte
FF	0		

supply. Under marginal conditions the system will *appear* to operate satisfactorily but it may crash or lock-up at some later time when the system temperature builds up or when one or more of the power rail voltages momentarily falls outside its tolerance limits.

PC video standards

The video capability of a PC depends not only on the display used but also on the type of 'graphics adapter' fitted. Most PCs will operate in a number of video modes which can be selected from DOS or from within an application.

The earliest PC display standards were those associated with the Monochrome Display Adapter (MDA) and Colour Graphics Adapter (CGA). Both of these standards are now obsolete, although they are both emulated in a number of laptop PCs that use LCD displays.

MDA and CGA were followed by a number of other much enhanced graphics standards. These include Enhanced Graphics Adapter (EGA), Multi-Colour Graphics Array (MCGA), Video Graphics Array (VGA), and the 8514 standard used on IBM PS/2 machines.

The EGA standard is fast becoming obsolete and all new PC systems are supplied with either VGA, 'super VGA', or displays which conform to IBM's 8514 standard. The characteristics of the most commonly used graphics standards are summarized in Table 1.20.

Table 1.20 Display adapter summary

Display standard	*Approx. year of introduction*	*Text capability (columns × lines)*	*Graphics capability (horizontal × vertical pixels)*
MDA	1981	80 × 25 monochrome	None
HGA	1982	80 × 25 monochrome	720 × 320 monochrome
CGA	1983	80 × 25 in 16 colours	320 × 200 in two sets of 4 colours
EGA	1984	80 × 40 in 16 colours	640 × 350 in 16 colours
MCGA	1987	80 × 30 in 16 colours	640 × 480 in 2 colours; 320 × 200 in 256 colours
8514/A (PS/2)	1987	80 × 60 in 16 colours	1024 × 768 in 256 colours
VGA	1987	80 × 50 in 16 colours	640 × 480 in 16 colours; 320 × 200 in 256 colours
SVGA (XGA)	1991	132 × 60 in 16 colours	1024 × 768 in 256 colours; 640 × 480 in 65 536 colours

Video modes

It is important to realize at the outset that graphics adapters normally operate in one of several different modes. A VGA card will, for example, operate in 'text mode' using either 80 or 40 columns, and in 'graphics mode' using 4, 16 or 256 colours.

The graphics adapter contains one or more VLSI devices that organize the data which produces the screen display. You should recall that a conventional cathode ray tube (CRT) display is essentially a serial device (screen data is built up using a beam of electrons which continuously scans the screen). Hence the graphics adapter must store the screen image while the scanning process takes place.

To determine the current video mode, you can simply read the machine's video mode byte stored in RAM at address 0449. This byte indicates the current video mode (using the hex. values shown in Table 1.21). You can examine the byte at this address using the DEBUG command D0:0449L1 (the equivalent SID command is D0:449,44A).

Graphics adapter memory

The amount of memory required to display a screen in text mode is determined by the number of character columns and lines and also by

Table 1.21 Video display modes and graphics adapter standards

Mode	*Display type*	*No. of colours*	*Screen resolution* (note1)*	*Display adapters supporting this mode* MDA	CGA	EGA	MCGA	VGA	HGA‡
00	Text	16	40 × 25		✓	✓	✓	✓	✓
01	Text	16	40 × 25		✓	✓	✓	✓	✓
02	Text	16	80 × 25		✓	✓	✓	✓	✓
03	Text	16	80 × 25		✓	✓	✓	✓	✓
04	Graphics	4	320 × 200		✓	✓	✓	✓	✓
05	Graphics	4	320 × 200		✓	✓	✓	✓	✓
06	Graphics	2	640 × 200		✓	✓	✓	✓	✓
07	Text	Mono	80 × 25	✓		✓		✓	✓
08	Graphics	16	160 × 200†						
09	Graphics	16	320 × 200†						
0A	Graphics	4	640 × 200†						
0B	§								
0C	§								
0D	Graphics	16	320 × 200			✓		✓	
0E	Graphics	16	640 × 200			✓		✓	
0F	Graphics	Mono	640 × 350			✓		✓	
10	Graphics	16	640 × 350			✓		✓	
11	Graphics	2	640 × 480				✓	✓	
12	Graphics	16	640 × 480					✓	
13	Graphics	256	320 × 200				✓	✓	

*Resolutions are quoted in (columns × lines) for text displays and (horizontal × vertical) pixels for graphics displays.
†Applies only to the PC Junior.
‡The Hercules Graphics Adapter card combines the graphics (but *not* the colour) capabilities of the CGA adapter with the high-quality text display of the MDA adapter.
§Reserved mode.

the number of colours displayed. In modes 0 to 6 and 8, a total of 16 kilobyte is reserved for display memory, whilst in mode 7 (monochrome 80 × 25 characters) the requirement is for only 4 kilobyte (colours are not displayed).

In modes 0 to 3, less than 16 kilobyte is used by the screen at any one time. For these modes, the available memory is divided into pages. Note that only one page can be displayed at any particular time. Displayed pages are numbered 0 to 7 in modes 0 and 1 and 0 to 3 in modes 2 and 3.

The extent of display memory required in a graphics mode depends upon the number of pixels displayed (horizontal × vertical) and also on the number of colours displayed. Provided that a display adapter has sufficient RAM fitted, the concept of screen pages also applies to graphics modes. Again, it is only possible to display one page at a time.

Colour

The basic 16-colour palette for a PC used in the vast majority of DOS applications is based on a 4-bit 'intensity plus RGB' code (see Table 1.22). This simple method generates colours by switching on and off the individual red, green and blue electron beams. The intensity signal simply serves to brighten up or darken the display at the particular screen location. The result is the 16 basic PC colours that we have all grown to know and love!

Table 1.22 The PC's 16-colour palette

Hex. code	*Binary code**				*Colour produced*
	I	*R*	*G*	*B*	
00	0	0	0	0	Black
01	0	0	0	1	Blue
02	0	0	1	0	Green
03	0	0	1	1	Cyan
04	0	1	0	0	Red
05	0	1	0	1	Magenta
06	0	1	1	0	Yellow
07	0	1	1	1	White
08	1	0	0	0	Grey
09	1	0	0	1	Bright blue
0A	1	0	1	0	Bright green
0B	1	0	1	1	Bright cyan
0C	1	1	0	0	Bright red
0D	1	1	0	1	Bright magenta
0E	1	1	1	0	Bright yellow
0F	1	1	1	1	Bright white

*I, intensity; R, red; B, blue; G, green.

The 16-colour palette is adequate for most text applications; however, to produce more intermediate shades of colour we need a larger palette. One way of doing this is to make use of a 6-bit code where *each* of the three basic colours (red, green and blue) is represented by 2 bits (one corresponding to bright and the other to normal). This allows each colour to have four levels and produces 64 possible colour combinations.

A better method (which generates a virtually unlimited colour palette) is to use 'analogue RGB' rather than 'digital RGB' signals. In this system (used in VGA, SVGA and XGA), the three basic colour signals (red, green and blue) are each represented by analogue voltages in the range 0–0.7 V (at the video connector). The number of colours displayed using such an arrangement depends upon the number of bits used to respresent the intensity of each colour before its conversion to an analogue signal.

Floppy disk drives

Floppy disk drives provide low-cost storage for data and programs. The first mini-floppy disk drive to gain popularity was the SA400 from Shugart Associates. This 35 (or 40) track drive provides an unformatted capacity of 125 kilobyte in single density (FM) or 250 kilobyte in double density (MFM), the data transfer rates being 125 and 250 kilobit/s, respectively. The drive provides an average latency of 100 ms, a stepping time (track to track) of 20 ms and an average access time of 280 ms. The drive requires d.c. supplies of +12 V ±5% at 0.9 A (typical) and +5 V ±5% at 0.5 A (typical).

In recent years, 'half-height' 5.25 and 3.5 inch drives have also become increasingly popular. Storage capacities of 360 kilobyte, 720 kilobyte, 1.44 megabyte and 2.88 megabyte are commonplace. The characteristics of various floppy drives are summarized in Table 1.23.

IBM-compatible floppy disk drives operate at 300 r.p.m. and use an 80-track format with 135 tracks per inch. The standard data transfer rate is 125 kilobit/s in single density (FM) and 250 kilobit/s in double density (MFM), while the unformatted storage capacity is 250 and 500 kilobyte, respectively.

Table 1.23 Characteristics of some typical floppy drives

Manufacturer	*Drive type*	*Media size (inch)*	*No. of tracks per side*	*No. of sides*	*Tracks per inch*
Mitsubishi	M4851	5.25	40	2	48
Mitsubishi	M4852	5.25	80	2	96
Mitsubishi	M4853	5.25	80	2	96
Mitsubishi	MF351	3.5	80	1	135
Shugart	SA300	3.5	80	1	135
Shugart	SA455	5.25	40	2	48
Shugart	SA465	5.25	80	2	96
Teac	FD-55A	5.25	40	1	48
Teac	FD-55B	5.25	40	2	48
Teac	FD-55E	5.25	80	1	96
Teac	FD-55F	5.25	80	2	96
Teac	FD-235F	3.5	80	2	135

Hard disk drives

Like floppy disks, the data stored on a hard disk takes the form of a magnetic pattern stored in the oxide coated surface of a disk. Unlike floppy disks, hard disk drives are sealed in order to prevent the ingress of dust, smoke and dirt particles. This is important since hard disks work to much finer tolerances (track spacing, etc.) than do floppy drives. Furthermore, the read/write heads of a hard disk 'fly' above the surface of the disk when the platters are turning. Due to the high speed of rotation (typically 3600 r.p.m.) it is essential that none of the read/write heads comes into direct contact with the area of the disk surface used for data storage.

A typical 120 megabyte IDE drive has two platters which provide four data surfaces. The drive is thus fitted with four read/write heads (one for each data surface). The read/write heads are all operated from the same 'voice coil' actuator so that they step in and out together across the surface of the disk. In addition, the innermost cylinder is designated as a 'landing zone'. No data is stored in this region and thus it provides a safe place for the heads to 'land' and make contact with the disk surface.

When the drive is static or coming up to speed, the heads remain in the landing zone. When sufficient rotational speed has been achieved, the heads leave the surface of the disk and are then stepped across to the active part of the disk surface where reading and writing takes place. Finally, when the disk becomes inactive, the motor ceases to rotate and the heads return to the landing zone where they are 'parked'.

Hard drive types

A variety of different hard disk types are supported by personal computers including ST506 (MFM), RLL, ESDI, and SCSI types. Since they all have different interfacing requirements, it is important to know which type you are dealing with. Typical data transfer rates and capacities for various types of hard disk drive are shown in Table 1.24.

The ST506 interface standard

The Shugart ST506 interface standard was popular in the late 1970s and early 1980s and it became the hard disk standard to be used in the first generations of IBM PCs that were supplied with hard disks. The ST506 uses 'modified frequency modulation' (MFM) to digitally encode its data. This method of recording digital information has now been superseded by 'run-length-limited' (RLL) encoding which uses the available disk space more efficiently. Note that MFM ST506 drives are

Table 1.24 Typical data transfer rates and capacities for various types of hard disk drive

Interface	*Data transfer rate* (megabit/s)	*Capacity* (megabyte)
ST506	0.1–1	10–20
ESDI	1.2–1.8	40–320
SCSI	1–2	80–680
IDE	1–7	80–340

normally formatted with 17 sectors per track, whilst RLL drives generally use 26 sectors per track.

When RLL encoding is used, a limit is imposed on the number of consecutive 0 bits that can be recorded before a 1 bit is included. '2,7 RLL' uses strings of between 2 and 7 consecutive 0 bits, whilst '3,9 RLL' is based on sequences of between 3 and 9 consecutive 0 bits before a 1 bit is inserted. 2,7 RLL and 3,9 RLL offer a 50% and 100% increase in drive capacity, respectively.

ST506 drives, whether MFM or RLL types, require a complex hard disk controller card. Drives are connected to the card by means of two separate ribbon cables: a 34-way cable for control signals and a 20-way cable for data.

ESDI drives

The 'enhanced small device interface' (ESDI) is an updated and improved standard based on the original ST506 interface. ESDI was first introduced by Maxtor in 1983 and its BIOS code is generally software-compatible with the earlier standard. Note, however, that most ESDI drives are formatted to 32 sectors per track. Like ST506 drives, ESDI units require the services of a separate hard disk controller card. Both the ESDI and ST506 interface standards support up to four physical drives, although usually no more than two drives are actually fitted.

IDE drives

Integrated Drive Electronics (IDE) drives are designed to interface very easily with the ISA bus. The interface can either make use of a simple adapter card (without the complex controller associated with ST506 and ESDI drives) or can be connected directly to the system motherboard where the requisite 40-way IDC connector has been made available by the manufacturer.

In either case, the 40-way ISA bus extension is sometimes known as an

'AT attachment'. This system interface is simply a subset of the standard ISA bus signals and it can support up to two IDE drives in a 'daisy-chain' fashion (i.e. similar to that used for floppy disk drives). This makes IDE drives extremely cost-effective since they dispense with the complex hard disk controller/adapter card required by their predecessors.

IDE drives are 'low-level formatted' with a pattern of tracks and sectors already in-place when they reach you. This allows them to be formatted much more efficiently; the actual *physical* layout of the disk is hidden from BIOS which only sees the *logical* format presented to it by the integrated electronics. This means that the disk can have a much larger number of sectors on the outer tracks than on the inner tracks. Consequently, a much greater proportion of the disk space is available for data storage (see Plate 1.2).

Plate 1.2 *Internal cabling of a modern tower system*

Local bus standards

Modern processors (such as the 486 and Pentium) are capable of operation at fast clock rates. Unfortunately, the PC's original bus (based on Industry Standard Architecture) is severely limited by virtue of its speed of operation (8 MHz) and limited bus width (16 bits). Windows and other modern applications packages (particularly those that are graphically intensive) require a much faster bus interface for optimum performance. For this reason, several manufacturers and other interested parties have developed improved local bus standards that can be instrumental in eliminating the bottleneck between processor and graphics peripherals.

The VL Bus (or VESA bus) was established by the Video Electronic Standards Association as a means of improving the graphics performance of modern PCs. The bus is useful in any graphics intensive environment (e.g. Windows) as it not only provides an extension to the standard 16-bit ISA bus so that 32-bit data transfers can take place, but is also capable of operating at speeds up to 33 MHz.

Intel's Peripheral Component Interconnect (PCI) bus is a somewhat more radical approach to the problem of making modern PCs run at the fastest possible speed. PCI offers a full 32-bit data path in order to take advantage of the full processing capability of modern 32-bit processors (such as the Pentium). PCI was introduced in 1992 as an intermediary (or 'mezzanine') bus in systems in which older ISA bus cards are required to co-exist with more modern PCI peripheral devices by means of a PCI-ISA bridge.

PCI requires a dedicated chip set for bus management. This chip set comprises a Data Patch Unit (DPU) and a Cache and DRAM Controller (CDC). The DPU manages the interface between the CPU's data bus and the PCI bus and system memory. The CDC provides a similar function for the interface between the CPU's address bus and the PCI bus and system memory. Note that, in a PCI system, the DPU and CDC are the only devices that are directly connected to the system's memory. Furthermore, the CDC can handle concurrent memory access requests, so the graphics adapter and processor can carry out memory transfers *at the same time*.

PCI supports both 5 V as well as the newer low-power 3.3 V chips, as well as an energy saving mode in which the bus can be powered down. PCI offers a measure of processor-independence and offers a 'plug-and-play' solution to PC bus design that is likely to see continuing support in the future.

2 PC expansion systems

The availability of a standard expansion bus system within the PC environment must surely be the single most crucial factor in harnessing the power of the machine. Not only does the expansion bus provide a means of adating a machine for a particular range of applications but it also provides a means of attaching a wide range of external hardware devices. Happily, a large number of manufacturers have recognized this fact and have developed expansion products specifically for control, data logging, and instrumentation applications.

An expansion bus should provide access to the full range of system bus signals. These signals can be divided into the following categories:

1 Address bus lines
2 Data bus lines
3 Read and write control signals
4 Interrupt request signals
5 DMA request and DMA acknowledge signals
6 Miscellaneous control signals
7 Clock signals
8 Power rails

PC expansion can be achieved by means of cards connected to the PC bus (using connectors fitted directly to the system motherboard) or by means of a *backplane bus* system (such as STE or VME). This chapter considers both methods of expansion, provides details of a variety of current PC bus products, and contains specific recommendations for those involved with the design and development of PC expansion cards.

PC expansion architecture

The most obvious method of expanding the PC bus is simply to provide a number of access points to the bus on the system motherboard. This approach was followed by IBM (and countless manufacturers of clones and compatibles) as a means of connecting essential items of peripheral hardware (such as displays and disk drives) via controllers fitted to *adapter* (or *option*) *cards*. This same method of connection can also be employed for more specialized applications such as analogue data acquisition, IEEE-488 bus control, etc.

Two basic standards are employed in conventional PC expansion bus schemes. The original and most widely used standard is based on *Industry Standard Architecture* (ISA) but is more commonly known simply as the *PC Expansion Bus*. This expansion scheme employes either one or two direct edge connectors.

The first connector (62-way) provides access to the 8-bit data bus and the majority of control bus signals and power rails whilst the second connector (36-way) gives access to the remaining data bus lines together with some additional control bus signals. Applications which require only an 8-bit data path and a subset of the PC's standard control signals can make use of *only* the first connector. Applications which require access to a full 16-bit data path (not available on the early PC and XT machines) must make use of *both* connectors.

With the advent of PS/2, a more advanced expansion scheme has become available. This expansion standard is known as *Micro Channel Architecture* (MCA) and it provides access to the 16-bit data bus in the IBM PS/2 Models 50 and 60 whereas access to a full 32-bit data bus is available in the Model 80 (fitted with an 80386 CPU).

An important advantage of MCA is that it permits data transfer at significantly faster rates than is possible with ISA. In fairness, the increase in data transfer rate may be unimportant in many applications and also tends to vary somewhat from machine to machine. As a rough guide, when a standard AT machine is compared with a PS/2 Model 50, data transfer rates can be expected to increase by around 25 per cent for conventional memory transfers and by 100 per cent (or more) for DMA transfers.

Since MCA interrupt signals are shared between expansion cards, MCA interrupt structure tends to differ from that employed within ISA where interrupt signals tend to remain exclusive to a particular expansion card. More importantly, MCA provides a scheme of bus arbitration in order to decide which of the 'feature cards' has rights to exercise control of the MCA bus at any particular time. The arbitration mechanism provides for up to 15 bus masters, each one able to exercise control of the bus. As a further bonus, MCA provides an auxiliary video connector and programmable option configuration to relieve the tedium of setting DIP switches on system boards and expansion cards.

PC expansion bus

The PC Expansion Bus is based upon a number of *expansion slots*, each of which is fitted with a 62-way direct edge connector together with an optional subsidiary 36-way direct edge connector. Expansion (or option) cards may be designed to connect only to the 62-way connector or may, alternatively, mate with both the 62-way and 36-way connectors. Since only the 62-way connector was fitted on early machines (which had an 8-bit data bus), cards designed for use with this connector are sometimes known as '8-bit expansion cards' or 'PC/XT expansion cards'. The AT machine, however, provides access to a full 16-bit data bus together with additional control signals and hence requires the additional 36-way connector. Cards which are designed to make use of *both* connectors are generally known as '16-bit expansion cards' or 'AT expansion cards'.

The original PC was fitted with only five expansion slots (spaced approximately 25 mm apart). The standard XT provided a further three slots to make a total of eight (spaced approximately 19 mm apart). Some cards, particularly those providing hard disk storage, require the width occupied by two expansion slot positions on the XT. This is unfortunate, particularly where the number of free slots may be at a premium!

All of the XT expansion slots provide identical signals with one

Plate 2.1 *Expansion bus cards fitted in a modern AT compatible PC*

notable exception; the slot nearest to the power supply was employed in a particular IBM configuration (the IBM 3270 PC) to accept a Keyboard/Timer Adapter. This particular configuration employs a dedicated 'card select' signal (B8 on the connector) which is required by the system motherboard. Other cards which *will* operate in this position include the IBM 3270 Asynchronous Communications Adapter.

Like the XT, the standard AT also provides eight expansion slots. Six of these slots are fitted with two connectors (62-way and 36-way) while two positions (slots 1 and 7) only have 62-way connectors. Slot positions 1 and 7 are designed to accept earlier 8-bit expansion cards which make use of the maximum allowable height throughout their length. If a 36-way connector *had* been fitted to the system motherboard, this would foul the lower edge of the card and prevent effective insertion of the card in question.

Finally, it should be noted that boards designed for AT systems (i.e. those specifically designed to take advantage of the availability of the full 16-bit data bus) will offer a considerable speed advantage over those which are based upon the 8-bit PC/XT data bus provided by the original XT expansion connector. In some applications, this speed advantage can be critical.

PC expansion cards

Expansion cards for PC systems tend to vary slightly in their outline and dimensions (see Figure 2.1). However, the *maximum* allowable dimensions for the adapter and expansion cards fitted to PC (and PS/2) equipment are as follows:

		Maximum card dimensions (approx.)					
Standard	*Bus*	*Height*		*Length*		*Width*	
		inches	*mm*	*inches*	*mm*	*inches*	*mm*
ISA	XT	4.2	107	13.3	335	0.5	12.7
ISA	AT	4.8	122	13.2	335	0.5	12.7
MCA	PS/2	3.8	96	13.2	335	0.5	12.7

It is important to note that, although the XT-286 is based on an AT motherboard, it is fitted in an XT enclosure and thus expansion and adapter cards used in this machine must conform with the general height restriction imposed on XT cards (i.e. 4.2 inches maximum). While many expansion card manufacturers are very conscious of this requirement, 4.8-inch height cards are still commonly available and XT

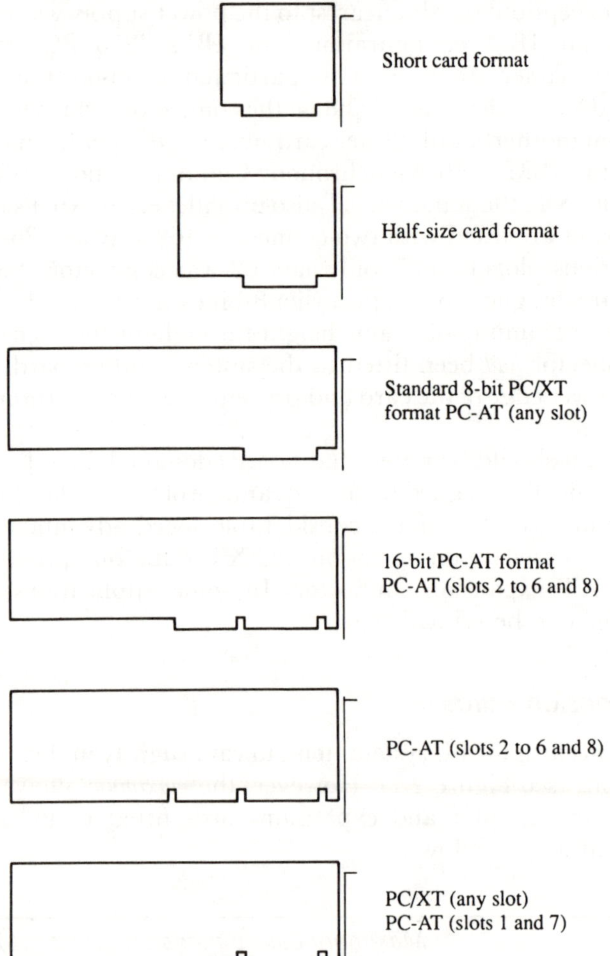

Figure 2.1 *Outlines for various types of PC expansion card*

users should thus exercise some caution when selecting expansion hardware.

Another difficulty is that some XT cards may fail to operate in AT equipment due to interrupt, DMA, addressing or other problems. In fairness, most manufacturers of adapters and expansion cards provide very clear indications of the systems with which their products are compatible. In any event, it is always wise to check with a manufacturer or distributor concerning compatibility of a card with a particular microcomputer system.

With the exception of slot eight in the XT, the position in which an adapter or expansion card is placed *should* be unimportant. In most cases, this does hold true however; in certain circumstances it is worth considering in which slot one should place a card.

The most important factor that should be taken into account is ventilation. Where cards are tightly packed together (particularly where ribbon cables may reduce airflow in the space between expansion cards) it is wise to optimize arrangements for cooling. Boards which are tightly packed with heat-producing components should be located in the positions around which airflow can be expected to be the greatest. This generally applies to the sixth, seventh, or eighth slot in a system fitted with a fan. Furthermore, when introducing a new card to a system, it may be worth re-arranging those cards which are already fitted in order to promote the unimpeded flow of air.

Accessibility of cards is also a point which is well worth considering. This is particularly important when the card in question is a prototype which may require adjustment or alignment when the system is running. The card placed in the first slot is usually very much more accessible than any of the others. Furthermore, measurements can easily be taken from the copper foil side of the board fitted in this slot position. This point is also worth bearing in mind when fault finding becomes necessary. It is, however, important to note that the first slot only provides access to the 8-bit data bus. Where a prototype card makes use of the full 16-bit bus, the card should be fitted in the second slot, rather than the first. The first slot position should then be left vacant in order to permit access to the rear of the PCB, and to allow voltage and waveform measurement to be made.

To avoid noise and glitches on the supply rails, it is usually beneficial to place boards which make large current demands or switch rapidly, in close proximity to the power supply (i.e. in slots six, seven and eight). This precaution can be instrumental in reducing supply borne disturbances (glitches) and can also help to improve overall system integrity and reliability. If, however, effective decoupling precautions have been observed, this precaution will be of minor importance.

Finally, whilst timing is not rarely a critical issue, some advantages can accrue from plaing cards as close to the CPU (and coprocessor) as possible. Expansion memory cards should, therefore, be fitted in slot positions six, seven and eight in preference to positions one, two, and three. In some cases this precaution may be instrumental in improving overall memory access times and avoiding parity errors.

The 62-way expansion bus connector

The 62-way PC expansion bus connector is a direct edge-type fitted to the system motherboard. One side of the connector is referred to as A

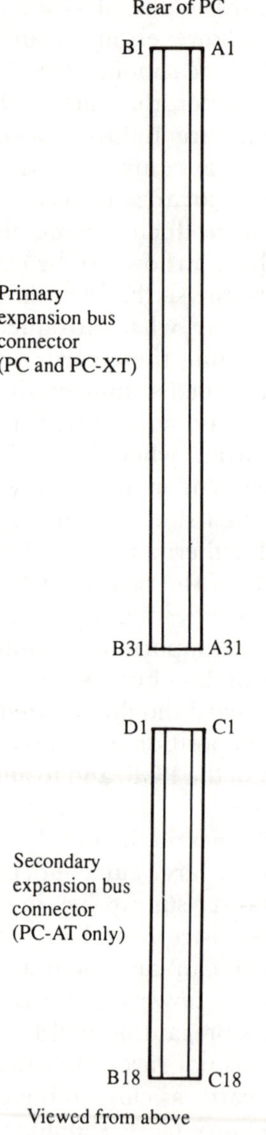

Figure 2.2 *PC expansion bus connectors*

(lines as numbered A1 to A31) while the other is referred to as B (lines are numbered B1 to B31). The address and data bus lines are grouped together on the A-side of the connector while the control bus and power rails occupy the B-side (see Figure 2.2).

It is, however, important to be aware that some early PC expansion bus pin-numbering systems did not use letters A and B to distinguish the two sides of the expansion bus connector. In such cases, odd-numbered lines (1 to 61) formed one side of the connector whilst even numbered lines (2 to 62) formed the other. Here we shall, however, adopt the more commonly used pin-numbering convention described earlier.

The following table describes each of the signals present on the 62-way expansion bus connector:

Pin no.	*Abbreviation*	*Direction*	*Signal*	*Function*
A1	$\overline{\text{IOCHK}}$	I	I/O Channel check	Taken low to indicate a parity error in a memory or I/O device
A2	D7	I/O	Data bus 7	Data bus line
A3	D6	I/O	Data bus 6	Data bus line
A4	D5	I/O	Data line 5	Data bus line
A5	D4	I/O	Data line 4	Data bus line
A6	D3	I/O	Data line 3	Data bus line
A7	D2	I/O	Data line 2	Data bus line
A8	D1	I/O	Data line 1	Data bus line
A9	D0	I/O	Data line 0	Data bus line
A10	$\overline{\text{IOCHRDY}}$	I	I/O channel ready	Pulsed low by a slow memory or I/O device to signal that it is not ready for data transfer
A11	AEN	O	Address enable	Issued by the DMA controller to indicate that a DMA cycle is in progress. Disables port I/O during a DMA operation in which $\overline{\text{IOR}}$ and $\overline{\text{IOW}}$ may be asserted
A12	A19	I/O	Address line 19	Address bus line
A13	A18	I/O	Address line 18	Address bus line
A14	A17	I/O	Address line 17	Address bus line
A15	A16	I/O	Address line 16	Address bus line
A16	A15	I/O	Address line 15	Address bus line
A17	A14	I/O	Address line 14	Address bus line
A18	A13	I/O	Address line 13	Address bus line
A19	A12	I/O	Address line 12	Address bus line
A20	A11	I/O	Address line 11	Address bus line
A21	A10	I/O	Address line 10	Address bus line
A22	A9	I/O	Address line 9	Address bus line
A23	A8	I/O	Address line 8	Address bus line
A24	A7	I/O	Address line 7	Address bus line
A25	A6	I/O	Address line 6	Address bus line
A26	A5	I/O	Address line 5	Address bus line
A27	A4	I/O	Address line 4	Address bus line

Pin no.	*Abbreviation*	*Direction*	*Signal*	*Function*
A28	A3	I/O	Address line 3	Address bus line
A29	A2	I/O	Address line 2	Address bus line
A30	A1	I/O	Address line 1	Address bus line
A31	A0	I/O	Address line 0	Address bus line
B1	GND	n.a.	Ground	Ground/common 0V
B2	RESET	O	Reset	When taken high this signal resets all expansion cards
B3	+5V	n.a.	+5V d.c.	Supply voltage rail
B4	IRQ2	I	Interrupt request level 2	Interrupt request (highest priority)
B5	−5V	n.a.	−5V d.c.	Supply voltage rail
B6	DRQ2	I	Direct memory access request level 2	Taken high when a DMA transfer is required. The signal remains high until the corresponding $\overline{\text{DACK}}$ line goes low
B7	−12V	n.a.	−12V d.c.	Supply voltage rail
B8	0WS	I	Zero wait state	Indicates to the microprocessor that the present bus cycle can be completed without inserting any additional wait cycles
B9	+12V	n.a.	+12V d.c.	Supply voltage rail
B10	GND	n.a.	Ground	Ground/common 0V
B11	$\overline{\text{MEMW}}$	O	Memory write	Taken low to signal a memory write operation
B12	$\overline{\text{MEMR}}$	O	Memory read	Taken low to signal a memory read operation
B13	$\overline{\text{IOW}}$	O	I/O write	Taken low to signal an I/O write operation
B14	$\overline{\text{IOR}}$	O	I/O read	Taken low to signal an I/O read operation
B15	$\overline{\text{DACK3}}$	O	Direct memory access acknowledge level 3	Taken low to acknowledge a DMA request on the corresponding level (see Notes below)
B16	DRQ3	I	Direct memory access request level 3	Taken high when a DMA transfer is required. The signal remains high until the corresponding $\overline{\text{DACK}}$ line goes low

Pin no.	*Abbreviation*	*Direction*	*Signal*	*Function*
B17	$\overline{\text{DACK1}}$	O	Direct memory access acknowledge level 1	Taken low to acknowledge a DMA request on the corresponding level (see Notes below)
B18	DRQ1	I	Direct memory access request level 1	Taken high when a DMA transfer is required. The signal remains high until the corresponding $\overline{\text{DACK}}$ line goes low
B19	$\overline{\text{DACK0}}$	O	Direct memory access acknowledge level 0	Taken low to acknowledge a DMA request on the corresponding level (see Notes on page 44)
B20	CLK4	O	4.77 MHz clock	CPU clock divided by 3, 210 ns period, 33 per cent duty cycle
B21	IRQ7	I	Interrupt request level 7	Asserted by an I/O device when it requires service (see Notes below)
B22	IRQ6	I	Interrupt request level 6	Asserted by an I/O device when it requires service (see Notes below)
B23	IRQ5	I	Interrupt request level 5	Asserted by an I/O device when it requires service (see Notes below)
B24	IRQ4	I	Interrupt request level 4	Asserted by an I/O device when it requires service (see Notes below)
B25	IRQ3	I	Interrupt request level 3	Asserted by an I/O device when it requires service (see Notes below)
B26	$\overline{\text{DACK2}}$	O	Direct memory access acknowledge level 2	Taken low to acknowledge a DMA request on the corresponding level (see Notes below)
B27	TC	O	Terminal count	Pulsed high to indicate that a DMA transfer terminal count has been reached

Pin no.	*Abbreviation*	*Direction*	*Signal*	*Function*
B28	ALE	O	Address latch enable	A falling edge indicates that the address latch is to be enabled. The signal is taken high during DMA transfers
B29	+5 V	n.a.	+5 V d.c.	Supply voltage rail
B30	OSC	O	14.31818 MHz clock	Fast clock with 70 ns period, 50 per cent duty cycle
B31	GND	n.a.	Ground	Ground/common 0 V

Notes:

1 Signal directions are quoted relative to the system motherboard; I represents input, O represents output, and I/O represents a bidirectional signal used both for input and also for output (n.a. indicates not applicable).

2 IRQ4, IRQ6 and IRQ7 are generated by the motherboard serial, disk and parallel interfaces, respectively.

3 $\overline{\text{DACK0}}$ (sometimes labelled REFRESH) is used to refresh dynamic memory whilst $\overline{\text{DACK1}}$ to $\overline{\text{DACK3}}$ are used to acknowledge other DMA requests.

The 36-way expansion bus connector

The PC–AT is fitted with an additional expansion bus connector which provides access to the upper eight data lines, D8 to D15, as well as further control bus lines. The AT-bus employs an additional 36-way direct edge-type connector. One side of the connector is referred to as C (lines are numbered C1 to C18) whilst the other is referred to as D (lines are numbered D1 to D18), as shown in Figure 2.2. The upper eight data bus lines and latched upper address lines are grouped together on the C-side of the connector (together with memory read and write lines) while additional interrupt request, DMA request, and DMA acknowledge lines occupy the D-side.

The following table describes each of the signals present on the 32-way expansion bus:

Pin no.	*Abbreviation*	*Direction*	*Signal*	*Function*
C1	SBHE	I/O	System bus high enable	When asserted this signal indicates that the high byte (D8 to D15) is present on the data bus
C2	LA23	I/O	Latched address line 23	Address bus line

Pin no.	*Abbreviation*	*Direction*	*Signal*	*Function*
C3	LA22	I/O	Latched address line 22	Address bus line
C4	LA21	I/O	Latched address line 21	Address bus line
C5	LA20	I/O	Latched address line 20	Address bus line
C6	LA23	I/O	Latched address line 19	Address bus line
C7	LA22	I/O	Latched address line 18	Address bus line
C8	LA23	I/O	Latched address line 17	Address bus line
C9	$\overline{\text{MEMW}}$	I/O	Memory write	Taken low to signal a memory write operation
C10	$\overline{\text{MEMR}}$	I/O	Memory read	Taken low to signal a memory read operation
C11	D8	I/O	Data line 1	Data bus line
C12	D9	I/O	Data line 1	Data bus line
C13	D10	I/O	Data line 1	Data bus line
C14	D11	I/O	Data line 1	Data bus line
C15	D12	I/O	Data line 1	Data bus line
C16	D13	I/O	Data line 1	Data bus line
C17	D14	I/O	Data line 1	Data bus line
C18	D15	I/O	Data line 1	Data bus line
D1	$\overline{\text{MEMCS16}}$	I	Memory chip-select 16	Taken low to indicate that the current data transfer is a 16-bit (single wait state) memory operation
D2	$\overline{\text{IOCS16}}$	I	I/O chip-select 16	Taken low to indicate that the current data transfer is a 16-bit (single wait state) I/O operation
D3	IRQ10	I	Interrupt request level 10	Asserted by an I/O device when it requires service
D4	IRQ11	I	Interrupt request level 11	Asserted by an I/O device when it requires service
D5	IRQ12	I	Interrupt request level 12	Asserted by an I/O device when it requires service

Pin no.	*Abbreviation*	*Direction*	*Signal*	*Function*
D6	IRQ13	I	Interrupt request level 13	Asserted by an I/O device when it requires service
D7	IRQ14	I	Interrupt request level 14	Asserted by an I/O device when it required service
D8	$\overline{\text{DACK0}}$	O	Direct memory access acknowledge level 0	Taken low to acknowledge a DMA request on the corresponding level
D9	DRQ0	I	Direct memory access request level 0	Taken high when a DMA transfer is required. The signal remains high until the corresponding DACK line goes low
D10	$\overline{\text{DACK5}}$	O	Direct memory access acknowledge level 5	Taken low to acknowledge a DMA request on the corresponding level
D11	DRQ5	I	Direct memory access request level 5	Taken high when a DMA transfer is required. The signal remains high until the corresponding DACK line goes low
D12	$\overline{\text{DACK6}}$	O	Direct memory access acknowledge level 6	Taken low to acknowledge a DMA request on the corresponding level
D13	DRQ6	I	Direct memory access request level 6	Taken high when a DMA transfer is required. The signal remains high until the corresponding DACK line goes low
D14	$\overline{\text{DACK7}}$	O	Direct memory access acknowledge level 7	Taken low to acknowledge a DMA request on the corresponding level
D15	DRQ7	I	Direct memory access request level 7	Taken high when a DMA transfer is required. The signal remains high until the

Pin no.	*Abbreviation*	*Direction*	*Signal*	*Function*
				corresponding DACK line goes low
D16	+5 V	n.a.	+5 V d.c.	Supply voltage rail
D17	$\overline{\text{MASTER}}$	I	Master	Taken low by the I/O processor when controlling the system address, data and control bus lines
D18	GND	n.a.	Ground	Ground/common 0 V

Electrical characteristics of the PC bus

All of the signals lines present on the expansion connector(s) are TTL-compatible. In the case of output signals from the system mother board, the maximum loading imposed by an expansion card adapter should be limited to no more than two low-power (LS) TTL devices. The following expansion bus lines are open-collector: $\overline{\text{MEMCS16}}$, $\overline{\text{IOCS16}}$ and 0WS.

The $\overline{\text{IOCHRDY}}$ line is available for interfacing slow memory or I/O devices. Normal processor generated read and write cycles use four clock (CLK) cycles per byte transferred. The standard PC clock frequency of 4.77 MHz results in a single clock cycle of 210 ns. Thus each processor read or write cycle requires 840 ns at the standard clock rate. DMA transfers, I/O read and write cycles, on the other hand, require five clock cycles (1050 μs). When the $\overline{\text{IOCHRDY}}$ line is asserted, the processor machine cycle is extended for an integral number of clock cycles.

Finally, when an I/O processor wishes to take control of the bus, it must assert the $\overline{\text{MASTER}}$ line. This signal should not be asserted for more than 15 μs as it may otherwise impair the refreshing of system memory.

Design of PC expansion cards

Several factors need to be taken into account when designing PC expansion cards. These include power-supply requirements, power-supply rail distribution and decoupling, and address decoding. In addition, access to the more specialized bus control signals (such as $\overline{\text{IOCHK}}$, $\overline{\text{IOCHRDY}}$, DRQ, IRQ) may be required in the case of cards which are fitted with slow I/O devices, require DMA transfer or need to

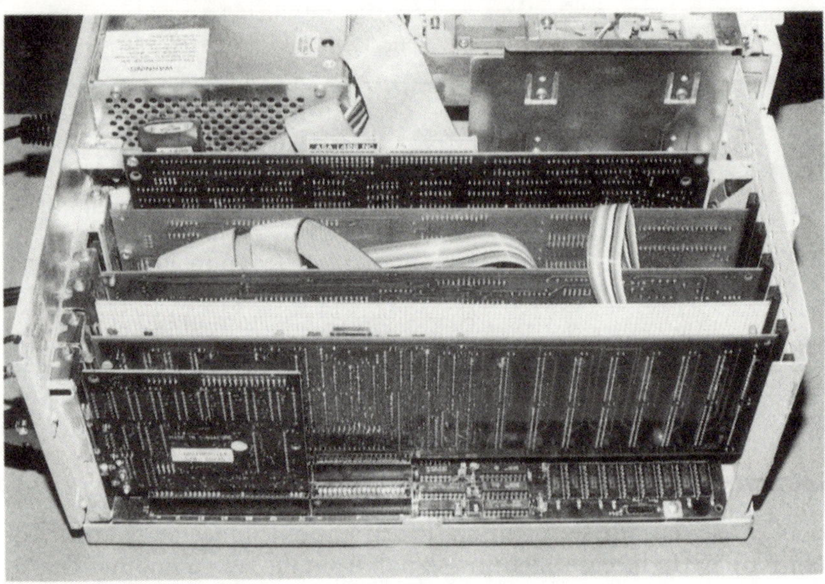

Plate 2.2 *Mixture of full-size, half-size and prototype cards fitted in an AT compatible system unit*

be interrupt driven (see Figure 2.3). The following pointers are given for the benefit of those involved with the design and development of PC expansion cards.

Power rails

The available power for additional expansion cards depends upon the rating of the system power supply, the requirements of the motherboard, and the demands of other adapter cards which may be fitted. When designing expansion cards, the recommended limit (per card) for each of the three power rails is as follows:

Voltage rail	*Connection*	*Maximum current*
+5 V	B3 and B29	1.5 A
−5 V	B5	100 mA
+12 V	B9	500 mA
−12 V	B7	100 mA

Where several adapter cards are fitted, the current demand for each supply rail should be estimated and the total power requirements

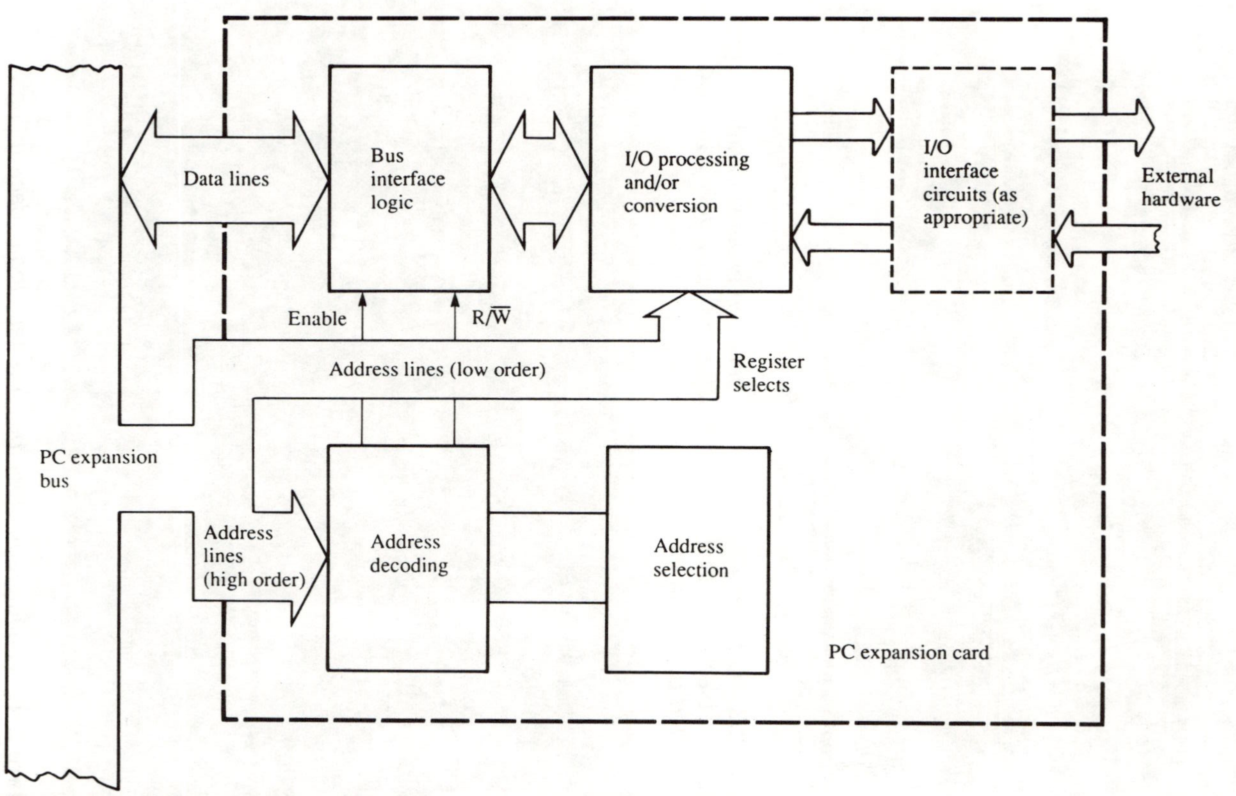

Figure 2.3 *Generalized block schematic for an expansion card*

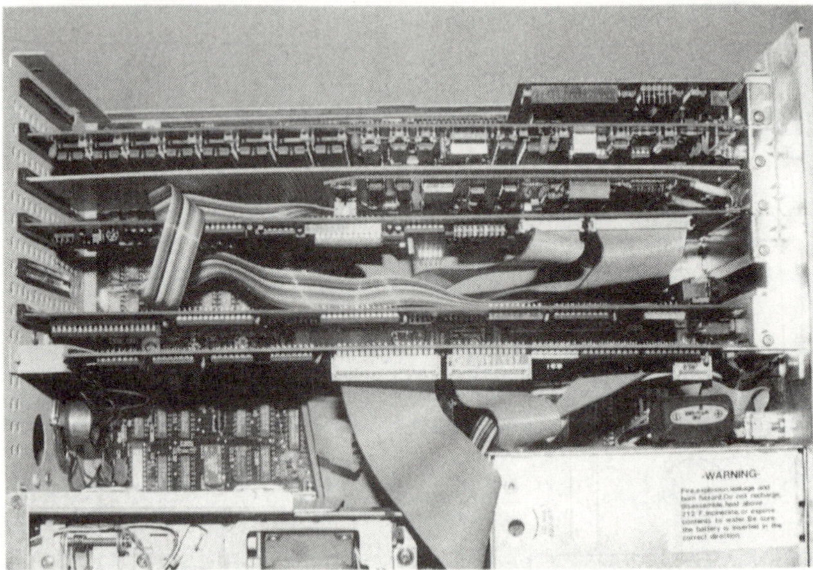

Plate 2.3 *Off-bus interconnection between expansion cards using ribbon cables (note that the fixed disk and floppy drives are also connected by ribbon cables)*

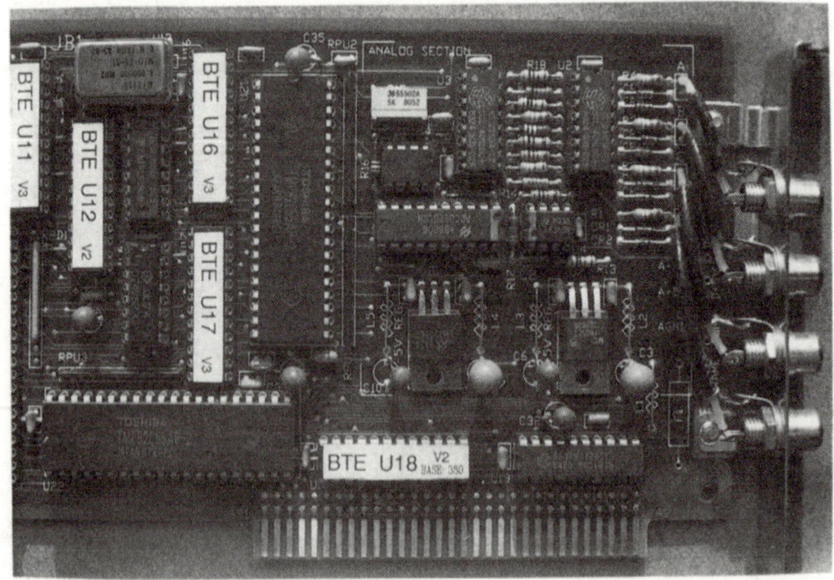

Plate 2.4 *Voltage regulators, bus interface logic, decoupling components and parallel I/O ports fitted to an analogue signal processing card (note the phono signal connectors used for analogue I/O)*

Plate 2.5 *Bus interface logic fitted to an XT compatible prototyping card (note the 40-way D-connector used for digital I/O)*

calculated. It should go without saying that the total demand should not exceed the spare capacity rating of the system power supply. In some cases this may be less than 15 W!

As a guide, the following data refers to the power supplies fitted as 'standard' on most PC, XT, XT–286 and AT systems:

System	*PC*	*XT*	*XT–286*	*AT*
Total power capability (W)	63.5	130	157	192
Maximum current rating (A)				
+5 V rail	7	15	20	20
−5 V rail	0.3	0.3	0.3	0.3
+12 V rail	2	4.2	4.2	7.3
−12 V rail	0.25	0.25	0.25	0.3

A fully populated system motherboard (including 80287 coprocessor) requires approximately 4 A and 2 A from the +5 V and +12 V rails

Plate 2.6 *Half-size expansion card (graphics adapter)*

respectively. An EGA graphics adapter and two standard floppy drives will demand an additional 2.4 A and 1.8 A from the +5 V and +12 V rails respectively. A cooling fan will require a further 0.3 A, or so, from the +12 V rail. The total load is thus 6.4 A from the +5 V rail and 4.1 A from the +12 V rail. With a standard XT power supply, reserves of 9.6 A from the +5 V supply and only 100 mA from the +12 V supply will be available!

Supply rail distribution

In order to minimize supply borne noise and glitches, the following recommendations should be observed when considering the design and layout of prototype expansion cards:

1 Ensure that the ground/common 0 V foil is adequate and that the three ground connections (B1, B10, and B31) are linked together via a substantial area of copper foil.

2 Include decoupling capacitors on each of the supply rails as follows:
   (a) 100 μF axial lead electrolytic to decouple the +5 V rail (locate close to pins B1 and B3 or B29 and B31).
   (b) 47 μF axial lead electrolytic to decouple the +12 V rail (locate close to pins B9 and B10).
   (c) 47 μ axial lead electrolytic to decouple the −12 V rail (locate close to pins B7 and B10).
   (d) 10 μF axial lead electrolytic to decouple the −5 V rail (locate close to pins B5 and B10).

   (NB: capacitors can be omitted when the relevant voltage rail is not used within the expansion card.)
3 Fit 10 μF 16 V radial lead decoupling capacitors to the +5 V rail at the rate of one capacitor for every eight to ten TTL or CMOS logic devices. Capacitors should be distributed at regular points along the supply rail.
4 Fit 100 nF 16 V disk ceramic capacitors to the +5V rail at the rate of one for every two to four TTL or CMOS logic devices. Capacitors should be placed at strategic points close to the supply pin connections of the integrated circuits.
5 Fit one 10 uF 16 V and one 100 nF 16 V capacitor to the +5 V rail for each VLSI device. Capacitors should be placed as close as possible to the supply pin connections of the devices in question.
6 Repeat (3), (4) and (5) for each of the other supply rails (where used).

Finally, it should go without saying that one should **never** *attempt to insert or remove an expansion or adapter card when the power is connected and the system is running*. Failure to observe this precaution may result in serious damage not only to the card in question but also to other cards which may be installed as well as to components on the system motherboard.

If this all sounds rather obvious, no apologies are made for repeating it. In the heat of the moment it is all too easy to forget that a system is 'live'. You are only likely to make this mistake once – but the cost and frustration are likely to have a long-lasting effect!

Address decoding

The I/O provided by an expansion card will be mapped into either address or I/O space (the latter being conventionally used for digital and analogue I/O cards). The expansion card must, therefore, contain some address decoding logic which must be configured to avoid conflicts with other system hardware. Figure 2.4 shows some representative address decoding logic which provides access to eight *base addresses* within I/O space. Address lines A0 and A1 may then be used as optional *register select* lines for connection to VLSI devices (e.g. and 8255 PPI).

The address decoder shown in Figure 2.4 employs a three-to-eight line decoder (74LS138) in which the enable lines (G2A, G2B and G1) are employed (note that G2A and G2B are active low, whilst G1 is an active high input). Outputs (Y0 to Y7) are active low and thus are ideal for use as chip select or enable signals. The truth table for the address decoder is as follows:

Address line											*Base address (hex)*	*Output selected (taken low)*
A12	*A11*	*A10*	*A9*	*A8*	*A7*	*A6*	*A5*	*A4*	*A3*	*A2*		
0	0	0	1	1	1	0	0	0	0	0	300	Y0
0	0	0	1	1	1	0	0	0	0	1	304	Y1
0	0	0	1	1	1	0	0	0	1	0	308	Y2
0	0	0	1	1	1	0	0	0	1	1	30C	Y3
0	0	0	1	1	1	0	0	1	0	0	310	Y4
0	0	0	1	1	1	0	0	1	0	1	314	Y5
0	0	0	1	1	1	0	0	1	1	0	318	Y6
0	0	0	1	1	1	0	0	1	1	1	31C	Y7
0	0	1	x	x	x	x	x	x	x	x	n/a	none
0	1	0	x	x	x	x	x	x	x	x	n/a	none
0	1	1	x	x	x	x	x	x	x	x	n/a	none
1	0	0	x	x	x	x	x	x	x	x	n/a	none
1	0	1	x	x	x	x	x	x	x	x	n/a	none
1	1	0	x	x	x	x	x	x	x	x	n/a	none
1	1	1	x	x	x	x	x	x	x	x	n/a	none

x = don't care.

The remaining address lines (A1 and A0) provide four *address offsets* from the base address, as follows:

Address lines		*Offset value*
A1	*A0*	
0	0	0
0	1	1
1	0	2
1	1	3

As an example, address 302 (hex) will be selected when the following address pattern appears:

A12	A11	A10	A9	A8	A7	A6	A5	A4	A3	A2	A1	A0
0	0	0	1	1	1	0	0	0	0	0	1	0

Base address = 300H   Offset = 2H

I/O address = 302H

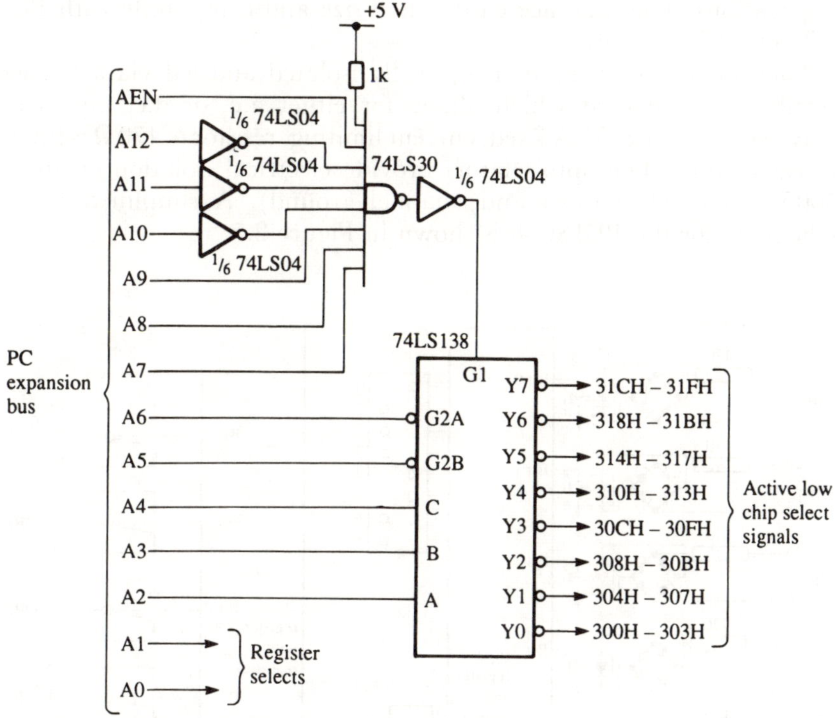

Figure 2.4 *Representative address decoder arrangement*

It is, of course, quite permissible to use the chip select lines without making use of the register select lines, A1 and A0. In such cases, it is important to remember that the I/O address will not be unique.

Representative I/O cards

The following information relates to a range of representative I/O interface cards for use with the PC expansion bus. Details have been included in order to provide readers with an insight into products which are available 'off-the-shelf'. Details of a popular prototyping card have also been included for those who may prefer to develop their own interface cards or for those with applications for which no suitable card is currently available.

MetraByte PDISO-8

The MetraByte Corporation PDISO-8 is an inexpensive eight-channel isolated input–output interface card designed for control and sensing applications. The interface card is half-size and compatible with PC, XT and AT systems.

Each of the eight inputs is optically isolated and fed via a bridge rectifier arrangement which allows for either a.c. or d.c. inputs of between 5 V and 24 V. A fixed, current limiting, resistor of 470 Ω is fitted to each input. The optoisolators provide electrical isolation of up to 500 V (channel–channel and channel–ground). A simplified block schematic for the PDISO-8 is shown in Figure 2.5.

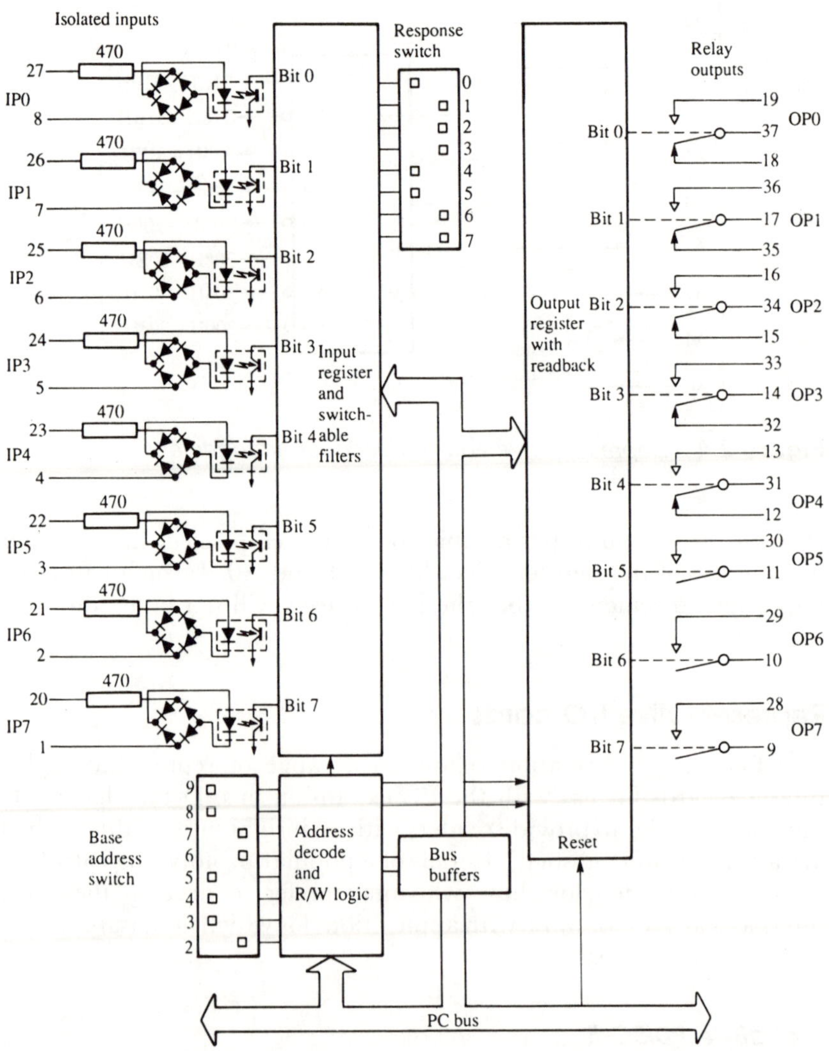

Figure 2.5 *Simplified block schematic for the PDISO-8*

The response time of each input may be individually selected using a dual-in-line switch; input response time is typically 20 μs without the filter and 10 ms when the filter is switched in (note that filters are normally required with a.c. inputs in order to avoid the digital input pulsing on and off at twice the a.c. input frequency!).

The eight relay outputs each have contacts rated at 3 A at 120 V a.c. or 28 V d.c. (resistive loads). The maximum contact resistance is 100 mΩ and both SPDT (Channels 0 to 4) and SPST (Channels 5 to 7) contacts are available. Relay operating time is 20 ms (max.) and release time 10 ms (max.).

The PDISO-8 uses only the +5 V power rail from the PC and requires a typical supply current of 1 A (all relays energized). The I/O lines from the board connect via a standard 37-pin D-type male connector fitted to the rear metal bracket. The I/O connector pin assignment is shown in Figure 2.6.

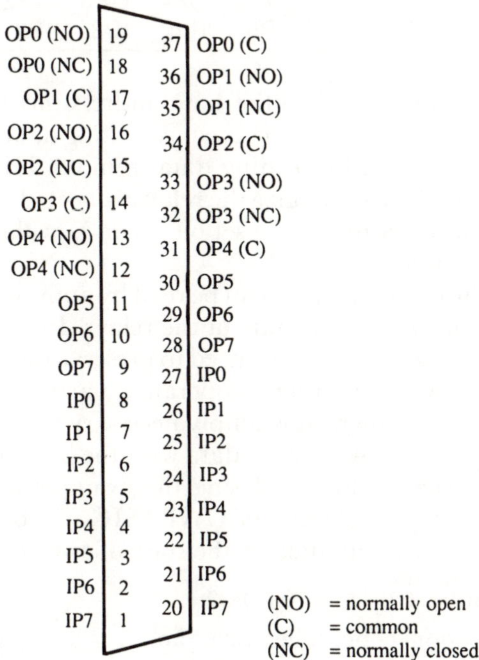

Figure 2.6 *Connector pin assignment for the PDISO-8*

The board address is selected by means of a dual-in-line switch. The PDISO-8 board occupies four consecutive addresses in the PC I/O address space of which only two addresses are actually used. The base address is selected by means of the dual-in-line switch and the two registers are located at (base address) and (base address + 1). The I/O map for the board is as follows:

I/O address	*Function*	*Mode*
Base address	Relay outputs	Read/write
Base address + 1	Isolated inputs	Read only

Each bit in the appropriate register corresponds to the equivalent I/O Channel Number. Bits are therefore allocated as follows:

Address	*Data bit*							
	D7	*D6*	*D5*	*D4*	*D3*	*D2*	*D1*	*D0*
Base	OP7	OP6	OP5	OP4	OP3	OP2	OP1	OP0
Base + 1	IP7	IP6	IP5	IP4	IP3	IP2	IP1	IP0

As an example, assuming that the base address has been set to 300 hexadecimal, the relays can be operated by writing data to 0300H while the inputs can be sensed by reading data from 0301H. In the former case, a set bit (logic 1) will energize the relay connected to the channel in question while in the latter case, a set bit (logic 1) will indicate that an input has been asserted.

The state of the output register can be read by appropriate software in order to ascertain the current state of the relays. In some applications this can be useful since it avoids the need to preserve the state of the relay port within a variable. In order to operate a particular relay without disturbing any of the others, it is simply necessary to first read the data from (base address), bit-wise *or* the data with the bit to be set, and then write it back to (base address). Using the example addresses quoted earlier, the following single line of GWBASIC will operate the relay connected to OP1 without altering the state of any of the other relays:

```
7010 OUT &H300, INP(&H300) OR 2
```

Further information concerning programming this type of interface appears in Chapter 11.

Blue chip technology AIP-24

The Blue Chip Technology AIP-24 analogue input card provides 24 channels of analogue input. The board is a 130 mm short format PC/XT/AT compatible card and its simplified block schematic is shown in Figure 2.7.

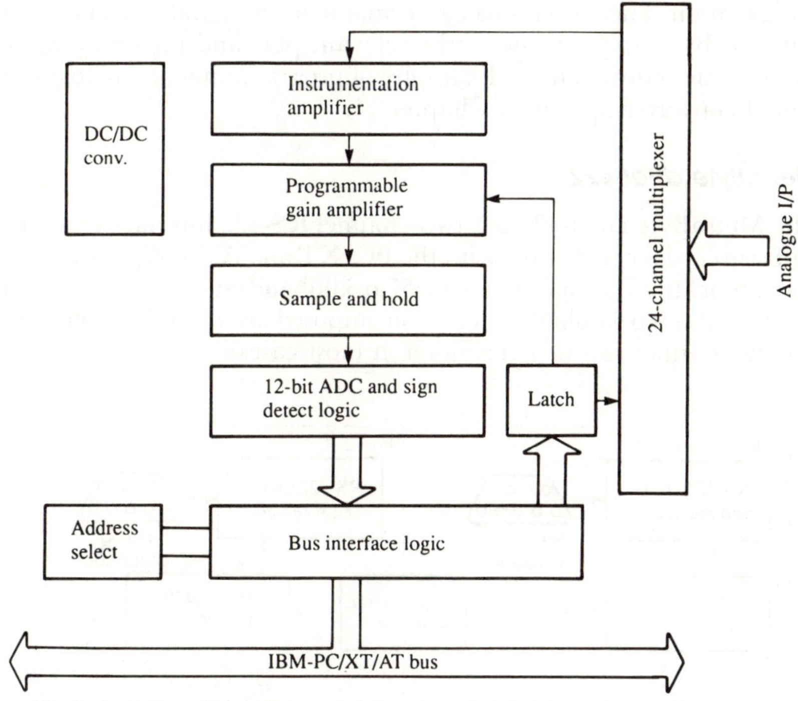

Figure 2.7 *Simplified block schematic of the AIP-24*

The AIP-24 uses a 12-bit analogue-to-digital converter which provides a resolution of 0.025 per cent. A sample and hold amplifier is used to capture fast moving analogue signals and freeze them in order to improve overall accuracy. The successive approximation ADC can operate in unipolar or bipolar modes and handles signals in the range 0 V to ± 10 V. In order to cope with low amplitude input signals, an on-board programmable gain amplifier can be used to provide input gains of 1, 10 or 100. Conversion time is 25 μs but faster ADC chips may be fitted where conversion speed is critical.

Input connection is made via a 50-way IDC connector attached to the metal rear bracket. A ribbon cable or screw terminal may be fitted directly to the 50-way connector. An on-board d.c.-to-d.c. converter provides power for the analogue circuitry of the ADC.

As with the PDISO-8, four addresses are used to set up and drive the card. These provide gain selection (write), initiate conversion (write), and converted data (read). The base address of the card is selected by means of PCB links.

Programming the card is reasonably straightforward. The gain of the analogue input will normally be set by writing appropriate bytes during initialization. Thereafter, successive analogue-to-digital conversions are initiated by simply writing to the relevant port and then reading the value of the returned data. Examples of programming an analogue-to-digital converter appear in Chapter 11.

MetraByte dual-422

The MetraByte Dual-422 is a two-channel RS-422 interface card. The half-size card is compatible with the PC, XT and AT and permits serial communications at speeds up to 57.6 kilobaud at distances of up to 1200 m (the 9.6 kilobaud limitation imposed by most PC communications routines can be overridden in most cases).

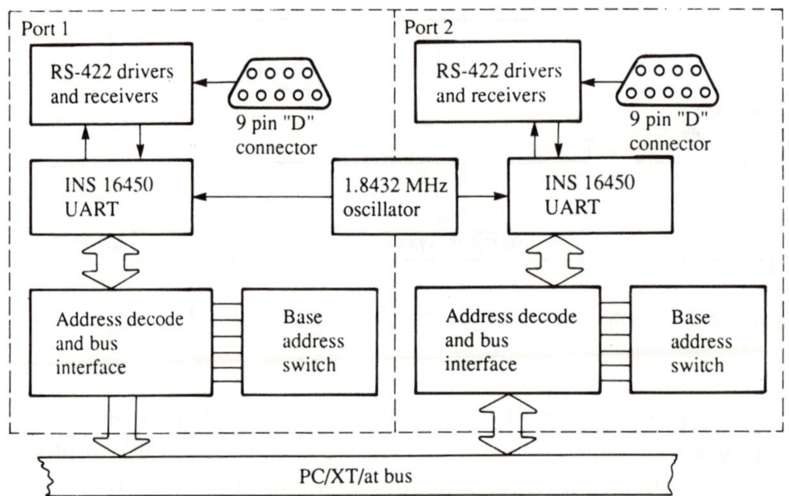

Figure 2.8 *Simplified block schematic of the Dual-422*

The simplified block schematic of the Dual-422 is shown in Figure 2.8. Both ports operate independently and each has its own case address and interrupt selection controls. A VLSI Universal Asynchronous Receiver/Transmitter (UART) device is used to form the basis of each channel and this device is augmented by external line drivers and receivers.

The UART employed is the National Semiconductor INS16450 (an improved device which is compatible with the original 6250 device employed in the PC). The INS16450 is fully programmable and offers a choice of serial data word length (5, 6, 7 or 8 data bits) with selectable even, odd, or no parity checking. Baud rates are also selectable in the range 120 baud to 57.6 kilobaud.

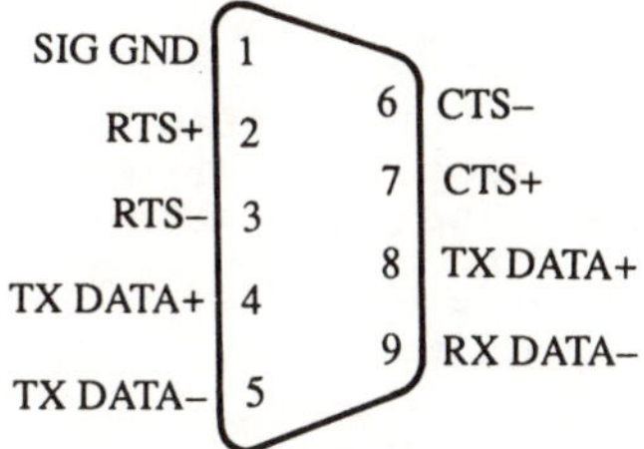

Figure 2.9 *I/O connector pin assignment of the Dual-422*

Base address selection (for each port) is obtained via a dual-in-line switch (see Figure 2.10). Links are used to select the desired interrupt level (either channel can be configured as MS–DOS serial port COM1: or COM2: or any other interrupt level may be selected) whilst a further link is provided in order to enable or disable CTS/RTS data transfer control.

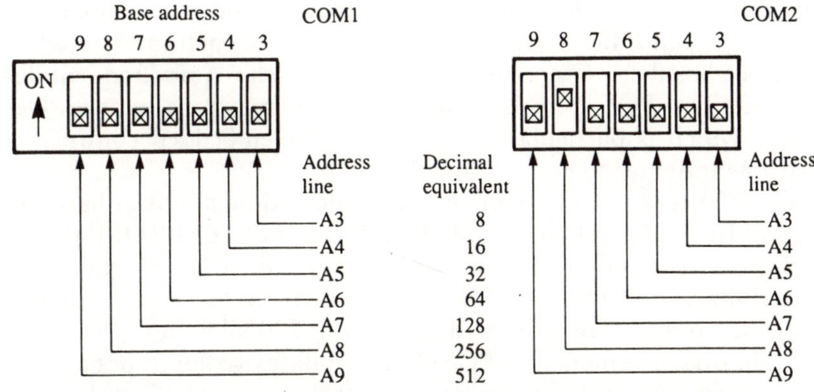

Note that a switch in the on position corresponds to a zero, and in the off position corresponds to the binary weight of the corresponding address bit (512, 256, 128, 64, 32, 16, 8). The base I/O address is the sum of all the off switch address bits.

Figure 2.10 *Base address selection of the Dual-422*

Programming the Dual-422 interface is extremely straightforward. Assuming that the ports have been configured as COM1: and COM2: (and that no other communication device has been configured to the same interrupt level), the following GWBASIC code transmits a test string (T$) output via COM1: for input via COM2: to the received string, R$:

```
110 REM Test string
120 T$="The quick brown fox jumps over the lazy dog"
130 REM Open the serial ports using 4800 baud
131 REM COM1 will be associated with channel 1
132 REM COM2 will be associated with channel 2
140 OPEN "COM1:4800" AS #1
150 OPEN "COM2:4800" AS #2
160 REM Enable COM1: RS-485 driver
170 OUT &H3FF,2
180 REM Enable COM2: RS-485 receiver
190 OUT &H2FF,1
200 REM Transmit data via COM1:
210 Print #1, T$
220 REM Receive data via COM2:
230 INPUT #2, R$
240 REM Close communication channels
250 CLOSE
260 END
```

The Dual-422 board uses only the + 5V supply rail and demands a current of 850 mA (typical).

Penny and Giles ROMDISK PCE/2

The Penny and Giles ROMDISK PCE provides solid-state emulation of a floppy or hard disk drive and is ideal for use in environments which would be hostile to disks or in applications which require frequent disk accesses. The PCE/2 card allows the designer to produce a diskless system which is particularly appropriate for embedded control and instrumentation systems.

The PCE/2 is capable of emulating single or dual floppy or hard disk drives. In single-disk mode, EPROMs or flash EEPROM devices emulate a read-only disk with capacities of up to 1.2 Megabytes. A battery-backed non-volatile SRAM daughterboard may be fitted in order to emulate a read/write disk of up to 770 kilobytes.

In the dual disk mode, the PCE/2 can emulate two floppy or hard disk drives. A primary or boot disk of up to 770 kilobytes of EPROM storage can be augmented (by means of the SRAM daughterboard) with an emulated read/write disk of 770 kilobytes.

The PC and AT disk standards which can be emulated include: 5.25" 180K SSDD, 360K DSDD, 770K and 1.2 megabytes or 3.5 × × 720K or a 1.3 megabyte portion of a 1.44 megabyte disk.

EPROMs are programmed by an on-board EPROM programmer which copies a disk (of the correct type and capacity) using a software utility (RDCOPY2). Other utilities are provided in order to compare the programmed EPROMs with a master disk and to format the SRAM devices fitted to the SRAM daughterboard.

The PCE/2 fits into a standard PC/XT/AT bus slot. An adapter ROM provides the facility for integration of the PCE/2 into the system by using the computer's BIOS and boot vectors to initiate operation and emulate a disk. The memory transfer mode is selectable as either DMA1 for transfer at RAM speeds or programmed I/O for use where there is a conflict with the DMA channel.

The PCE/2 card automatically configures the system for four floppy drives in the Fixed Drive designation mode or to the number of ROMDISKs plus the number of physical drives in the Autoset mode. A program mode (to program EPROMs) and three read modes are provided. The read modes include two autoboot modes and a non-system file mode. One of the autoboot modes boots from ROMDISK whilst the other allows for disk booting. All PC and MS–DOS commands operate normally except for those which require the physical presence of a disk drive adapter.

Amplicon PC-35 prototyping card

The Amplicon PC-35 prototyping card provides a means of assembling and testing prototype circuits as well building 'one-off' designs. The card is a full-size board which uses the original (8-bit) PC expansion bus and is supplied already populated with the bus interface circuitry (comprising seven dual-in-line devices). In order to facilitate I/O connections, a 37-way D connector may be fitted to the rear metal bracket.

Plate 2.7 *Bus interface circuitry on the PC-35 prototyping card*

More than 60 14-pin dual-in-line devices can be accommodated within the board area which incorporates a matrix of 3000 holes. Soldered or wire-wrapped sockets can be used and power supply rails are distributed for easy access across the whole board area. The manufacturer recommends the following maximum current loading:

Supply rail	*Maximum loading*
+5 V	1.5 A
+12 A	0.5 A
−12 V	−0.1 A

Eight address decoding lines are provided for direct interfacing of peripheral chips such as the 8253, 8254 and 8255. The address range conforms to that which is normally associated with IBM prototyping cards, i.e. 300H to 31FH. Address lines A0 and A1 provide selection of four address offsets from the base address according to the following scheme:

Chip select line	*Address range (hex)*	*Address selected (hex)*			
		A1 A0 *1 1*	*A1 A0* *1 0*	*A1 A0* *0 1*	*A1 A0* *1 0*
CS0	300–303	303	302	301	300
CS1	304–307	307	306	305	304
CS2	308–30B	30B	30A	309	308
CS3	30C–30F	30F	30E	30D	30C
CS4	310–313	313	312	311	310
CS5	314–317	317	316	315	314
CS6	318–31B	31B	31A	319	318
CS7	31C–31F	31F	31E	31D	31C

The A0, A1, RESET, IOW, IOR, MEMW, MEMR, and D0–D7 bus lines are all buffered and can be loaded with up to ten LS-TTL loads. All other lines may be accessed subject to a maximum loading equivalent to two LS-TTL devices.

The data bus is buffered by means of a bidirectional bus transceiver (74LS245). The direction of this device may be controlled from either the IOW/IOR or MEMW/MEMR lines (the former option is used for port I/O whilst the latter is employed when the card is to be memory mapped). An example of using this card appears in Chapter 11.

Backplane bus systems

While the market for PCs was in its infancy, many electronics manufacturers active in the control and instrumentation sector were moving towards the development of products suitable for modern backplane bus systems (such as STE and/or VME). Boards designed for such systems can be designed to offer almost identical functionality to those associated with the PC expansion bus. However, both STE and VME offer an environment which is more rugged and reliable (both mechanically and electrically) and therefore offer considerable superiority when employed in the vast majority of industrial applications.

Despite this, the deciding factor which brought about a widespread availability of a 'PC engine' for the STE and VME backplane bus systems was simply the demand for software compatibility; end-users required a hardware configuration which would support a PC compatible processor board in order to be able to run MS-DOS software.

Several manufacturers rose to the challenge and the designer now has the option of building a PC-based system using off-the-shelf products designed for the STE or VME bus.

The STE bus

The STE bus was originally conceived as a Eurocard replacement for an earlier 8-bit microprocessor bus standard, the STD bus. The system was first specified in 1982 and the first STE cards appeared in 1984. Since 1985 the standard has attracted considerable support and many users of earlier 8-bit bus systems have changed to STE. In 1987 the system was defined under the IEEE-1000 specification which establishes it as an internationally recognized standard.

The link between the STE bus standard and the PC microcomputer may at first sight appear to be somewhat tenuous. However, the appearance of complete PC microcomputers based on Standard Eurocards has made it eminently possible to implement an industrial PC at low cost using the STE bus. Furthermore, the availability of a wide range of STE bus products has made it possible to solve an enormous variety of interfacing problems using off-the-shelf parts. All of this is good news to the engineer; a PC based on the STE bus offers considerable flexibility coupled with inherent reliability.

The STE bus provides for an 8-bit data path and 20 address lines (permitting 1 megabyte of directly addressable memory space). The system provides for an I/O space of 4 kilobytes and up to three potential bus masters may be present within a system. Furthermore, while the backplane provides only for an 8-bit data path, this does not mean that the system cannot support 16-bit processors (such as the 8086 and 80286). The restriction is simply that data transfers between the bus master (i.e. the PC CPU) and bus slaves (e.g. I/O cards, memory cards, etc.) must take place on a byte-by-byte basis.

STE bears more than a passing resemblance to the VME bus and some enthusiastic users have likened it to an 8-bit implementation of the VME standard. One of the beauties of the STE bus is that, whilst the system is flexible enough to permit mixing of a wide variety of processors from different manufacturers, it is definitive in terms of bus signals and protocol. System designers can thus have every confidence that STE Bus products obtained from a variety of sources can be interconnected and the whole system will operate as planned.

STE boards measure 160 × 100 mm (i.e. Standard Eurocard size) and the connector specified is a 64-pin male DIN 41612 type utilizing rows (a) and (c) (the inside row is not used). This connector is specified in IEC 603-2 and the corresponding female connector is specified for the STE backplane. The function of the signals present (see Figure 2.11) are as follows:

D0 to D7	Eight data lines
A0 to A19	Twenty address lines
ADRSTB*	Address strobe. This line is taken low to indicate that a valid address has been placed on the bus
DATSTB*	Data strobe. This line is taken low to indicate that valid data has been placed on the bus
CM0 to CM2	Command modifiers which indicate the current type of bus cycle
BUSRQ0* and BUSRQ1*	Bus request lines. These lines are taken low when a potential bus master wishes to gain access to the bus
BUSAK0* and BUSAK1*	Bus acknowledge lines. These lines are taken low to indicate that the bus request has been granted. A potential bus master may only drive the bus when it has received an acknowledge signal on the bus request line
DATACK*	This handsake line is asserted by a bus slave on a write cycle in order to indicate that it has accepted data or, on a read cycle, to indicate that its data is valid
TRFERR*	A bus slave asserts this signal instead of DATACK* if an error is detected
ATNRQ0* to ATNRQ7*	Attention request/interrupt lines. (ATNRQ0* has the highest priority)
SYSCLK	16-MHz system clock
SYSRST*	System reset

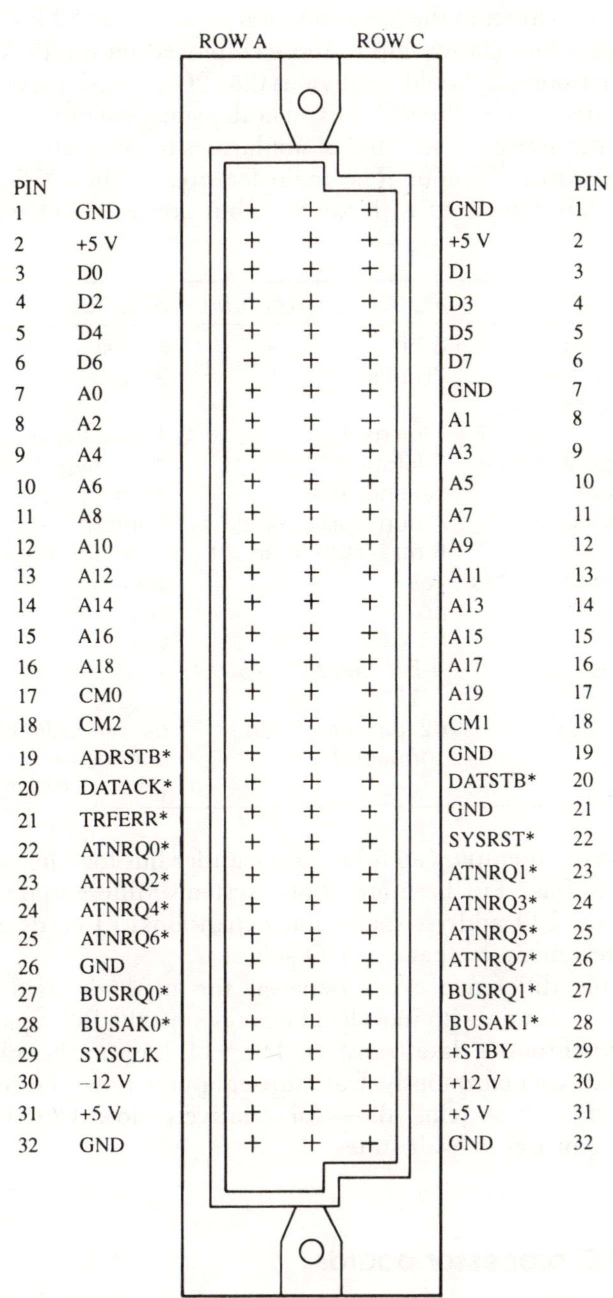
ROW A
ROW C
PIN
PIN
1 GND GND 1
2 +5 V +5 V 2
3 D0 D1 3
4 D2 D3 4
5 D4 D5 5
6 D6 D7 6
7 A0 GND 7
8 A2 A1 8
9 A4 A3 9
10 A6 A5 10
11 A8 A7 11
12 A10 A9 12
13 A12 A11 13
14 A14 A13 14
15 A16 A15 15
16 A18 A17 16
17 CM0 A19 17
18 CM2 CM1 18
19 ADRSTB* GND 19
20 DATACK* DATSTB* 20
21 TRFERR* GND 21
22 ATNRQ0* SYSRST* 22
23 ATNRQ2* ATNRQ1* 23
24 ATNRQ4* ATNRQ3* 24
25 ATNRQ6* ATNRQ5* 25
26 GND ATNRQ7* 26
27 BUSRQ0* BUSRQ1* 27
28 BUSAK0* BUSAK1* 28
29 SYSCLK +STBY 29
30 –12 V +12 V 30
31 +5 V +5 V 31
32 GND GND 32

Figure 2.11 *STE bus connector*

Comparison of STE and PC bus systems

It should be clear from the foregoing that many of the STE bus control signals bear little relationship to those employed on the PC expansion bus. Furthermore, it should be obvious that PC expansion cards will not interface directly with the STE bus on both logical *and* physical grounds! Despite this, there are a number of similarities between the two systems which are worthy of note. The main features of the (XT 8-bit) PC expansion bus compared with the STE bus are listed below:

Feature	*PC/XT expansion bus*	*STE bus*
Data bus width	8 bits	8 bits
Address bus width	20 bits	20 bits
Memory addressing range	1 megabyte	1 megabyte
I/O addressing range	1 kilobyte	4 kilobyte
Data transfer	synchronous	asynchronous
Interrupt provision	1 non-maskable plus 6 maskable interrupt	8 attention/ interrupt request
Direct memory access	yes	yes
Multiple master	no	yes
Vector fetch	no	yes
Supply rails	+5 V, +12 V, −5 V, −12 V	+5 V, +12 V, −12 V
Bus connector	62-pin direct edge connector	64-pin indirect connector (DIN 41612 type C)

Both bus systems employ an 8-bit data transfer bus together with a 20-bit address bus. Furthermore, both systems differentiate between memory and I/O address space and expansion I/O cards are either mapped into memory space or I/O space.

A notable difference exists between the methods used for data transfer. In common with most local bus systems, the PC expansion bus employs synchronous data transfers. The STE bus, on the other hand, transfers data asynchronously. Data throughput is, however, reasonably similar and both systems allow for relatively slow I/O devices by introducing processor wait states.

STE bus PC processor boards

The following information relates to three popular PC-compatible processor boards which are available in the form of standard STE bus cards. All three of these cards are proving to be immensely popular with designers who wish to implement PC-based systems within a backplane bus environment. STE I/O cards will not, however, be described in detail as details are available in *Bus-based Industrial Process Control* (see Appendix E for details).

DSP design eurocard personal computer

The DSP Design Eurocard Personal Computer (ECPC) is an IBM PC-compatible processor card available for the STE bus. The board represents a hitherto unprecedented level of miniaturization in that it packs the complete functionality of a PC into an STE bus single Eurocard measuring a mere 100 × 160 mm.

The ECPC employs an NEC V40 processor (enhanced 8086) running at 7.12 MHz. The on-board memory may be populated with 256K, 512K or 768K of dynamic RAM. In addition, two sockets are provided for EPROM or static RAM devices. A memory map for the ECPC is shown in Figure 2.12.

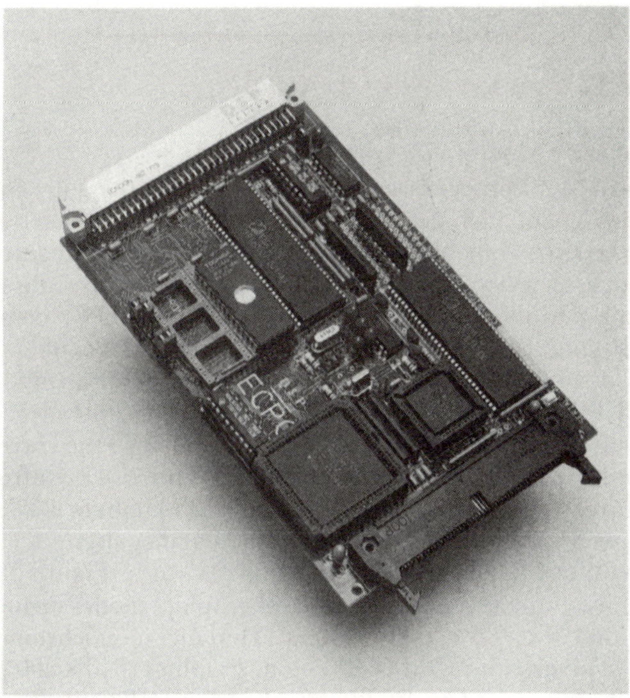

Plate 2.8 *DSP Eurocard Personal Computer (photo courtesy of DSP Design)*

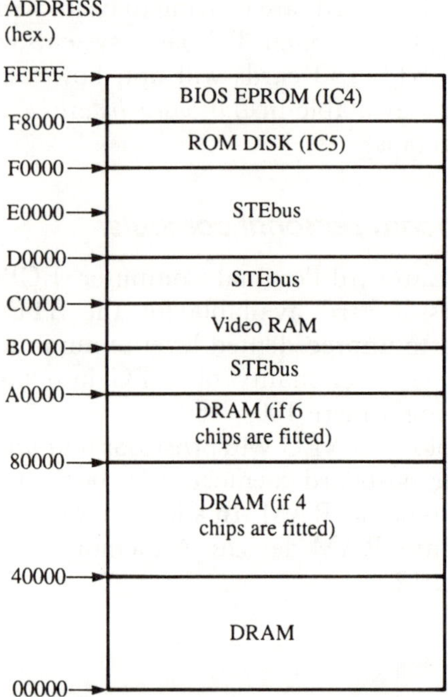

Figure 2.12 *Memory map of the ECPC*

The NEC V40 processor executes a superset of the Intel 8086/80186 instruction set and includes on-chip peripherals such as timers, interrupt controller, DMA controller, and serial communications controller and, whilst such devices appear to the programmer as subsets of the equivalent peripheral chips on the original IBM PC, compatibility extends beyond BIOS level to register level. The ECPC offers a Norton Speed Index of 2.6 (i.e. it runs 2.6 times faster than a standard PC).

The ECPC is fitted with an on-board floppy disk controller as well as a multimode graphics controller. These correspond to the corresponding components used on the IBM PC/XT. The disk controller will, however, read and write 360 kilobyte and 720 kilobyte 3.5″ and 5.25″ disks as well as modern high-capacity 1.44 megabyte 3.5″ disks. A simplified block diagram of the ECPC is shown in Figure 2.13.

The video controller offers a choice of graphics modes including IBM monochrome text (80 × 25 characters), Hercules monochrome graphics (720 × 348 pixels) and IBM CGA colour graphics (640 × 200 pixels and 80 × 25 characters). The ECPC graphics controller also caters for a double scan CGA mode which provides a superior display to that which is normally associated with the conventional CGA graphics mode.

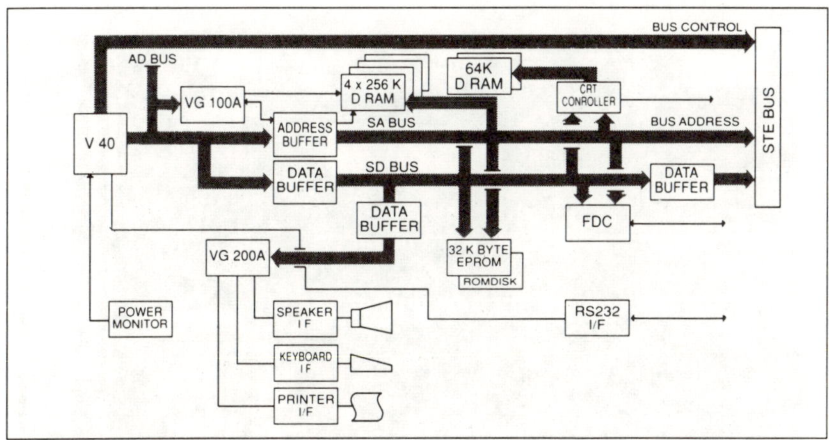

Figure 2.13 *Simplified block schematic of the ECPC*

Control Universal CeleSTE PC

The Control Universal CeleSTE PC-compatible processor provides full PC/XT hardware compatability on a standard STE bus Eurocard. The system can be configured with a variety of option daughterboards which provide memory expansion, system utilities, etc.

Plate 2.9 *Control Universal CeleSTE PC Processor Board*

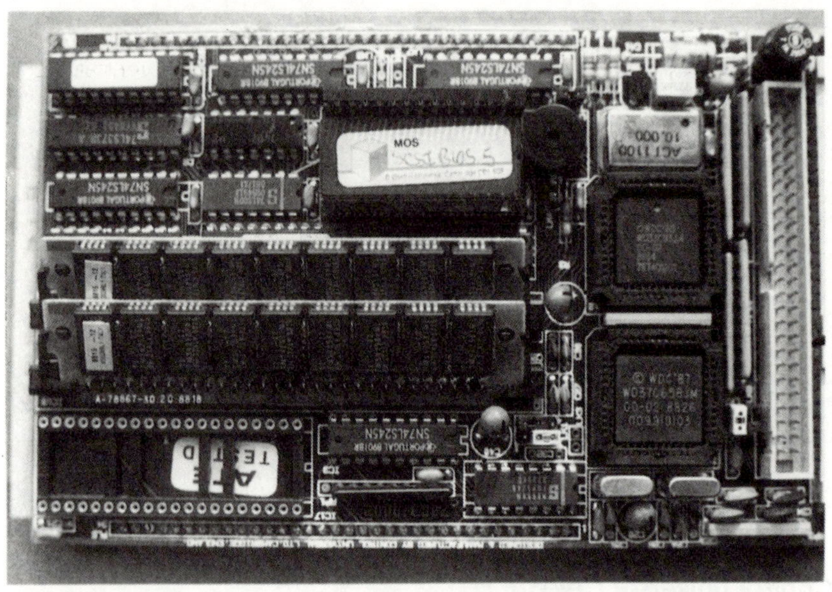

Plate 2.10 *Control Universal CeleSTE PC Options Board*

The CeleSTE PC employs either an NEC V20 (enhanced 8088) or an Intel 8088 processor. Software selectable clock frequencies are 4.77 MHz and 8 MHz (8088) or 4.77 MHz and 10 MHz (V20). The Norton Speed Index is 1.6 and 3.0 or 1.0 and 2.0, respectively. The processor board offers 128 kilobytes of ROM space for Digital Research's DR-DOS, BIOS and a ROM disk together with 256 kilobytes of zero wait state RAM. The memory map for the CeleSTE PC is shown in Figure 2.14.

An in-built LIM V4.0 EMS-compatible memory controller allows the addition of a further 1.25 megabytes of zero wait state DRAM using an 'Options Board'. This allows the system memory to be extended to the DOS maximum of 640 kilobytes and provides paged access to the remainder.

The simplified block schematic of the CeleSTE PC is shown in Figure 2.15. XT functionality is provided by means of Chips and Technology's 82C100 *XT Controller* whilst standard PC I/O is provided by an 82C606 *Multi I/O Controller*.

An NCR 72C81 *CGMA Controller* provides CGA, MDA and Hercules graphics modes offering 720 × 350 pixels in monochrome, 640 × 200 pixels in two colours, 320 × 200 pixels in four colours, 160 × 200 pixels in 16 colours and 80/40 column text in 16 colours.

The 82C100 XT Controller is a highly integrated VLSI device which

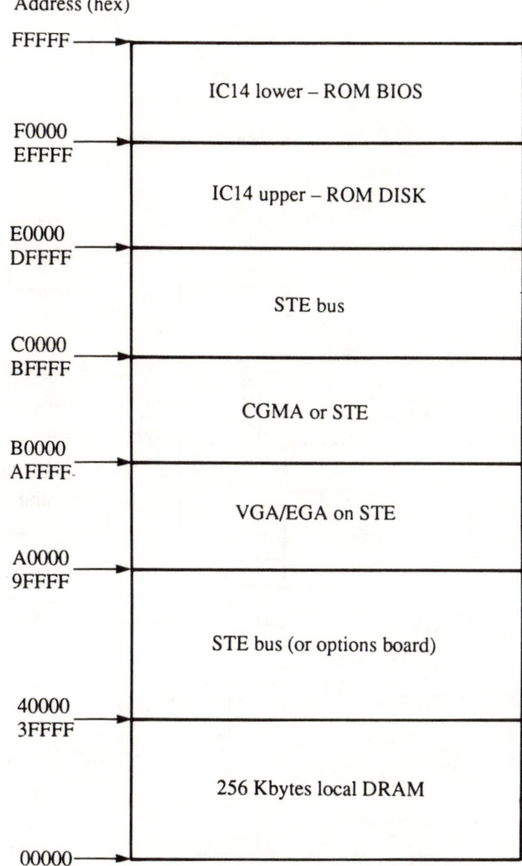

Figure 2.14 *Memory map of the CeleSTE PC*

replaces the following devices found in the original PC/XT:

8237	DMA Controller
8253	Counter/Timer
8255	Peripheral Interface
8259	Interrupt Controller
8284	Clock Generator
8288	Bus Controller
DRAM	Control Logic

The 82C606 Multi I/O Controller (also from Chips and Technology) provides further PC/XT hardware functionality. This device provides the following facilities:

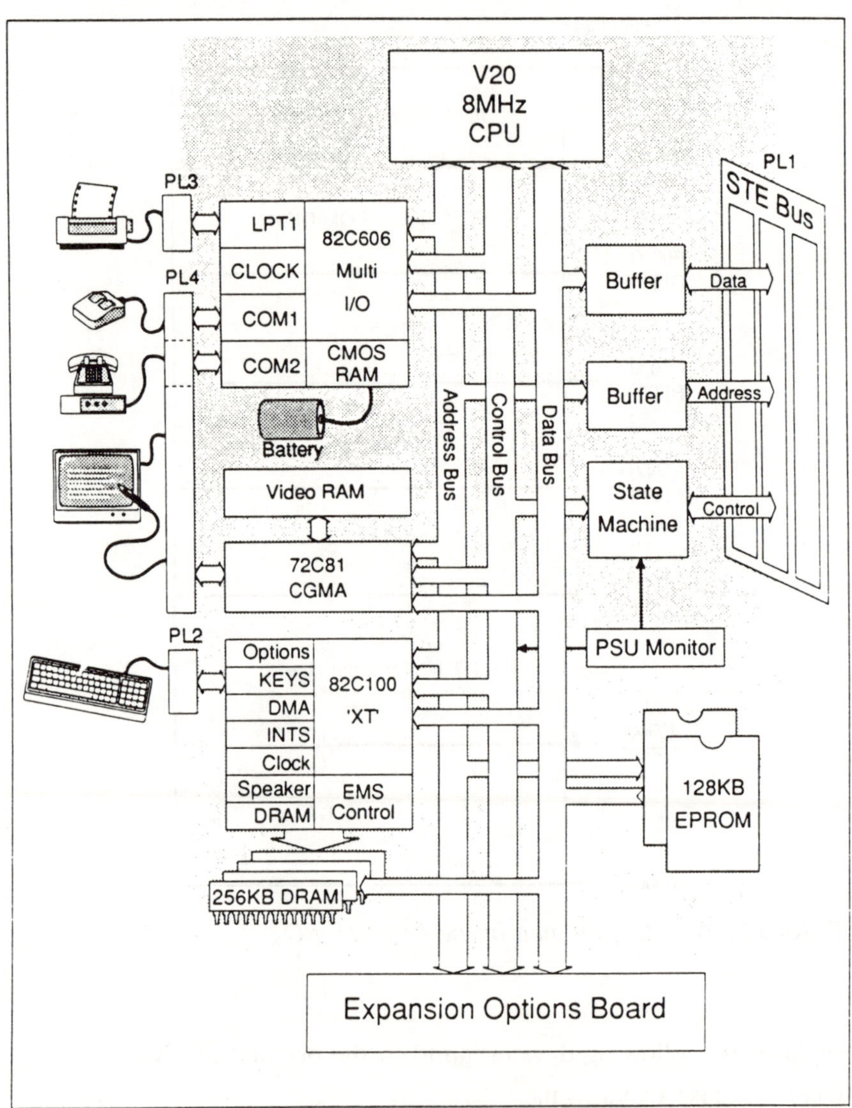

Figure 2.15 *Simplified block schematic of the CeleSTE PC*

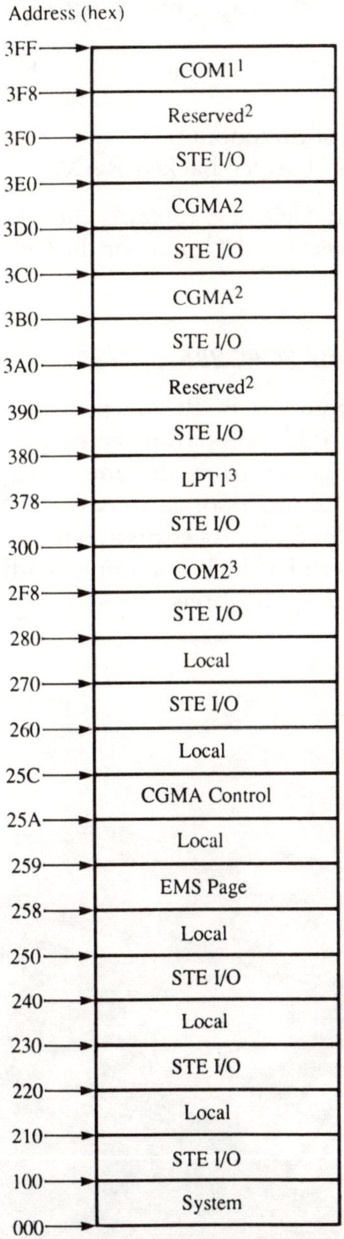

Address (hex)	
3FF	
	COM1[1]
3F8	
	Reserved[2]
3F0	
	STE I/O
3E0	
	CGMA2
3D0	
	STE I/O
3C0	
	CGMA[2]
3B0	
	STE I/O
3A0	
	Reserved[2]
390	
	STE I/O
380	
	LPT1[3]
378	
	STE I/O
300	
	COM2[3]
2F8	
	STE I/O
280	
	Local
270	
	STE I/O
260	
	Local
25C	
	CGMA Control
25A	
	Local
259	
	EMS Page
258	
	Local
250	
	STE I/O
240	
	Local
230	
	STE I/O
220	
	Local
210	
	STE I/O
100	
	System
000	

Notes: 1 If disabled by software, the location is reserved.
2 These locations will be STE I/O if the CGMA is not fitted or is disabled.
3 If disabled by software, these locations will be STE I/O.

Figure 2.16 *I/O map for the CeleSTE PC*

- Dual INS16450 (8250 compatible) UART (with modem control signals).
- MC146818 compatible real-time clock (which may be battery-backed).
- Parallel interface (bi-directional).
- Limited (114 bytes) battery-backed RAM.

As with the ECPC, the CeleSTE replaces the PC/XT I/O channel is replaced by the STE bus. The I/O map for the CeleSTE PC is shown in Figure 2.16.

Arcom control systems SCPC286

The Arcom Control Systems SCPC286 provides all of the standard functions of an AT or PS/2-compatible computer within an STE bus environment. Following on from the immensely popular SCPC88 PC/XT-compatible STE processor card, Arcom's SCPC286 represents current state of the art for STE processors and provides the designer with an exceptionally powerful vehicle for implementing a range of data acquisition and control applications.

Plate 2.11 *Arcom Control Systems SCPC286 (photo courtesy of Arcom Control Systems)*

The SCPC286 employs an 80286 CPU running at 12 MHz (Norton Speed Index = 11.5). The standard Eurocard is fitted with 0.5 megabyte of RAM (expandable to 1 megabyte), 64 kilobytes of EPROM, a socket for an 80287 mathematics coprocessor, a battery-backed real-time clock and a loudspeaker.

A fast hard-disk BIOS is supplied with the board which is designed to run standard XT/AT or PS/2 MS-DOS software. Further utilities include EMS providing extended memory support to LIM specifications, a disk PROM and RAM disk for implementing diskless systems, and a shadow RAM to speed BIOS execution of real-time software. Expansion options (available through the STE bus) include CGA/EGA graphics controllers and disk controllers.

3 The operating system

Anyone who has made passing use of a microcomputer system will be aware of at least some of the facilities offered by its operating system. Such an awareness is developed by means of the interface between the operating system and the user; the system generates prompts and messages, and the user makes an appropriate response.

Many of the functions of an operating system (like those associated with disk filing) are obvious. Others, however, are so closely related to the machine's hardware that the average user remains blissfully unaware of them. This, of course, is as it should be. As far as most end-users of computer systems are concerned, the operating system provides an environment from which it is possible to launch and run applications programs and to carry out elementary maintenance of disk files. Here, the operating system is perhaps better described as a 'microcomputer resource manager'.

The operating system provides an essential bridge between the user's application programs and the system hardware. In order to provide a standardized environment (which will cater for a variety of different hardware configurations) and ensure a high degree of software portability, part of the operating system is hardware independent whilst the hardware dependent remainder provides the individual low-level routines required by the machine in question.

An applications program will primarily interact with the hardware independent (DOS) routines. These, in turn, will interact with the lower-level hardware dependent (BIOS) routines. Figure 3.1 illustrates this important point.

In the context of software development, the operating system takes on a new significance. The software developer needs to interact with the operating system at a much lower level than the applications user. The operating system has a vital role to play in making effective use of the hardware resource. A range of utility programs are required (including

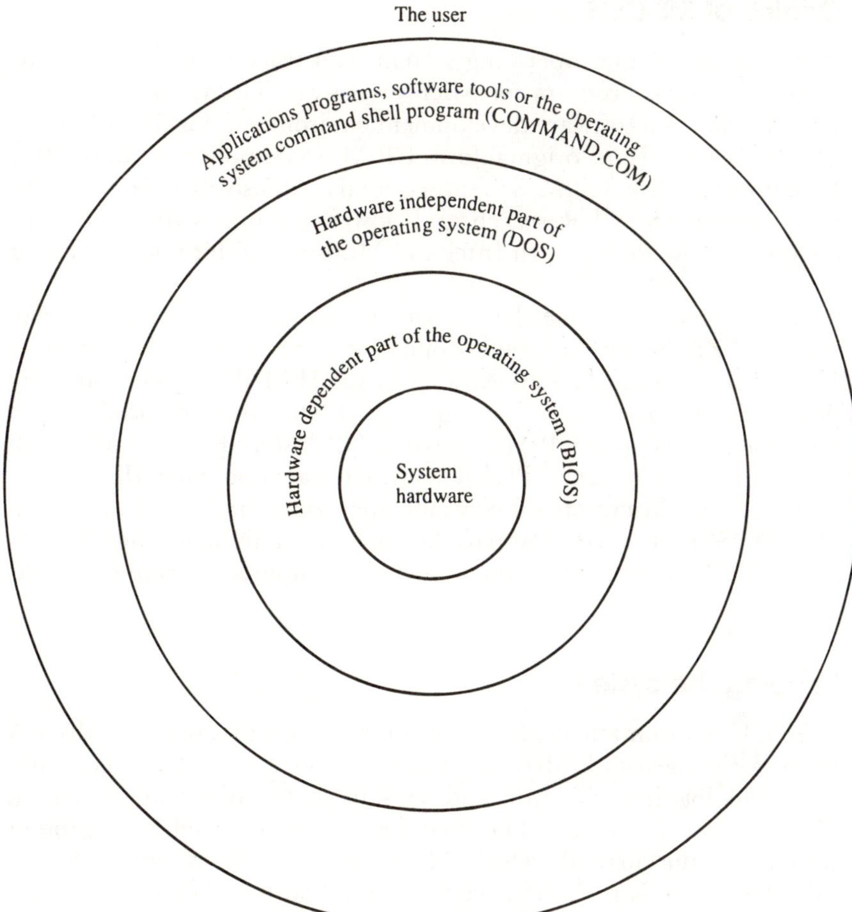

Figure 3.1 *Model showing the relationship between the different layers of the operating system (DOS and BIOS)*

such items as editors, assemblers, linkers and debuggers). These software tools work together with the operating system to provide an environment which facilitates effective software development.

This chapter outlines the facilities provided by the ever-popular MS-DOS operating system. Particular emphasis has been placed on those features which are of particular relevance to the engineer user and the software developer. Time spent in getting to know an operating system (including its pecularities and foibles) can be instrumental in saving considerable time and avoiding a variety of pitfalls. Readers who require further information of a general nature are advised to refer to one, or more, of the texts listed in Appendix E.

Origins of MS-DOS

Most microcomputer operating systems can trace their origins to the original Control Program for Microprocessors (CP/M) developed by Gary Kildall as a software development environment for the Intel 8080 microprocessor. In its original form CP/M was supplied on an 8″ IBM format floppy disk. CP/M was subsequently extensively developed and marketed by Digital Research and now exists in a variety of forms including those for use with Intel and Motorola 16-bit microprocessor families.

CP/M rapidly became the *de facto* operating system for most 8080, 8085 and Z80 based 8-bit microcomputer systems. With the advent of 16-bit machines and the appearance of the IBM PC, a new and more powerful operating system was required. Digital Research produced an 8086-based version of CP/M (known as CP/M-86); however Microsoft produced a rival product (PC-DOS) which was adopted by IBM for use with the PC. Microsoft quickly also developed an operating system (MS-DOS) for use with compatible machines. This operating system is currently the world's most popular microcomputer operating system.

Loading the system

MS-DOS is automatically loaded from the floppy disk placed in drive A or from the hard disk (drive C) whenever the system is booted. After successful loading, the title and version of the operating system is displayed on the screen. The message is then followed by a prompt which gives the currently selected drive (A or C). This prompt indicates that the system is ready to accept a command from the user.

If an AUTOEXEC batch file is present, the commands which it contains are executed before control is passed to the user. Furthermore, if such a file contains the name of an executable program (i.e. a file with a COM or EXE extension) then this program will be loaded from disk and executed. The program may take one of several forms including a program which simply performs its function and is then cleared from memory, a 'terminate and stay resident' (TSR) program, or a fully-blown application (such as a text editor or BASIC compiler).

In any event, it is important to remember that the currently selected drive remains the default drive unless explicitly changed by the user. As an example, consider a system which is booted with a system disk (floppy) placed in drive A. The default drive will then be A (unless an AUTOEXEC file is present which contains commands to change the current drive). The system prompt will indicate that A is the current drive. Thereafter, it is implicit that all commands which do not specify a drive refer, by default, to that drive. The SET PATH command (see page 120) can, however, be used to specify a directory path which will be searched if a command of filename does not appear in the current directory.

Making back-up copies of disks

It is often necessary to make back-up copies of software supplied on distribution disks. Indeed, it should be considered essential to make at least one back-up copy of *every* disk in current use. This simple precaution can help to save much agonizing when a disk becomes corrupt or is inadvertently subjected to a FORMAT command!

Having made a back-up copy, the distribution or master disk should be safely stored away and the working copy clearly labelled with the program name, version number and creation date, where appropriate. Assuming that a hard disk-based system is in use and that the DOS command utilities are placed in a sub-directory named DOS on drive C, the procedure for backing-up a floppy disk is as follows:

1 Boot the system from the hard disk in the normal way.
2 When the system prompt (C>) appears, enter the command:

```
SET PATH=C:\DOS
```

(the command may be entered in either upper or lower case and should be immediately followed by the <ENTER> key). An alternative to using the SET PATH command is to make the DOS directory by entering the command:

```
CD DOS
```

This step can be omitted if the DOS command utilities are present in the root directory (see page 352) of drive C.
3 Now enter the command:

```
DISKCOPY A: A:
```

The system will respond with a message of the form:

```
Insert SOURCE diskette in drive A:
Press any key when ready . . .
```

4 Now insert the distribution or master disk in drive A. Close the drive door and press any key (e.g. <SPACE>). The system will read information from the master disk and transfer the contents of the disk to memory. At the start of this process a message of the form:

```
Copying 80 tracks
9 Sectors/Track, 2 Side(s)
```

will be displayed.

5 When all data has been transferred from the disk to memory, the system will prompt for insertion of the destination or target disk. The following message will appear:

```
Insert TARGET diskette in drive A:
Press any key when ready . . .
```

The destination disk should then be inserted. This disk may be a blank (unformatted) disk or may be a disk which has been previously written to. In the latter case, the disk write protection should be removed.

The drive door should then be closed and a key pressed. A blank (unformatted) disk will be formatted during the process, in which case the following message will appear:

```
Formatting while copying
```

6 When the copying process has been completed, the user will be prompted with the following message:

```
Copy another diskette (Y/N)?
```

Further disks may then be copied or the user may choose to exit from the DISKCOPY utility and return to the command prompt. In the latter case, the contents of the target disk may be checked by issuing the following command from the system prompt:

```
DIR A:
```

7 If it is necessary to abort the copying process at any stage, the user should use the <CTRL-C> key combination. It should also be noted that earlier versions of MS-DOS require the user to format disks before using DISKCOPY.

Devices

The MS-DOS operating system can be configured for operation with a wide variety of peripheral devices including various types of monitor (CGA, HGA, EGA, VGA etc.), serial and parallel printers and modems. Each individual hardware configuration requires its own particular I/O provision and this is achieved by means of a piece of software known as a *device driver*. A number of device drivers (e.g. those which deal with the standard serial and parallel ports) are resident within the BIOS ROM. Others which may be required must be loaded into RAM during system initialization.

Input and output channels

In order to simplify the way in which MS-DOS handles input and output, the system recognizes the names of its various I/O devices. This may, at first, appear to be unnecessarily cumbersome but it is instrumental in allowing MS-DOS to redirect data. This feature can be extremely useful when, for example, output normally destined for the printer is to be redirected to an auxiliary serial port.

The following I/O channels are recognised by MS-DOS:

Channel	*Meaning*	*Function*	*Notes*
COM1: and COM2:	Communications	Serial I/O	Via RS-232 ports
CON:	Console	Keyboard (input) and screen (output)	This channel is equivalent to that which would be associated with a computer *terminal*
LPT1: LPT2: and LPT3:	Line printer	Parallel printer (output)	This interface conforms to the Centronics standard
PRN:	Printer	Serial or parallel printer (output)	
NUL:	Null device	Simulated I/O	Provides a means of simulating a physical I/O channel without data transfer taking place

The COPY command (see page 118) can be used to transfer data from one device to another. As an example, the command:

```
COPY CON: PRN:
```

copies data from the keyboard (consule input device) to the printer. Similarly, the command:

```
COPY CON: COM1:
```

copies data from the keyboard to the serial port (COM1). In either case, the end-of-file character, <CTRL-Z> or <F6>, must be entered to abort the command.

Redirection

The ability to redirect data is an extremely useful facility. The < and > characters are used within certain MS-DOS commands to indicate redirection of input and output data. Hence:

```
TYPE A:README.DOC > PRN:
```

can be used to redirect the normal screen output produced by the TYPE command to the printer. In this case, the content of a file named README.DOC (present in the root directory of the disk in drive A:) is sent to the printer.

Finally, the MODE command can be used to establish redirection of printer output from the (default) parallel port to a serial port. This facility is extremely useful where a serial printer has to be used in place of the (more usual) parallel printer. For further information see page 126.

MS-DOS commands

MS-DOS responds to command lines typed at the console and terminated with a <RETURN> or <ENTER> keystroke. A command line is thus composed of a command keyword, an optional command tail, and <RETURN>. The command keyword identifies the command (or program) to be executed. The command tail can contain extra information relevant to the command, such as a filename or other parameters. Each command line must be terminated using <RETURN> or <ENTER> (not shown in the examples which follow).

As an example, the following command can be used to display a directory of all assembly language source code files (i.e. those with an ASM extension) within a directory named MIC in drive A, indicating the size of each:

```
DIR A:\MIC\*.ASM
```

(see Figure 3.2)

Note that, in this example and the examples which follow, we have omitted the prompt generated by the system (indicating the current drive).

It should be noted that the command line can be entered in any combination of upper-case or lower-case characters. MS-DOS converts all letters in the command line to upper-case before interpreting them. Furthermore, whilst a command line generally immediately follows the system prompt, MS-DOS permits spaces between the prompt (>) and the command keyword.

As characters are typed at the keyboard, the cursor moves to the right in order to indicate the position of the next character to be typed.

```
 Volume in drive C is DISK1_VOL1
 Directory of  C:\MIC

DI        ASM       89  28-01-89  12:26a
EN        ASM       86  28-01-89  12:47a
TEST      ASM      323  24-01-89  11:52p
PAGERR    ASM    11192   1-02-88   1:00p
SHOWR     ASM     8944   1-02-88   1:00p
RECORD    ASM     3059   9-04-89   7:04p
PAGERP    ASM    11504   1-02-88   1:00p
SHOWP     ASM    15079   1-02-88   1:00p
PTEST     ASM      811  29-10-89   9:07p
          9 File(s)    5228544 bytes free
```

Figure 3.2 *Typical response to the MS-DOS DIR command*

Depending upon the keyboard used, a <BACKSPACE>, or <DELETE> key can be used to delete the last entered character and move the cursor backwards one character position. Alternatively, a combination of the CONTROL and H keys (i.e. <CTRL-H> may be used instead. Various other control characters are significant in MS-DOS and these have been shown in Table 3.1.

If it is necessary to repeat or edit the previous command, the <F1> (or right-arrow) key may be used to reproduce the command line, character by character, on the screen. The left-arrow key permits backwards movement through the command line for editing purposes. The <F3> key simply repeats the last command in its entirety.

File specifications

Many of the MS-DOS commands make explicit reference to files. A file is simply a collection of related information stored on a disk. Program files comprise a series of instructions to be executed by the processor whereas data files simply contain a collection of records. A complete file specification has three distinct parts: a drive specifier, a filename and a filetype. The drive specifier is a single letter followed by a colon which separates it from the filename which follows. The filename comprises one to eight characters while the filetype is usually specified in a one- to three-character extension. The filetype extension is separated from the filename by means of a full stop. A complete file specification (or *filespec*) thus takes the form:

[drive specifier]:[filename].[filetype]

Table 3.1 MS-DOS control characters

Control character	*Hex.*	*Function*
<CTRL-C>	03	Terminates the current program (if possible) and returns control to the user.
<CTRL-G>	07	Sounds the audible warning device (bell). Can only be used as part of a program of batch file.
<CTRL-H>	08	Moves the cursor back by one space (i.e. the same as the <BACKSPACE> key) and deletes the character present at that position.
<CTRL-I>	09	Tabs the cursor right by a fixed number of columns (usually eight). Performs the same function as the <TAB> key.
<CTRL-J>	10	Issues a line feed and carriage return, effectively moving the cursor to the start of the next line.
<CTRL-L>	12	Issues a form feed instruction to the printer.
<CTRL-M>	13	Produces a carriage return (i.e. has the same effect as <RETURN>).
<CTRL-P>	16	Toggles screen output to the printer (i.e. after the first <CTRL-P> is issued, all screen output will be simultaneously echoed to the printer. A subsequent <CTRL-P> will disable the simultaneous printing of the screen output). Note that <CTRL-PRT.SC.> has the same effect as <CTRL-P>.
<CTRL-S>	19	Pauses screen output during execution of the TYPE command (<CTRL-NUM.LOCK > has the same effect).
<CTRL-Z>	26	Indicates the end of a file (can also be entered using <F6>).

Note that, while not a control character, <CTRL-ALT-DEL> can be used to terminate the current program and perform a system reset. This particular combination of keys should only be used in the last resort as it will clear the system memory. Any program or data present in RAM will be lost!

As an example, the following file specification refers to a file named PROCESS and having a COM filetype stored on the disk on drive A:

```
A:PROCESS.COM
```

MS-DOS allows files to be grouped together within directories and sub-directories. Directory and sub-directory names are separated by means of the backslash (\) character. Directories and sub-directories are organized in an hierarchical (tree) structure and thus complete file specifications must include directory information. The following general format is used:

[drive specifier]:\[directory name(s)]\[filename].[filetype]

The base directory (i.e. that which exists at the lowest level in the hierarchical structure) is known as the *root directory*. The root directory is accessed by default when we simply specify a drive name without further reference to a directory. Thus:

```
A:PROCESS.COM
```

refers to a file in the root directory whilst:

```
A:\APPS\PROCESS.COM
```

refers to an identically named file resident in the APPS sub-directory.

Sub-directories can be extended to any practicable level. As an example:

```
A:\APPS\DATA\PROCESS.DAT
```

refers to a file named PROCESS.DAT present in the DATA sub-directory within a directory which is itself named APPS.

When it is necessary to make explicit reference to the root directory, we can simply use a single backslash character as follows:

```
A:\
```

Figure 3.3 shows a typical hierarchical directory structure.

File extensions

The filetype extension provides a convenient mechanism for distinguishing different types of file and MS-DOS provides various methods for manipulating groups of files having the same filetype extension. We could, for example, delete all of the assembly language source code files present in the root directory of the disk in drive A using a single command of the form:

```
ERA A:\*.ASM
```

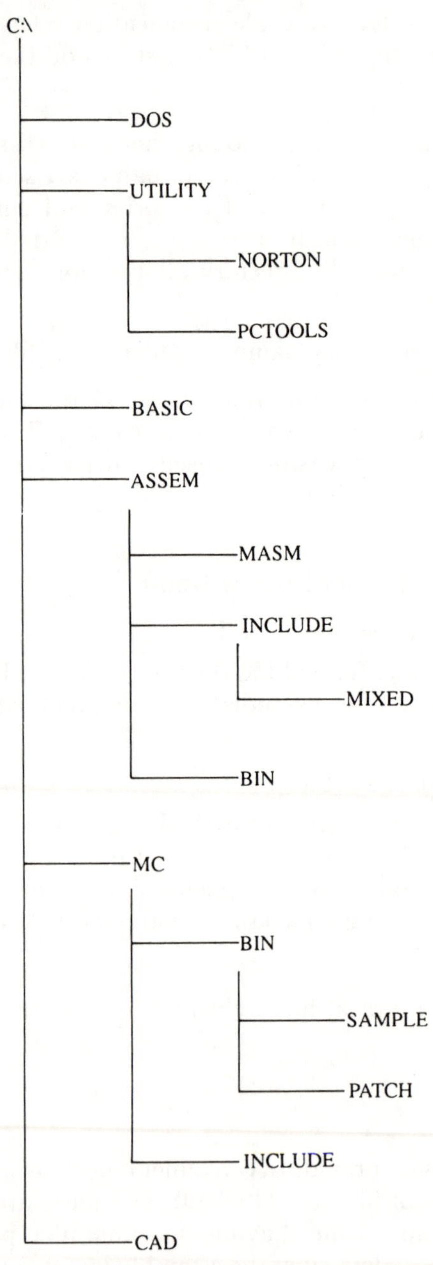

Figure 3.3 *Typical directory tree structure*

Alternatively, we could copy all of the executable (EXE) files from one disk in drive A to another in drive B using the command:

```
COPY A:\*.EXE B:\
```

Commonly used filetype extensions are shown in Table 3.2.

Wildcard characters

MS-DOS allows the user to employ wildcard characters when specifying files. The characters, * and ?, can be used to replace complete fields and individual characters respectively within a file specification. MS-DOS will search, then carry out the required operation on all files for which a match is obtained.

The following examples illustrate the use of wildcard characters:

`A:\*.ASM`	refers to *all* files having an ASM extension present in the root directory of drive A.
`C:\MC\*.*`	refers to *all* files (regardless of name or extension) present in the directory named MC on drive C.
`B:\PROC?.C`	refers to *all* files having a C extension present in the TURBO directory on the disk in drive B which have PROC as their first three letters and any alphanumeric character in the fourth character place. A match will occur for *each* of the following files: PROC1.C PROC2.C PROC3.C PROCA.C PROCB.C etc

Internal and external commands

A distinction must be made between MS-DOS commands which relate to the resident part of the operating system (internal commands) and those which involve other utility programs (external commands). Intrinsic commands are executed immediately whereas extrinsic commands require the loading of transient utility programs from disk and hence there is a short delay before the command is acted upon.

In the case of external commands, MS-DOS checks only the

Table 3.2 Common MS-DOS file types

Extension	*Type of file*
ASC	An ASCII text file.
ASM	An assembly language source code file.
BAK	A back-up file (often created automatically by a text editor which renames the source file with this extension and the revised file assumes the original file specification).
BAS	A BASIC program source file.
BAT	A batch file which contains a sequence of operating system commands.
BIN	A binary file (comprising instructions and data in binary format).
C	A source code file written in the C language.
COM	An executable program file in small memory format (i.e. confined to a single 64-kilobyte memory segment).
CRF	A cross reference file (for processing by a utility which will produce a cross-reference listing of symbols for debugging purposes).
DAT	A data file (usually presented in either binary or ASCII format).
DOC	A document file (not necessarily presented in standard ASCII format).
EXE	An executable program file in large memory format (i.e. not confined to a 64-kilobyte memory model).
HEX	A file presented in hexadecimal (an intermediate format sometimes used for object code).
INI	An initialization file which may contain a set of inference rules and/or environment variables.
LIB	A library file (containing multiple object code files).
LST	A listing file (usually showing the assembly code corresponding to each source code instruction together with a complete list of symbols).
MAK	A make file generated by a make utility (i.e. MAKE). A make file contains information which allows a compiler to identify together the various modules and include files required in the compilation process.
MAP	A map file generated by a linker utility (i.e. LINK). A map file consists of a list of symbols and their corresponding addresses.
OBJ	An object code file.
SYS	A system file.
TXT	A text file (usually in ASCII format).

command keyword. Any parameters which follow are passed to the utility program without checking.

At this point we should perhaps mention that MS-DOS only recognizes command keywords which are correctly spelled! Even an obvious typing error will result in the non-acceptance of the command and the system will respond with an appropriate error message.

As an example, suppose the user is attempting to format a disk but types FORMATT instead of FORMAT. The system responds with the message:

```
Bad command or file name
```

indicating that the command is unknown and that no file of that name (with a COM, BAT or EXE extension) is present in the current directory.

Internal MS-DOS commands

We shall now briefly examine the function of each of the most commonly used internal MS-DOS commands. Examples have been included wherever they can help to clarify the action of a particular command. Readers intending to make extensive use of MS-DOS should refer on or more of the texts referred to in Appendix E.

Command	*Function*
BREAK	The BREAK command disables the means by which it is possible to abort a running program. This facility is provided by means of the <CTRL-C> or <CTRL-BREAK> key combinations and it normally only occurs when output is being directed to the screen or the printer. BREAK accepts two parameters, ON and OFF. *Examples:* `BREAK ON` enables full <CTRL-C> or <CTRL-BREAK> key checking (it is important to note that this will normally produce a dramatic reduction in the speed of execution of a program). `BREAK OFF` restores normal <CTRL-C> or <CTRL-BREAK> operation (i.e. the default condition).
CD	*See* CHDIR.
CHDIR	The CHDIR command allows users to display or change the current directory. CHDIR may be abbreviated to CD.

Command	*Function*
	Examples: `CHDIR A:` displays the current directory path for the disk in drive A. `CHDIR C:\APPS` changes the directory path to APPS on drive C. `CD D:\DEV\PROCESS` changes the directory path to the sub-directory PROCESS within the directory named DEV on drive D. `CD\` changes the directory path to the root directory of the current drive.
CLS	The screen may be cleared using the CLS command.
COPY	The COPY command can be used to transfer a file from one disk to another using the same or a different filename. The COPY command is effective when the user has only a single drive. The COPY command must be followed by one or two file specifications. When only a single file specification is given, the command makes a single drive copy of a file. The copied file takes the same filename as the original and the user is prompted to insert the source and destination disks at the appropriate point. Where both source and destination file specifications are included, the file is copied to the specified drive and the copy takes the specified name. Where only a destination drive is specified (i.e. the destination filename is omitted) the COPY command copies the file to the specified drive without altering the filename. COPY may be used with the * and ? wildcard characters in order to copy all files for which a match is found (see page 115). *Examples:* `COPY A:ED.COM` copies the file ED.COM present on the disk in drive A to another disk inserted in the same drive. The COPY utility generates prompts during the process. `COPY A: ED.COM B:` copies the file ED.COM present on the disk in drive A to the disk present in drive B. The copy will be given the name ED.COM.
DATE	The DATE command allows the date to be set or displayed. *Examples:* `DATE` displays the date on the screen and also prompts the user to make any desired changes. The

Command	*Function*
	user may press <RETURN> to leave the settings unchanged. `DATE 12-27-90` sets the date to 27th December 1990.
DEL	*See* ERASE.
DIR	The DIR command displays the names of all non-system files in the directory. Variations of the command allow the user to specify the drive to be searched and the types of files to be displayed. Further options govern the format of the directory display. *Examples:* `DIR` displays all files in the current default directory. `DIR A:` displays all files on the disk in drive B (but without changing the default drive to B). `A: DIR` changes the default drive to A (root directory) and then displays the contents of the root directory of the disk in drive A. `DIR *.BAS` displays all files with a BAS extension present in the current default directory drive. `DIR C:\DEV.*` displays all files named DEV (regardless of their type or extension) present in the root directory of drive C (the hard disk). `DIR C:\MC\*.BIN` displays all files having a BIN extension present in the sub-directory named MC on drive C (the hard disk). `DIR/W` displays a directory listing in 'wide' format (excluding size and creation date/time information) of the current default directory.
ERASE	The ERASE command is used to erase a filename from the directory and release the storage space occupied by a file. The ERASE command is identical to the DEL command and the two may be used interchangeably. ERASE may be used with the * and ? wildcard characters in order to erase all files for which a match occurs. *Examples:* `ERASE PROG1.ASM` erases the file named PROG1.ASM from the disk placed in the current default drive. `ERASE B:TEMP.DAT` erases the file named TEMP.DAT from the disk placed in drive B. `ERASE C:*.COM` erases all files having a COM

Command	*Function*
	extension present on the disk in drive C. `ERASE A:PROG1.*` erases all files named PROG1 (regardless of their type extension) present on the disk currently in drive A.
MKDIR	The MKDIR command is used to make a new directory or sub-directory. The command may be abbreviated to MD. *Examples:* `MKDIR APPS` creates a sub-directory named APPS within the current directory (note that the MKDIR command is often used after a CHDIR).
PATH	The PATH command may be used to display the current directory path. Alternatively, a new directory path may be established using the SET PATH command. *Examples:* `PATH` displays the current directory path (a typical response would be PATH=c:\pctools). `SET PATH=C:\DOS` makes the directory path C:\DOS.
PROMPT	The PROMPT command allows the user to change the system prompt. The PROMPT command is followed by a text string which replaces the system prompt. Special characters may be inserted within the string, as follows: $d current date $e escape character $g > $h backspace and erase $l < $n current drive $p current directory path $q = $t current time $v DOS version number $$ = $_ newline *Examples:* `PROMPT $t$g` changes the prompt to the current time followed by a >. `PROMPT Howard Associates PLC $-?` changes the prompt to Howard Associates PLC followed by a

Command	*Function*
	carriage return and newline on which a ? is displayed. `PROMPT` restores the normal system prompt (e.g. C>).
RD	*See* RMDIR.
RENAME	The RENAME command allows the user to rename a disk file. RENAME may be used with the * and ? wildcard characters in order to rename all files for which a match occurs. RENAME may be abbreviated to REN.
TIME	The TIME command allows the time to be set or displayed. *Examples:* `TIME` displays the time on the screen and also prompts the user to make any desired changes. The user may press <RETURN> to leave the settings unchanged. `TIME 14:30` sets the time to 2.30 p.m.
TYPE	The TYPE command allows the user to display the contents of an ASCII (text) file on the console screen. The TYPE command can be used with options which enable or disable paged mode displays. The <CTRL-S> key combination may be used to halt the display and <CTRL-Q> used to restart. <CTRL-C> may be used to abort the execution of the TYPE command and exit to the system. The file may be simultaneously echoed to the printer by means of <CTRL-P> which should be used before issuing the TYPE command. A second <CTRL-P> can be issued to disable the echo facility. *Example:* `TYPE B:PROG1.ASM` will display the contents of a file called PROG1.ASM stored on the disk placed in drive B. The file will be sent to the screen.
VER	The VER command displays the current DOS version.
VERIFY	The VERIFY command can be used to enable or disable disk file verification. VERIFY ON enables verification while VERIFY OFF disables verification. If VERIFY is used without ON or OFF, the system will display the state of verification (on or off).
VOL	The VOL command may be used to display the volume label of a disk.

External MS-DOS commands

Unlike internal commands, these commands will not function unless the appropriate MS-DOS utility program is resident in the current (default) directory. External commands are simply the names of utility programs (normally resident in the DOS sub-directory). If you need to gain access to these utilities from any directory or sub-directory, then the following lines should be included in your AUTOEXEC.BAT file (see page 340:

```
SET PATH=C:\DOS
```

The foregoing assumes that you have created a sub-directory called DOS on the hard disk and that this sub-directory contains the MS-DOS utility programs.

Command	*Function*
APPEND	The APPEND command allows the user to specify drives, directories and sub-directories which will be searched through when a reference is made to a particular data file. The APPEND command follows the same syntax as the PATH command.
ASSIGN	The ASSIGN command allows users to re-direct files between drives. ASSIGN is particularly useful when a RAM disk is used to replace a conventional disk drive. *Examples:* `ASSIGN A=D` results in drive D being searched for a file whenever a reference is made to drive A. The command may be countermanded by issuing a command of the form: `ASSIGN A=A` Alternatively, all current drive assignments may be overridden by simply using: `ASSIGN`
ATTRIB	The ATTRIB command allows the user to examine and/or set the attributes of a single file or a group of files. The ATTRIB command alters the file attribute byte (which appears within a disk directory) and which determines the status of the file (e.g. read-only). *Examples:* `ATTRIB A:\PROCESS.DOC` displays the attribute status of copies the file PROCESS.DOC contained in the root directory of the disk in drive A. `ATTRIB +R A:\PROCESS.DOC` changes the status of the file PROCESS.DOC contained in the root directory

Command	Function
	of the disk in drive A so that it is a read-only file. This command may be countermanded by issuing a command of the form: `ATTRIB -R A:\PROCESS.DOC`
BACKUP	The BACKUP command may be used to copy one or more files present on a hard disk to a number of floppy disks for security purposes. It is important to note that the BACKUP command stores files in a compressed format (i.e. not in the same format as that used by the COPY command). The BACKUP command may be used selectively with various options including those which allow files to be archived by date. The BACKUP command usually requires that the target disks have been previously formatted however, from MS-DOS V3.3 onwards, an option to format disks has been included. *Examples:* `BACKUP C:*.* A:` backs up all of the files present on the hard disk. This command usually requires that a large number of (formatted) disks are available for use in drive A. Disks should be numbered so that the data can later by RESTOREd in the correct sequence. `BACKUP C:\DEV\*.C A:` Backs up all of the files with a C extension present within the DEV sub-directory on drive C. `BACKUP C:\PROCESS\*.BAS A:/D:01-01-89` Backs up all of the files with a BAS extension present within the PROCESS sub-directory of drive C that were created or altered on or after 1st January 1990. `BACKUP C:\COMMS\*.* A:/F` Backs up all of the files present in the COMMS sub-directory of drive C and formats each disk as it is used.
CHKDSK	The CHKDSK command reports on disk utilization and provides information on total disk space, hidden files, directories, and user files. CHKDSK also gives the total memory and free memory available. CHKDSK incorporates options which can be used to enable reporting and to repair damaged files. *Example:* `CHKDSK C:\DEV\*.ASM/F/V` checks the specified disk and directory, examining all files with an ASM extension, reporting errors and attempting to correct them.

Command	*Function*
COMP	The COMP command may be used to compare two files on a line by line or character by character basis. The following options are available: /A use . . . to indicate differences /B perform comparison on a character basis /C do not report character differences /L perform line comparison for program files /N add line numbers /T leave tab characters /W ignore white space at beginning and end of lines *Example:* `COMP /B PROC1.ASM PROC2.ASM` carries out a comparison of the files PROC1.ASM and PROC2.ASM on a character by character basis.
DISKCOMP	The DISKCOMP command provides a means of comparing two (floppy) disks. DISKCOMP accepts drive names as parameters and the necessary prompts are generated when a single-drive disk comparison is made. *Example:* `DISKCOMP A: B:` compares the disk in drive A with that placed in drive B.
EXE2BIN	The EXE2BIN utility converts, where possible, an EXE program file to a COM program file (which loads faster and makes less demands on memory space). The command will *not* operate on EXE files that require more than 64 kilobytes of memory space (including space for the stack and data storage) and/or those that make reference to other memory segments (i.e. CS, DS, ES, and SS must all remain the same within the program). *Example:* `EXE2BIN PROCESS` will search for the program PROCESS.EXE and generate a program PROCESS.COM.
FASTOPEN	The FASTOPEN command provides a means of rapidly accessing files. The command is only effective when a hard disk is fitted and should ideally be used when the system is initialized (e.g. from within the AUTOEXEC.BAT file). The command retains details of files within memory and must not be used concurrently with the commands ASSIGN, JOIN, and/or SUBST.

Command	*Function*
	Example: `FASTOPEN C:32` enables fast opening of files and provides for the details of up to 32 files to be retained in RAM.
FDISK	The FDISK utility allows users to format a hard (fixed) disk. Since the command will render any existing data stored on the disk inaccessible, FDISK should be used with *extreme* caution. Furthermore, improved hard disk partitioning and formatting utilities are normally supplied when a hard disk is purchased. These should be used in preference to FDISK whenever possible.
FIND	The FIND command can be used to search for a character string within a file. Options include: /C display the line number(s) where the search string has been located /N number the lines to show the position within the file /V display all lines which do *not* contain the search string *Example:* `FIND/C "outport" C:/DEV/PROCESS.C` searches the file PROCESS.C present in the DEV sub-directory for occurrences of outport. When the search string is located, the command displays the appropriate line number.
FORMAT	The FORMAT command is used to initialize a floppy or hard disk. The command should be used with caution since it will generally not be possible to recover any data which was previously present. Various options are available including: /1 single-sided format /8 format with 8 sectors per track /B leave space for system tracks to be added (using the SYS command) /N:8 format with 8 sectors per track /S write system tracks during formatting (note that this must be the last option specified when more than one option is required) /T:80 format with 80 tracks /V format and then prompt for a volume label *Examples:* `FORMAT A:` formats the disk placed in drive A. `FORMAT B:/S` formats the disk placed in drive B as a system disk.

Command	*Function*
JOIN	The JOIN command provides a means of associating a drive with a particular directory path. The command must be used with care and must *not* be used with ASSIGN, BACKUP, DISKCOPY, FORMAT etc.
KEYB	The KEYB command invokes the MS-DOS keyboard driver. KEYB replaces earlier utilities (such as KEYBUK) which were provided with MS-DOS versions prior to V3.3. The command is usually incorporated in an AUTOEXEC.BAT file and must specify the country letters required. *Example:* `KEYB UK` selects the UK keyboard layout.
LABEL	The LABEL command allows a volume label (maximum 11 characters) to be placed in the disk directory. *Example:* `LABEL A: PROCESS` will label the disk present in drive A as PROCESS. This label will subsequently appear when the directory is displayed.
MODE	The MODE command can be used to select from a range of screen and printer options. MODE is an extremely versatile command and offers a wide variety of options. *Examples:* `MODE LPT1: 120,6` initializes the parallel printer LPT1 for printing 120 columns at 6 lines per inch. `MODE LPT2: 60,8` initializes the parallel printer LPT2 for printing 60 columns at 8 lines per inch. `MODE COM1: 1200,N,8,1` initializes the COM1 serial port for 1200 baud operation with no parity, eight data bits and one stop bit. `MODE COM2: 9600,N,7,2` initializes the COM2 serial port for 9600 baud operation with no parity, seven data bits and two stop bits. `MODE 40` sets the screen to 40 column text mode. `MODE 80` sets the screen to 80 column mode. `MODE BW80` sets the screen to monochrome 40 column text mode. `MODE CO80` sets the screen to colour 80 column mode. The MODE command can also be used to permit redirection of printer output.

Command	*Function*
	Example: `MODE LPT1:=COM1:` redirects output from the (default) parallel port to the RS-232 serial port. Normal operation can be restored by the command: `MODE LPT1:`
PRINT	The PRINT command sends the contents of an ASCII text file to the printer. Printing is carried out as a background operation and data is buffered in memory. The default buffer size is 512 bytes; however the size of the buffer can be specified using /B: (followed by required buffer size in bytes). When the utility is first entered, the user is presented with the opportunity to redirect printing to the serial port (COM1:). A list of files (held in a queue) can also be specified. *Examples:* `PRINT README.DOC` prints the file README.DOC from the current directory. `PRINT /B:4096 HELP1.TXT HELP2.TXT HELP3.TXT` establishes a print queue with the files HELP1.TXT, HELP2.TXT, and HELP3.TXT and also sets the print buffer to 4 kilobytes. The files are sent to the printer in the specified sequence.
RESTORE	The RESTORE command is used to replace files on the hard disk which were previously saved on floppy disk(s) using the BACKUP command. Various options are provided (including restoration of files created before or after a specified date. *Examples:* `RESTORE C:\DEV\PROCESS.COM` restores the files PROCESS.COM in the sub-directory named DEV on the hard disk partition, C. The user is prompted to insert the appropriate floppy disk (in drive A). `RESTORE C:\BASIC /M` restores all modified (altered or deleted) files present in the sub-directory named BASIC on the hard disk partition, C.
SYS	The SYS command creates a new boot disk by copying the hidden MS-DOS system files. SYS is normally used to transfer system files to a disk which has been formatted with the /S or /B option. SYS cannot be used on a disk which has had data written to it after initial formatting.

Command	*Function*
TREE	The TREE command may be used to display a complete directory listing for a given drive. The listing starts with the root directory.
XCOPY	The XCOPY utility provides a means of selectively copying files. The utility creates a copy which has the same directory structure as the original. Various options are provided: `/A` only copy files which have their archive bit set (but do *not* reset the archive bits) `/D` only files which have been created (or that have been changed) after the specified date `/M` copy files which have their archive bit set but reset the archive bits (to avoid copying files unnecessarily at a later date) `/P` prompt for confirmation of each copy `/S` copy files from sub-directories `/V` verify each copy `/W` prompt for disk swaps when using a single drive machine *Example:* `XCOPY C:\DEV\*.* A:/M` copy all files present in the DEV sub-directory of drive C:. Files will be copied to the disk in drive A:. Only those files which have been modified (i.e. had their archive bits set) will be copied.

Batch files

Batch files provide a means of avoiding the tedium of repeating a sequence of operating system commands many times over. Batch files are nothing more than straightforward ASCII text files which contain the commands which are to be executed when the name of the batch is entered. Execution of a batch file is automatic; the commands are executed just as if they had been typed in at the keyboard. Batch files may also contain the names of executable program files (i.e. those with a COM or EXE extension), in which case the specified program is executed and, provided the program makes a conventional exit to DOS upon termination, execution of the batch file will resume upon termination.

Batch file commands

MS-DOS provides a number of commands which are specifically intended for inclusion within batch files.

Command	*Function*
ECHO	The ECHO command may be used to control screen output during execution of a batch file. ECHO may be followed by ON or OFF or by a text string which will be displayed when the command line is executed. *Examples:* `ECHO OFF` disables the echoing (to the screen) of commands contained within the batch file. `ECHO ON` re-enables the echoing (to the screen) of commands contained within the batch file. (Note that there is no need to use this command at the end of a batch file as the reinstatement of screen echo of keyboard generated commands is automatic.) `ECHO Sorting data - please wait!` displays the message: Sorting data – please wait! on the screen.

FOR	FOR is used with IN and DO to implement a series of repeated commands.
	
	Examples:
	`FOR %A IN (IN.ASM OUT.ASM MAIN.ASM) DO COPY %A LPT1:`
	copies the files IN.ASM, OUT.ASM, and MAIN.ASM in the current directory to the printer.
	`FOR %A IN (*.ASM) DO COPY %A LPT1:`
	copies all the files having an ASM extension in the current directory to the printer. The command has the same effect as COPY *.ASM LPT1:.

IF	If is used with GOTO to provide a means of branching within a batch file. GOTO must be followed by a label (which must begin with :). *Example:* `IF NOT EXIST INPUT.DAT GOTO :EXIT` transfers control to the label :EXIT if the file INPUT.DAT cannot be found in the current directory.
PAUSE	the pause command suspends execution of a batch file until the user presses any key. The message: `Press any key when ready . . .` is displayed on the screen.
REM	The REM command is used to precede lines of text which will constitute remarks. *Example:* `REM Check that the file exists before copying`

Creating batch files

Batch files may be created using an ASCII text editor or a word processor (operating in ASCII mode). Alternatively, if the batch file comprises only a few lines, the file may be created using the MS-DOS COPY command. As an example, let us suppose that we wish to create a batch file which will:

1 Erase all of the files present on the disk placed in drive B.
2 Copy all of the files in drive A having a TXT extension to produce an identically named set of files on the disk placed in drive B.
3 Rename all of the files having a TXT extension in drive A so that they have a BAK extension.

The required operating system commands are thus:

```
ERASE B:\*.*
COPY A:\*.TXT B:\
RENAME A:\*.TXT A:\*.BAK
```

The following *keystrokes* may be used to create a batch file named ARCHIVE.BAT containing the above commands:

```
COPY CON: ARCHIVE.BAT<ENTER>
ERASE B:\*.*<ENTER>
COPY A:\*.TXT B:\<ENTER>
RENAME A:\*.TXT A:\*.BAK<ENTER>
<CTRL-Z><ENTER>
```

If you wish to view the batch file which you have just created simply enter the command:

```
TYPE ARCHIVE.BAT
```

Whenever you wish to execute the batch file simply type:

```
ARCHIVE
```

Note that, if necessary, the sequence of commands contained within a batch file may be interrupted by typing:

```
<CTRL-C>
```

The system will respond by asking you to confirm that you wish to terminate the batch job. Respond with Y to terminate the batch process or N if you wish to continue with it.

Additional commands can be easily appended to an existing batch file. Assume that we wish to view the directory of the disk in drive A after running the archive batch file. We can simply append the extra commands to the batch files by entering:

```
COPY ARCHIVE.BAT + CON:
```

The system displays the filename followed by the CON prompt. The extra line of text can now be entered using the following keystrokes:

```
DIR A:\<ENTER>
<CTRL-Z><ENTER>
```

Passing parameters

Parameters may be passed to batch files by including the % character to act as a place holder for each parameter passed. The parameters are numbered strictly in the sequence in which they appear after the name of the batch file. As an example, suppose that we have created a batch file called REBUILD, and this file requires two file specifications to be passed as parameters. Within the text of the batch file, these parameters will be represented by %1 and %2. The first file specification following the name of the batch file will be %1 and the second will be %2. Hence, if we enter the command:

```
REBUILD PROC1.DAT PROC2.DAT<ENTER>
```

During execution of the batch file, %1 will be replaced by PROC1.DAT whilst %2 will be replaced by PROC2.DAT.

It is also possible to implement simple testing and branching within a batch file. Labels used for branching should preferably be stated in lower case (to avoid confusion with operating systems commands) and should be preceded by a colon when they are the first (or only) statement in a line. The following example which produces a sorted list of directories illustrates these points:

```
ECHO OFF
IF EXIST %1 GOTO valid
ECHO Missing or invalid parameter
GOTO end
:valid
ECHO Index of Directories in %1
DIR %1 : FIND ''<DIR>'' : SORT
:end
```

The first line disables the echoing of subsequent commands contained within the batch file. The second line determines whether, or not, a valid parameter has been entered. If the parameter is invalid (or missing) the ECHO command is used to print an error message on the screen.

The MS-DOS debugger

One of the most useful tools available to the software developer within the MS-DOS environment is the debugger, DEBUG.COM. This program provides a variety of facilities including single-stepping a program to permit examination of the CPU registers and the contents of memory after execution of each instruction.

The debug command line can accept several arguments. Its syntax is as follows:

```
DEBUG [filespec] [parm1] [parm2]
```

where [filespec] is the specification of the file to be loaded into memory, [parm1] and [parm2] are optional parameters for the specified file.

As an example, the following MS-DOS command will load debug along with the file PROCESS.COM (taken from the disk in drive B) ready for debugging:

```
DEBUG B:PROCESS.COM
```

When debug has been loaded, the familiar MS-DOS prompt is replaced by a hyphen (-). This indicates that DEBUG is awaiting a command from the user. Commands comprise single letter (in either upper or lower case). Delimiters are optional between commands and parameters. They must, however, be used to separate adjacent hexadecimal values.

<CTRL-BREAK> can be used to abort a DEBUG command while <CTRL-NUM.LOCK> can be used to pause the display (any other keystroke restarts the output). Commands may be edited using the keys available for normal MS-DOS command editing.

The following commands are available:

Command	*Meaning*	*Function*
A [addr]	Assemble	Assemble mnemonics into memory from the specified address. If no address is specified, the code will be assembled into memory from address CS:0100. The <ENTER> key is used to terminate assembly and return to the debug prompt. *Examples:* `A 200` starts assembly from address CS:0200. `A 4E0:100` starts assembly from address 04E0:0100 (equivalent to a physical address of 04F00).
C range addr	Compare	Compare memory in the specified range with memory starting at the specified address.
D [addr]	Dump	Dump (display) memory from the given starting address. If no start address is specified, the dump will commence at DS:0100.

Command	*Meaning*	*Function*
		Examples: `D 400` dumps memory from address DS:0400. `D CS:0` dumps memory from address CS:0000.
D [range]		Dump (display) memory within the specified range. *Example:* `D DS:200 20F` displays 16 bytes of memory from DS:0200 to DS:0210 inclusive.
E addr [list]	Enter	Enter (edit) bytes into memory starting at the given address. If no list of data bytes is specified, byte values are displayed and may be sequentially overwritten. <SPACE> may be used to advance, and <–> may be used to reverse the memory pointer. *Example:* `E 200, 3C, FF,1A,FE` places byte values of 3C, FF, 1A and FE into four consecutive memory locations commencing at DS:0200.
F range list	Fill	Fills memory in the given range with data in the list. The list is repeated until all memory locations have been filled. *Examples:* `F 100,10F,FF` fills 16 bytes of memory with FF commencing at address DS:0100. `F 0,FFFF,AA,FF` fills 65536 bytes of memory with alternate bytes of AA and FF.
G [=addr]	Go	Execute the code starting at the given address. If no address is specified, execution commences at address CS:IP. *Example:* `G=100` executes the code starting at address CS:0100.

Command	*Meaning*	*Function*
G [=addr]	[addr] [addr] ...	Executes the code starting at the given address with the specified breakpoints. *Example:* `G=100 104 10B` executes the code starting at address CS:0100 and with breakpoints at addresses CS:0104 and CS:010B.
H value value	Hexadecimal	Calculates the sum and difference of two hexadecimal values.
I port	Port input	Inputs a byte value from the specified I/O port address and display the value. *Example:* `I 302` inputs the byte value from I/O port address 302 and displays the value returned.
L [addr]	Load	Loads the file previously specified by the Name (N) command. The file specification is held at address CS:0080. If no load address is specified, the file is loaded from address CS:0100.
M range addr	Move	Moves (replicates) memory in the given range so that it is replicated starting at the specified address.
N filespec	Name	Names a file to be used for a subsequent Load (L) or Write (W) command. *Example:* `N B:PROCESS.COM` names the file PROCESS.COM stored in the root directory of drive B for a subsequent load or write command.
O port byte	Port output	Output a given byte value to the specified I/O port address. *Example:* `O 303 FE` outputs a byte value of FE from I/O port address 303.

Command	*Meaning*	*Function*
P [=addr] [instr]	Proceed	Executes a subroutine, interrupt, loop or string operation and resumes control at the next instruction. Execution starts at the specified address and continues for the specified number of instructions. If no address is specified, execution commences at the address given by CS:IP.
Q	Quit	Exits debug and return control to the current MS-DOS shell.
R [regname]	Register	Displays the contents of the specified register and allows the contents to be modified. If a name is not specified, the contents of all of the CPU registers (including flags) is displayed together with the next instruction to be executed (in hexadecimal and in mnemonic format).
S range list		Search memory within the specified range for the listed data bytes. *Example:* `S 0100 0800 20, 1B` searches memory between address DS: 0100 and DS:0800 for consecutive data values of 20 and 1B.
T [=addr] [instr]	Trace	Traces the execution of a program from the specified address and executing the given number of instructions. If no address is specified, the execution starts at address CS:IP. If a number of instructions is not specified then only a single instruction is executed. A register dump (together with a disassembly of the next instruction to be executed) is displayed at each step. *Examples:* `T` traces the execution of the single instruction referenced by CS:IP. `T =200,4` traces the execution of four instructions commencing at address CS:0200.

Command	*Meaning*	*Function*
U [addr]	Unassemble	Unassemble (disassemble) code into mnemonic instructions starting at the specified address. If no address is specified, disassembly starts from the address given by CS:IP. *Examples:* `U` disassembles code starting at address CS:IP. `U 200` disassembles code starting at address CS:0200.
U [range]		Unassemble (disassemble) code into mnemonic instructions within the specified range of addresses. *Example:* `U 200 400` disassembles the code starting at address CS:0200 and ending at address CS:0400.
W [addr]	Write	Writes data to disk from the specified address. The file specification is taken from a previous Name (N) command. If the address is not specified, the address defaults to that specified by CS:IP. The file specification is located at CS:0080.

Notes:

1. Parameters enclosed in square brackets ([and]) are optional.
2. The equal sign (=) *must* precede the start address used by the following commands; Go (G), Proceed (P), and Trace (T).
3. Parameters have the following meanings:

Parameter	*Meaning*

addr	address (which may be quoted as an offset or as the contents of a segment register or segment address followed by an offset). The following are examples of acceptable addresses:
	
	`CS:0100`
	`04C0:0100`
	`0200`

byte	A byte of data (i.e. a value in the range 0 to FF). The following are examples of acceptable data bytes:
	
	`0`
	`1F`
	`FE`

filespec	A file specification (which may include a drive letter and sub-directory, etc.). The following are examples of acceptable file specifications:
	`PROCESS.COM`
	`B:PROCESS.COM`
	`B:\PROGS\PROCESS.COM`

instr	The number of instructions to be executed within a Trace (T) or Proceed (P) command.

list	A list of data bytes, ASCII characters (which must be enclosed in single quotes), or strings (which must be enclosed in double quotes). The following examples are all acceptable data lists:
	`3C,2F,C2,00,10`
	`'A','.','/'`
	`"Insert disk and press ENTER"`

port	A port address. The following are acceptable examples of port addresses: E (the DMA controller) 30C (within the prototype range) 378 (the parallel printer)
range	A range of addresses which may be expressed as an address and offset (e.g. CS:100,100) or as an address followed by a size (e.g. DS:100 L 20).
regname	A register name (see 5). The following are acceptable examples of register names: AX DS IP
value	A hexadecimal value in the range 0 to FFFF.

4 The following register and flag names are used within debug:

AX, BX, CX, DX	16-bit general purpose registers
CS, DS, ES, SS	Code, data, extra and stack segment registers
SP, BP, IP	Stack, base and instruction pointers
SI, DI	Source and destination index registers
F	Flag register

5 The following abbreviations are used to denote the state of the flags in conjunction with the Register (R) and Trace (T) commands:

Flag	*Abbreviation*	*Meaning/status*
Overflow	OV	Overflow
	NV	No overflow
Carry	CY	Carry
	NC	No carry
Zero	ZR	Zero
	NZ	Non-zero

Direction	DN	Down
	UP	Up
Interrupt	EI	Interrupts enabled
	DI	Interrupts disabled
Parity	PE	Parity even
	PO	Parity odd
Sign	NG	Negative
	PL	Positive
Auxiliary carry	AC	
	NC	

6 All numerical values within Debug are in hexadecimal.

Using Debug

The following 'walkthrough' has been provided in order to give readers an insight into the range of facilities offered by Debug. We shall assume that a short program TEST.COM has been written to test a printer connected to the parallel port. The program is designed to generate a single line of upper and lower case characters but, since an error is present, the compiled program prints only a single character. The source code for the program (TEST.ASM) is shown in Figure 3.4.

```
        ; This program outputs ASCII data
        ; in the range 41H to 7FH to the
        ; parallel printer port
        ; Registers used: AX,CL,DL
        ; Parameters passed:   none
        TITLE   ptest
_TEXT   SEGMENT          ; Define code segment
        ASSUME cs:_TEXT,ds:_TEXT,ss:_TEXT
        ORG     100H     ; Normal for COM program
start:  MOV     AH,05H   ; Printer output function code
        MOV     DL,0AH   ; First generate a
        INT     21H      ; line feed
        MOV     DL,0DH   ; Next generate a
        INT     21H      ; carriage return
        MOV     DL,41H   ; First character to print is A
        MOV     CL,3EH   ; Number of characters to print
        MOV     AH,05H   ; Set up the function code
        INT     21H      ; and print the character
prch:   INC     DL       ; Get the next character
        LOOP    prch     ; and go round again
        MOV     AL,00H   ; Set up the return code
        MOV     AH,4CH   ; and the function code
        INT     21H      ; for an exit to DOS
_TEXT   ENDS
        END     start
```

Figure 3.4 *Source code for the program* TEST.ASM

The first stage in the debugging process is to invoke Debug from MS-DOS using the command:

```
DEBUG TEST.COM
```

The command assumes that TEST.COM is present in the current directory and that DEBUG.COM is accessible either directly or via previous use of the SET PATH command.

After the Debug hyphen prompt appears, we can check that our code has loaded, we use the Dump (D) command. There is, however, no need to specify an address parameter since the code for TEST.COM will have been loaded at the default address (CS:0100). The resulting display has been shown in Figure 3.5 (CS will almost certainly be different on a different system).

```
1685:0100  B4 05 B2 0A CD 21 B2 0D-CD 21 B2 41 B1 3E B4 05   .....!...!.A.>..
1685:0110  CD 21 FE C2 E2 FC B0 00-B4 4C CD 21 89 46 F8 89   .!.......L.!.F..
1685:0120  56 FA 2B C0 50 FF 36 46-43 8B 46 FC 8B 56 FE B1   V.+.P.6FC.F..V..
1685:0130  04 D1 E0 D1 D2 FE C9 75-F8 52 50 8B 46 F8 8B 56   .......ų.RP.F..V
1685:0140  FA 05 0C 00 52 50 E8 7F-69 83 C4 04 50 E8 28 1F   ....RP..i...P.(.
1685:0150  83 C4 0A 0A C0 74 1A 80-3E B2 10 00 75 13 A1 06   .....t..>...u...
1685:0160  3C 8B 16 0B 3C A3 1A 41-89 16 1C 41 C6 06 B2 10   <...<..A...A....
1685:0170  FF 5E 8B E5 5D C3 55 8B-EC 56 8B 5E 04 8B 76 06   .^..].U..V.^..v.
```

Figure 3.5 *Using Debug's Dump (D) command to display* TEST.COM *in memory*

The extreme left-hand column of Figure 3.5 gives the address (in segment register:offset format). The next 16 columns comprise hexadecimal data showing the bytes stored at the 16 address locations starting at the address shown in the left-hand column. The first byte in the block (address 1685:0100) has a value of B4H, while the second (address 1685:0101) has a value of 05H, and so on. The last byte in the block (address 1685:017F) has the value 06H.

An ASCII representation of the data is shown in the right-hand column of Figure 3.5. Byte values that do not correspond to printable ASCII characters are shown simply as a full-stop. Hence B4H and 05H (which are both non-printable characters) are shown by full-stops whilst 21H appears as !, and 41H as A.

The hexadecimal/ASCII dump shown in Figure 3.5 is not particularly useful and a more meaningful representation can be achieved by using the Unassemble (U) command. Again, no parameters are required since the default CS:IP address will point to the start of our program code. The Unassemble command yields the display shown in Figure 3.6 which comprises a disassembly of 16 instructions starting from address 1685:0100.

The first instruction occupies 2 bytes of memory (addresses 1685:0100 and 1685:0101) and it comprises a move of a byte of immediate data

```
1685:0100 B405        MOV     AH,05
1685:0102 B20A        MOV     DL,0A
1685:0104 CD21        INT     21
1685:0106 B20D        MOV     DL,0D
1685:0108 CD21        INT     21
1685:010A B241        MOV     DL,41
1685:010C B13E        MOV     CL,3E
1685:010E B405        MOV     AH,05
1685:0110 CD21        INT     21
1685:0112 FEC2        INC     DL
1685:0114 E2FC        LOOP    0112
1685:0116 B000        MOV     AL,00
1685:0118 B44C        MOV     AH,4C
1685:011A CD21        INT     21
1685:011C 8946F8      MOV     [BP-08],AX
1685:011F 8956FA      MOV     [BP-06],DX
```

Figure 3.6 *Using Debug's Unassemble (U) command to disassemble* TEST.COM

(05H) into the AH register. The last program instruction is at address 1685:011A and is a software interrupt relating to an address in the interrupt vector table of 21H. The (unusual) instructions at addresses 1685:011C and 1685:011F do not form part of our program code and can be ignored. Furthermore, it is important to note that Unassemble command cannot distinguish program code from data and will consequently produce some meaningless instructions when it does attempt to disassemble data!

Having disassembled the program code resident in memory we can check it against the original source code file. Normally, however, this will not be necessary unless the object code file has become changed or corrupted in some way.

The next stage is that of tracing program execution. The Debug Trace (T) command could be employed for this function. However we can use the Proceed (P) command to avoid tracing execution of the DOS interrupt routines in order to keep the amount of traced code manageable.

The Proceed command expects its first parameter to be the address of the first instruction to be executed (in this case we can simply express the address offset as 100H). Furthermore, since our program terminates normally, we can supply any sufficiently large number of instructions as the second parameter to the Proceed command (in this case we shall again use 100H in the safe knowledge that we shall be returned to the debugger once our code terminates). The requisite command is thus:

```
AX=0541  BX=0000  CX=0000  DX=007F  SP=0000  BP=0000  SI=0000  DI=0000
DS=1675  ES=1675  SS=1685  CS=1685  IP=0102   NV UP EI PL NZ NA PO NC
1685:0102 B20A          MOV     DL,0A

AX=0541  BX=0000  CX=0000  DX=000A  SP=0000  BP=0000  SI=0000  DI=0000
DS=1675  ES=1675  SS=1685  CS=1685  IP=0104   NV UP EI PL NZ NA PO NC
1685:0104 CD21          INT     21

AX=050A  BX=0000  CX=0000  DX=000A  SP=0000  BP=0000  SI=0000  DI=0000
DS=1675  ES=1675  SS=1685  CS=1685  IP=0106   NV UP EI PL NZ NA PO NC
1685:0106 B20D          MOV     DL,0D

AX=050A  BX=0000  CX=0000  DX=000D  SP=0000  BP=0000  SI=0000  DI=0000
DS=1675  ES=1675  SS=1685  CS=1685  IP=0108   NV UP EI PL NZ NA PO NC
1685:0108 CD21          INT     21

AX=050D  BX=0000  CX=0000  DX=000D  SP=0000  BP=0000  SI=0000  DI=0000
DS=1675  ES=1675  SS=1685  CS=1685  IP=010A   NV UP EI PL NZ NA PO NC
1685:010A B241          MOV     DL,41

AX=050D  BX=0000  CX=0000  DX=0041  SP=0000  BP=0000  SI=0000  DI=0000
DS=1675  ES=1675  SS=1685  CS=1685  IP=010C   NV UP EI PL NZ NA PO NC
1685:010C B13E          MOV     CL,3E

AX=050D  BX=0000  CX=003E  DX=0041  SP=0000  BP=0000  SI=0000  DI=0000
DS=1675  ES=1675  SS=1685  CS=1685  IP=010E   NV UP EI PL NZ NA PO NC
1685:010E B405          MOV     AH,05

AX=050D  BX=0000  CX=003E  DX=0041  SP=0000  BP=0000  SI=0000  DI=0000
DS=1675  ES=1675  SS=1685  CS=1685  IP=0110   NV UP EI PL NZ NA PO NC
1685:0110 CD21          INT     21
A
AX=0541  BX=0000  CX=003E  DX=0041  SP=0000  BP=0000  SI=0000  DI=0000
DS=1675  ES=1675  SS=1685  CS=1685  IP=0112   NV UP EI PL NZ NA PO NC
1685:0112 FEC2          INC     DL

AX=0541  BX=0000  CX=003E  DX=0042  SP=0000  BP=0000  SI=0000  DI=0000
DS=1675  ES=1675  SS=1685  CS=1685  IP=0114   NV UP EI PL NZ NA PE NC
1685:0114 E2FC          LOOP    0112

AX=0541  BX=0000  CX=0000  DX=007F  SP=0000  BP=0000  SI=0000  DI=0000
DS=1675  ES=1675  SS=1685  CS=1685  IP=0116   NV UP EI PL NZ NA PO NC
1685:0116 B000          MOV     AL,00

AX=0500  BX=0000  CX=0000  DX=007F  SP=0000  BP=0000  SI=0000  DI=0000
DS=1675  ES=1675  SS=1685  CS=1685  IP=0118   NV UP EI PL NZ NA PO NC
1685:0118 B44C          MOV     AH,4C

AX=4C00  BX=0000  CX=0000  DX=007F  SP=0000  BP=0000  SI=0000  DI=0000
DS=1675  ES=1675  SS=1685  CS=1685  IP=011A   NV UP EI PL NZ NA PO NC
1685:011A CD21          INT     21
```

Figure 3.7 *Debug program trace of* TEST.COM

p=100,100

The resulting program trace is shown in Figure 3.7. The state of the CPU registers is displayed as each instruction is executed together with the *next* instruction in disassembled format. Taking the results of executing the first instruction (MOV AH,05H) as an example, we see that 05H has appeared in the upper byte of AX and the instruction pointer (IP) has moved on to offset address 0102H. The next instruction to be executed (located at the address which IP is pointing to) is MOV DL,0AH. The state of the CPU flags is also shown within the register dump. In this particular case, none of the flags has been changed as a result of executing the instruction.

In order to obtain a hard copy of the program trace, a <CTRL-P> command was issued immediately before issuing the Proceed (P) command. From that point onwards, screen output was echoed to the printer. Since the program directs its own output to the printer, this also

```
AX=0500  BX=0000  CX=0000  DX=007F  SP=0000  BP=0000  SI=0000  DI=0000
DS=1675  ES=1675  SS=1685  CS=1685  IP=0102   NV UP EI PL NZ NA PO NC
1685:0102 B20A          MOV     DL,0A

AX=0500  BX=0000  CX=0000  DX=000A  SP=0000  BP=0000  SI=0000  DI=0000
DS=1675  ES=1675  SS=1685  CS=1685  IP=0104   NV UP EI PL NZ NA PO NC
1685:0104 CD21          INT     21


AX=050A  BX=0000  CX=0000  DX=000A  SP=0000  BP=0000  SI=0000  DI=0000
DS=1675  ES=1675  SS=1685  CS=1685  IP=0106   NV UP EI PL NZ NA PO NC
1685:0106 B20D          MOV     DL,0D

AX=050A  BX=0000  CX=0000  DX=000D  SP=0000  BP=0000  SI=0000  DI=0000
DS=1675  ES=1675  SS=1685  CS=1685  IP=0108   NV UP EI PL NZ NA PO NC
1685:0108 CD21          INT     21

AX=050D  BX=0000  CX=0000  DX=000D  SP=0000  BP=0000  SI=0000  DI=0000
DS=1675  ES=1675  SS=1685  CS=1685  IP=010A   NV UP EI PL NZ NA PO NC
1685:010A B241          MOV     DL,41

AX=050D  BX=0000  CX=0000  DX=0041  SP=0000  BP=0000  SI=0000  DI=0000
DS=1675  ES=1675  SS=1685  CS=1685  IP=010C   NV UP EI PL NZ NA PO NC
1685:010C B13E          MOV     CL,3E

AX=050D  BX=0000  CX=003E  DX=0041  SP=0000  BP=0000  SI=0000  DI=0000
DS=1675  ES=1675  SS=1685  CS=1685  IP=010E   NV UP EI PL NZ NA PO NC
1685:010E B405          MOV     AH,05

AX=050D  BX=0000  CX=003E  DX=0041  SP=0000  BP=0000  SI=0000  DI=0000
DS=1675  ES=1675  SS=1685  CS=1685  IP=0110   NV UP EI PL NZ NA PO NC
1685:0110 CD21          INT     21
A
AX=0541  BX=0000  CX=003E  DX=0041  SP=0000  BP=0000  SI=0000  DI=0000
DS=1675  ES=1675  SS=1685  CS=1685  IP=0112   NV UP EI PL NZ NA PO NC
1685:0112 FEC2          INC     DL

AX=0541  BX=0000  CX=003E  DX=0042  SP=0000  BP=0000  SI=0000  DI=0000
DS=1675  ES=1675  SS=1685  CS=1685  IP=0114   NV UP EI PL NZ NA PE NC
1685:0114 E2FA          LOOP    0110
BCDEFGHIJKLMNOPQRSTUVWXYZ[\]^_`abcdefghijklmnopqrstuvwxyz{|}~
AX=057E  BX=0000  CX=0000  DX=007F  SP=0000  BP=0000  SI=0000  DI=0000
DS=1675  ES=1675  SS=1685  CS=1685  IP=0116   NV UP EI PL NZ NA PO NC
1685:0116 B000          MOV     AL,00

AX=0500  BX=0000  CX=0000  DX=007F  SP=0000  BP=0000  SI=0000  DI=0000
DS=1675  ES=1675  SS=1685  CS=1685  IP=0118   NV UP EI PL NZ NA PO NC
1685:0118 B44C          MOV     AH,4C

AX=4C00  BX=0000  CX=0000  DX=007F  SP=0000  BP=0000  SI=0000  DI=0000
DS=1675  ES=1675  SS=1685  CS=1685  IP=011A   NV UP EI PL NZ NA PO NC
1685:011A CD21          INT     21
```

Figure 3.8 *Program trace of the corrected* TEST.COM *code*

appears amidst the traced output. A single line feed has been generated after execution of the third instruction (INT 21H), a carriage return after the fifth instruction (INT 21H), and a single character, A, after the ninth instruction. Thereafter, the program executes the loop formed by the instructions at offset addresses 0110H and 0112H. No printing takes place within this loop even though the DL register is incremented through the required range of ASCII codes. Clearly the loop is not returning to the INT 21H instruction responsible for making the DOS call which generates character output to the printer!

Fortunately, we can easily overcome this problem from within the debugger without returning to the macro assembler. We simply need to modify the LOOP instruction at offset address 0114H. To do this we can make use of the Assemble (A) command to over-write the existing instruction. The required command is:

```
A114
```

The CS:IP prompt is then displayed (in this case it shows 1685:0114) after which we simply enter:

```
LOOP 110
```

The CS:IP prompt is incremented however; since we need to make no

```
        ; This program outputs ASCII data
        ; in the range 41H to 7FH to the
        ; parallel printer port
        ; Registers used: AX,CL,DL
        ; Parameters passed:   none
        TITLE   ptest
_TEXT   SEGMENT             ; Define code segment
        ASSUME cs:_TEXT,ds:_TEXT,ss:_TEXT
        ORG     100H        ; Normal for COM program
start:  MOV     AH,05H      ; Printer output function code
        MOV     DL,0AH      ; First generate a
        INT     21H         ; line feed
        MOV     DL,0DH      ; Next generate a
        INT     21H         ; carriage return
        MOV     DL,41H      ; First character to print is A
        MOV     CL,3EH      ; Number of characters to print
        MOV     AH,05H      ; Set up the function code
prch:   INT     21H         ; and print the character
        INC     DL          ; Get the next character
        LOOP    prch        ; and go round again
        MOV     AL,00H      ; Set up the return code
        MOV     AH,4CH      ; and the function code
        INT     21H         ; for an exit to DOS
_TEXT   ENDS
        END     start
```

Figure 3.9 *Corrected source code for* TEST.ASM

further changes to the code, we can simply escape from the Debug line assembler by simply pressing <RETURN>.

Having modified our code, we can again trace the program using the Proceed (P) command exactly as before. The traced output produced by the modified program is shown in Figure 3.8. Note that we have now succeeded in producing a line of printed output!

Since no further errors have been found, we can exit from Debug, load the macro assembler, make the necessary changes to our source code, assemble and link to produce a modified COM program file. Alternatively, we could simply write the modified code back to TEST.COM using the debugger. If this expedient course of action *is* followed, it is important to remember that the modification will not be reflected in our source code. Failure to update the source code may, in some cases, cause confusion at some future time! The correctly modified source code has been shown in Figure 3.9.

Windows

Microsoft Windows is a graphical user interface (GUI) based on icons and pull-down menus. Current versions of Windows (3.0 and 3.1) run under DOS; however, it now appears that future versions of Windows will not require DOS to be present. The core of Windows lies in two important programs: Program Manager and File Manager. Program Manager provides a means of locating and running programs within the Windows multi-tasking environment. Program selection is made easy with the use of icons that indicate a program's name or its function. It is simply necessary to point and click with the mouse in order to launch a program. Figure 3.10 illustrates the use of icons within the Windows 'Control Panel'.

Programs can be collected together in 'groups'. This helps to keep the system tidy and makes it easy to locate a particular program without having to search through a cluttered screen full of icons. Typical names for groups include 'Accessories', 'Applications', 'Network', 'Start Up', 'Tools', 'Utilities', etc.

File Manager provides the necessary file-oriented housekeeping functions. From within File Manager you can copy, move, and delete files. File Manager will also allow you to create and rename directories and format disks.

Windows is a substantial software product, requiring in excess of 4 megabyte of disk space and at least 640 kilobyte of RAM, even in its most basic configuration. During installation, Windows will modify your CONFIG.SYS and AUTOEXEC.BAT files (preserving copies of the old versions under the names CONFIG.OLD and

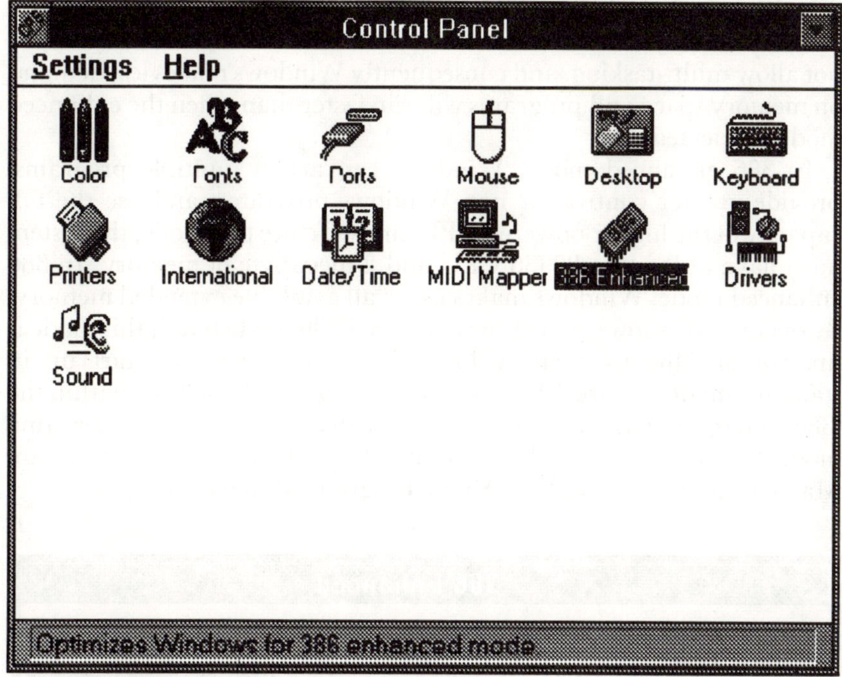

Figure 3.10 *Windows' 'Control Panel'*

AUTOEXEC.OLD, respectively). Windows will also create temporary or permanent swap files on your hard disk. Optimum performance will result from a reasonably large (e.g. 4 or 8 megabyte) permanent swap file. These files should be reserved for Windows use and should not be deleted or moved.

Starting Windows

Systems can be easily configured to run Windows whenever they start up, simply by including an appropriate command at the end of the AUTOEXEC.BAT file. Alternatively, Windows can be started when necessary by issuing a commend of the form:

```
WIN /mode command
```

where WIN is the name of the program that contains the Windows code, /mode is an optional mode switch, and command is an optional start-up command.

The mode switch can be used to indicate a specific operating mode. The two mode switches are /S which forces Windows to operate in 'standard' mode and /3 which runs Windows in 'enhanced' mode.

Standard mode is designed for systems that have a limited amount of memory (e.g. systems based on an 80286 CPU). Standard mode does not allow multi-tasking, and consequently Windows makes less demand on memory space and programs will run faster than when the enhanced mode is selected.

In 386 enhanced mode, Windows can access multiple programs, provide greater control for non-Windows programs, and use the full capacity of the higher powered CPU chips. To use this mode, the system must have at least an 80386 CPU and 2 megabyte of memory. In 386 enhanced mode, Windows makes use of all available extended memory. By default, Windows adopts the enhanced 386 mode but, if this mode is unavailable due to memory limitations, Windows will start up in standard mode. Figure 3.11 shows the settings available from within the '386 Enhanced' window. Note that you can determine which operating mode Windows is using by selecting the Help menu from Program Manager and choosing the 'About Program Manager' option.

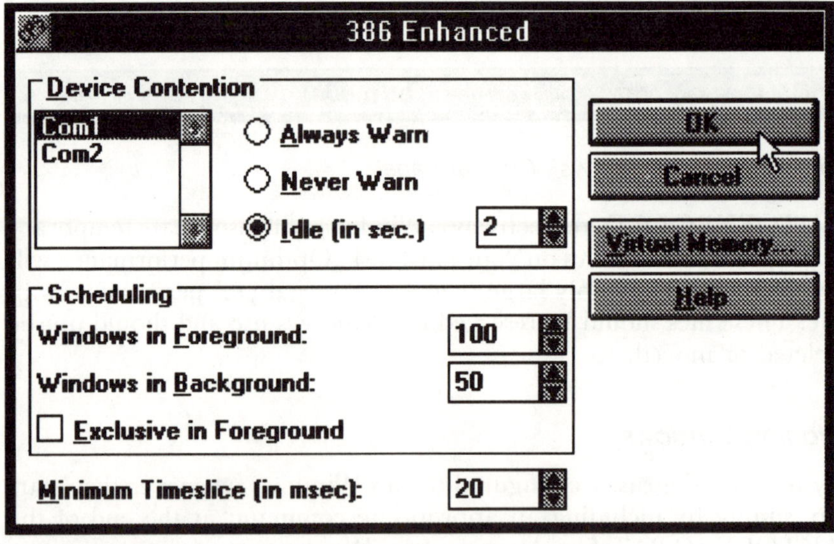

Figure 3.11 *The '386 Enhanced' selection window*

The start-up command

The start-up command allows you to execute a particular application whenever Windows is run. Suppose that you always use a program called PROCESS.EXE, the start-up command for Windows would then be:

```
WIN PROCESS
```

Note that the foregoing command assumes that PROCESS.EXE is present within the directory from which you are running Windows or that it has been included in your AUTOEXEC.BAT PATH statement. As an alternative, the command:

```
WIN C:\APPS\PROCESS
```

refers to a program called PROCESS.EXE which is contained within the APPS directory.

An alternative method of automatically starting a windows program is that of including the program within the StartUp program group. This is a powerful new feature available from Version 3.1 onwards. Any program icons residing in this program group will be automatically opened and started when Windows starts. Thus, if PROCESS is required every time your system is used, you can simply place its icon within the Window for the StartUp program group.

Changing system settings

The display, keyboard, mouse and network configuration can be changed using the Change System Settings window (shown in Figure 3.12). This window will allow you to change the display mode (resolution and colours), keyboard and mouse type by selecting from the options available within the drop-down selections.

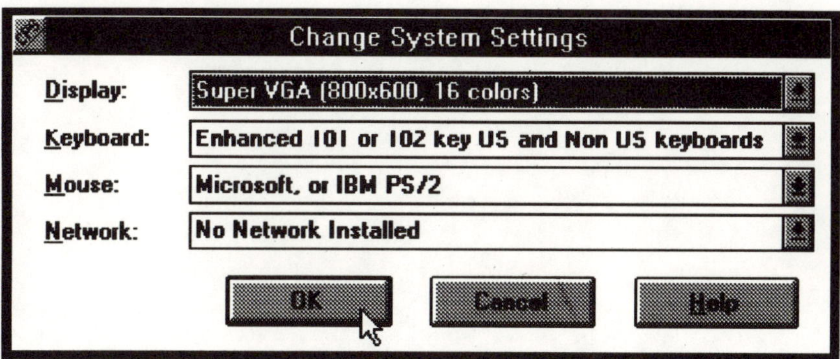

Figure 3.12 *Changing Windows' System Settings*

Windows for Workgroups

Microsoft Windows for Workgroups is an enhancement of Windows 3.1 which incorporates networking capability so that two or more computers can share a common set of resources. If the computers are already on a network, Windows for Workgroups will enhance the network by allowing users to make use of the Windows' graphical user interface.

Windows NT

Windows NT is a multiprocessing operating system that conforms with the US Government's POSIX procurement standard (POSIX is an acronym for Portable Operating System Interface for UNIX), and offers further significant advantages over earlier versions of Windows. Windows NT is also downward-compatible and will operate on any ISA type machine from a 386 up to a Pentium. When installed as server software on a network, Windows NT will allow lowly XT and AT workstations to co-exist with their more powerful 32-bit successors.

Windows NT will run existing applications written for DOS or Windows, but new applications can be written to take advantage of the 32-bit system in which the old 640 kilobyte barrier no longer exists. Windows NT can access up to 4 gigabyte of RAM, and a single application can address up to 2 gigabyte of memory. Windows NT software is compressed so that it can be distributed on floppy disks. It is also available on a CD-ROM. A hard disk with 80 megabyte of free disk space is required for installation (20 megabyte of this disk is set aside for virtual memory). A minimum of 8 megabyte of RAM is required, while 12 to 16 megabyte is recommended for optimum performance.

4 Programming

Whilst many users of PC-based instrumentation and control systems will be able to make use of off-the-shelf software packages, others may have specific applications for which there is no existing software package available. This is often the case with dedicated process control systems where a particular operational configuration is unique to the system concerned or where an existing software package is limited in some way.

The control engineer should be perfectly capable of developing simple, robust and efficient control programs without the assistance of a programmer or software engineer. However, where the software is complex, sophisticated or requires a high degree of optimization, then the services of a software engineer/programmer will almost certainly be required.

At the outset, it should be stated that there is a great deal more to programming than simply entering code. Programming benefits from a disciplined approach and this is absolutely essential when developing software which must operate reliably and be easy to maintain.

Experience shows that electronic engineers (particularly those involved with control systems) generally make excellent software engineers. They have usually developed a high degree of familiarity with hardware (microprocessors and support devices) and will be only too well aware of the characteristics and constraints of such devices.

Software engineering should not be confused with programming. A programmer is not necessarily a software engineer neither is a software engineer necessarily a programmer. In fairness, a software engineer will normally be proficient in several computer languages; however, such proficiency will be relatively unimportant if the software he or she produces behaves erratically or is impossible to maintain.

This chapter introduces some of the basic concepts associated with the production of structured code which is both predictable and reliable and easy to maintain. This information should be invaluable to the electronic or control system engineer who may be increasingly involved with the development of programs to control PC-based systems.

Choice of language

Sooner or later, the software developer must make some decisions concerning the choice of language used for software development. To some extent this decision will be crucial to the success of a project. The essential features to consider when selecting a language for software development in PC-based instrumentation and control applications are as follows:

- *What control flow structures are provided to facilitate the development of structured code?*

  Such control structures may take several forms but should ideally include the ability to handle user-defined functions and procedures (with or without local variables) and such control structures as IF . . . ELSE . . . ENDIF, DO WHILE . . . LOOP, SELECT . . . CASE . . . END SELECT, and WHILE . . . WEND.

- *What provision is there for handling I/O?*

  Most languages provide functions and statements (e.g. BASIC's PEEK and POKE) which facilitate direct access to memory. A language for PC-based instrumentation and control applications should have statements that allow reading from and writing to I/O port addresses. Taking BASIC as an example, functions such as:

  INP(port)

  and statements such as:

  OUT port, data

  make writing I/O routines extremely easy.

- *How easy is it to combine/interface modules written in the same or a different language?*

  A facility for combining/interfacing modules written in the same or a different language will be essential in any other than the simplest of applications. As an example, it may be convenient to develop an assembly language routine to handle some critical I/O process and then interface this to a high-level language program which deals with more mundane processes, such as keyboard input, display output, and disk filing. In such a case, it will generally be necessary to have some mechanism for passing parameters between the main program and the code generated by the assembly language module.

- *What, if any, provision is there for handling interrupts?*

  Some mechanism for allowing the user to incorporate his or her own interrupt handling routines will be essential in most control applications.

- *What provision is there for event/error trapping?*

  The ability to include specific event/error trapping routines can be important in making the program robust and suitable for non-technical users. Error handling routines should permit meaningful error reporting as well as the ability to retain control of the program with an orderly shutdown when operation cannot continue.

- *Finally, will the language allow multi-tasking?*

  In control applications, the ability to support multi-tasking is a highly desirable feature. In addition to the main program, the programmer will then be able to define one, or more, background tasks to run concurrently with the main program. These tasks will be switched to repeatedly during program execution and thus effectively run in parallel with the main program.

  Unfortunately, true multi-tasking is a problem within an 8088/8086 based MS-DOS environment as the Real Mode provided by the 8088/8086 microprocessor in the original PC, PC-XT, and PC-AT employs straightforward addressing and no interprocess protection. The limitation in available memory (640 kilobytes under PC-DOS or MS-DOS) further mitigates against true multi-tasking applications.

  The *Protected Mode* environment available within PS/2 (Models 50, 60 and 80) is more appropriate for applications which require multi-tasking. Furthermore, it is important to note that DOS applications originally written for a PC will not run in Protected Mode on a PS/2 system unless they have been specifically modified by the programmer.

If you can answer with an unqualified 'yes' to the majority of the foregoing questions, you can be assured that language under consideration is an ideal candidate for software development in the control and instrumentation field. Coupled with an integrated development environment (editor and debugger) it should be able to cope with almost anything!

Software development

Software development should normally be a top–down process in which one moves from the general to the specific. The process can be divided into a number of identifiable phases which generally include:

1 Problem *analysis*, leading to
2 a *software specification*.
3 Development of an *algorithm* and
4 a *program definition*.
5 *Coding* and
6 *testing* (against the specification) and
7 *debugging*.
8 *Implementation* and
9 *evaluation*.

In practice, Steps (5), (6) and (7) will invariably be repeated a number of times in order to refine the software and eliminate errors made during the coding phase. At this stage, it is perhaps worth examining each of the phases in the *software development cycle* in a little more detail.

The first two stages (problem analysis and the production of a software specification) involve first determining the user's requirements and then itemizing the functions and facilities expected of the software. The specification should, of course, be agreed with the user. Furthermore the initial stages will normally require a dialogue with the user in order to establish the parameters within which the system should operate. Very few users are able to give a precise definition of their requirements and, since it is important to consider all eventualities, it is important to explore with the user what should happen in abnormal circumstances as well as in routine situations.

As an example, consider the case of the operator of an aggregates processing plant which comprises several conveyor belts, processing drums, and a washing plant. The problem essentially involves delivering various grades of aggregate at rates which are sufficient to ensure that the capacity of the stockpile is not exceeded and that a certain minimum amount of each grade of material is always available. The software specification (agreed with the operator) will involve delivery rates and volumes. However, the plant operator may forget to mention that, in the event of an interruption of the water supply, part of the plant must shut down with a consequent and drastic change in delivery rates elsewhere.

In extreme cases, a problem of this type may only come to light when the system is commissioned. Clearly, this would not have happened if the initial stages of the development model had been rigorously followed.

Steps (3) and (4) can be considered to be the 'design' phases. The first of these (development of an algorithm) involves conceptualizing the means of solving the problem. This is often done with the aid of a flowchart or a data flow diagram and usually involves breaking down the problem into a number of smaller steps (processes). Figure 4.1 shows the set of standard symbols which are used in flowcharts.

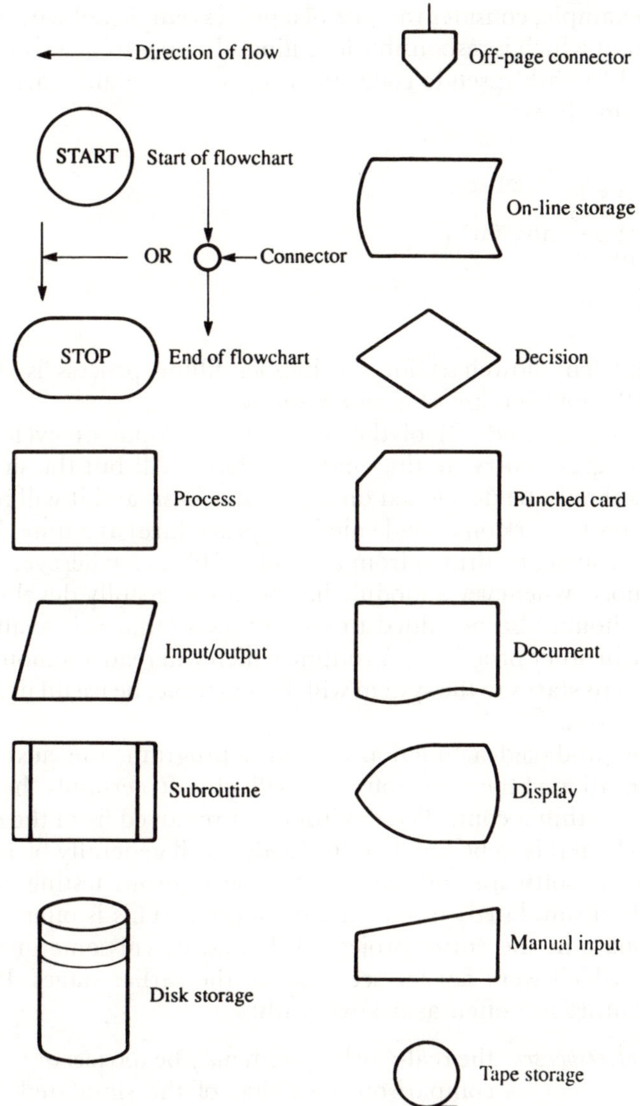

Figure 4.1 *Flowchart symbols*

The second of the design phases involves defining the various program modules and procedures. These will often be associated with the individual stages of the flowchart model (or its equivalent) and may be separately documented. The action of each module can be summarized using *structured English* (or *pseudo code*). Each line of *pseudo code* will generally correspond to one, or more, lines of program code.

As an example, consider the case of a process employed within a grain drying plant which is responsible for filling a hopper from a conveyor. In structured English (pseudo code) the process can be summarized along the following lines:

```
Begin
Close hopper outlet
Start conveyor
While hopper not full
  Run conveyor
EndWhile
Stop conveyor
End
```

The equivalent flowchart for the hopper filling process is shown in Figure 4.2 (note the use of a *conditional loop*).

Steps (5), (6) and (7) of the software development cycle involve routine program entry, testing and debugging. All but the simplest of programs should be developed on a modular basis and it will normally be necessary to work on a single module (procedure) at a time. Modules should, of course, be drawn from a standard library wherever possible. Furthermore, whenever a module has been successfully developed and tested, it should also be added to the library so that it is available for future use in other programs. A routine which will read a remote keypad and return its status to the system will, for example, be useful in a variety of applications.

Having produced a functional control program, the next stage is implementation. Since the software will almost certainly have been developed within a controlled environment removed from the environment in which it is to be finally imbedded, it will generally be necessary to install the software and carry out some rigorous testing with real (rather than simulated) inputs and/or outputs. This is often the most critical phase in the entire project cycle and it will sometimes reveal problems which were not foreseen during the earlier stages. Problems and difficulties are often associated with:

- *speed of response* the real-world system may be too fast or too slow in comparison with that of the simulated development environment
- *noise* signals in the real-world environment are rarely ideal and often contain a significant amount of noise.

As an example, a system installed to monitor the flow of gas along a pipeline behaved erratically when an apparently functional (and fully debugged) program was installed within its industrial PC-based controller. Sixteen remote sensors (based on rotating vanes) were used to sample the flow rate at various points. Each sensor was connected, via

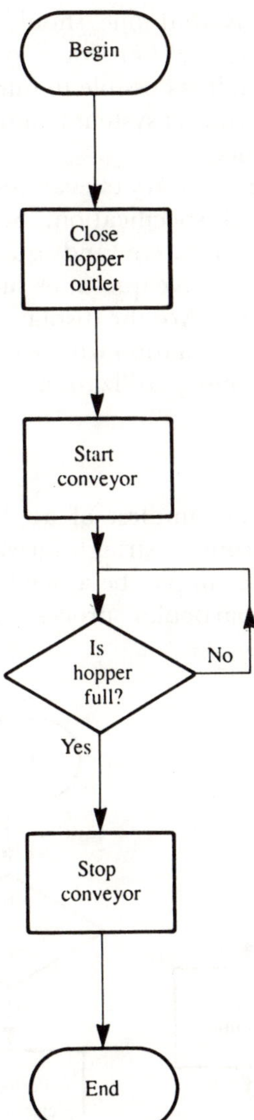

Figure 4.2 *Flowchart for the hopper filling process*

an asynchronous serial data link, to the controller. Under certain conditions, the PC indicated that the flow rates were well outside the prescribed limits for the system. However, upon examination it was found that, not only did the sensors exhibit a reluctance to respond to very low flow rates but the signals from the furthermost sensors were regularly erroneous due to power-line induced switching transients and lack of RS-232 parity checking.

The moral, of course, is that one should attempt to anticipate problems at the earliest stages of hardware/software development. By planning for the unforeseen, it is possible to minimize the time taken to imbed the software into the target system to a bare minimum, reducing both costs and inconvenience.

Finally, it will usually be necessary to evaluate the performance of the system against the original specification. Such an evaluation will generally involve both qualitative and quantitative aspects. The qualitative evaluation will involve questions such as 'Does the user feel at ease with the system?' and 'Are the displays and prompts meaningful?' while the quantitative evaluation will be concerned with collecting data on response times, memory utilization, disk space, etc.

Control structures

In anything other than the simplest of applications, programs will involve some deviation from a straightforward linear sequence of processes. There may, for example, be a need for conditional forward branching (bypassing a particular process) depending upon some

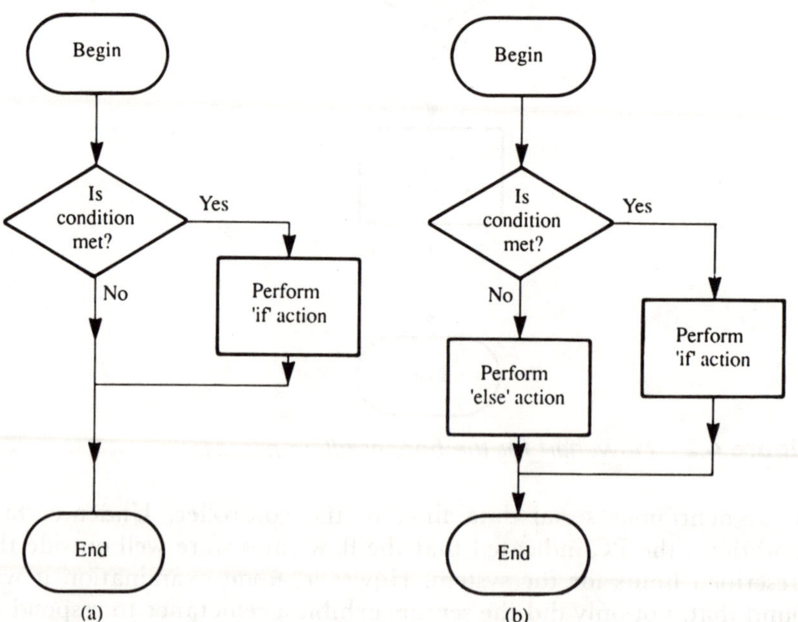

Figure 4.3 *Commonly available control structures: (a) simple branch (IF . . . EndIf); (b) binary branch (If . . . Else . . . EndIf); (c) multiple branch (Select . . . Case . . . EndSelect)*

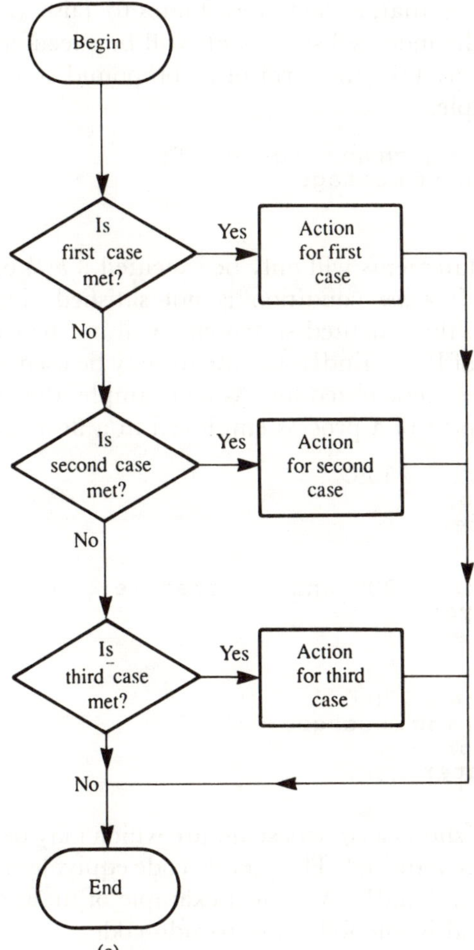

(c)

particular outcome, or for a certain process to be repeated a number of times until a particular result is obtained.

Several common control structures (available within the majority of today's programming languages) are illustrated in Figure 4.3. The first of these (Figure 4.3(a)) involves a simple branch forwards depending upon the outcome of the conditional test. A typical example of this control structure, expressed in pseudo code, is:

```
If tank empty
  Open valve
  Operate pump
EndIf
```

It should be noted that, if the test evaluates to 'false' (i.e. if the valve is open) none of the indented statements will be executed. Furthermore, the condition may take the form of a compound statement, as in the following example:

```
If temperature high and coolant off
  Display warning message
  Turn heat off
Endif
```

The indented statements will only be executed if *both* of the conditions evaluate true. If *either* condition is not satisfied (i.e. one or other evaluates false), the indented statements will not be executed.

A succession of If . . . EndIf statements may be used where a number of outcomes need to be tested for. As an example, the following pseudo code describes part of a process employed in a flow soldering plant:

```
If temperature < 230°C
  Stop conveyor
  Turn on heater
EndIf

If temperature > 230°C and temperature < 270°C
  Start conveyor
  Turn on heater
EndIf

If temperature > 270°C
  Display warning message
  Stop conveyor
  Turn off heater
EndIf
```

Figure 4.3(b) shows a control structure which may be adopted where two outcomes are required. The pseudo code equivalent of this is known as IF . . . Else . . . EndIf. A typical example of the use of this control structure is found in the following pseudo code:

```
If light level low
  Lights on
Else
  Lights off
EndIf
```

A further control structure provides for multiple branching (rather than binary branching, as in the case of If . . . Else . . . EndIf). This structure is illustrated in the flowchart of Figure 4.3(c) and a typical application might be in the selection of a main menu option, as described by the following representative pseudo code:

```
Select Case
  1, Input new data
  2, Get old data
  3, Sort data
  4, Print data
```

```
  5, Exit
  Else warn user
EndSelect
```

This (apparently complex) pseudo code can quite easily be implemented in both BASIC and C. A typical BASIC routine to satisfy the pseudo code would take the form:

```
SELECT CASE R$
  CASE "1"
    CALL NewData
  CASE "2"
    CALL OldData
  CASE "3"
    CALL SortData
  CASE "4"
    CALL PrintData
  CASE "5"
    CALL UpdateFile
  CASE "6"
    EXIT DO
  CASE ELSE
    BEEP
    PRINT "Input not valid!"
END SELECT
```

while its equivalent in C would be:

```
switch(c)
{
  case '1':
    newdata();
    break;
  case '2':
    olddata();
    break;
  case '3':
    sortdata();
    break;
  case '4':
    printdata();
    break;
  case '5':
    exit();
  default
    beep();
    printf ("Input not valid!\n");
}
```

Loops

A loop structure (backwards branch) may be used in order to avoid the need to repeat blocks of code several times over whenever a process is to be repeated more than once. Various types of loop are possible (both conditional and unconditional) and these are supported by pseudo code statements such as Do . . . Loop While, Do . . . Loop Until, Do While

. . . Loop, and Do Until . . . Loop. These structures are explained in detail in Chapter 6.

Error checking and input validation

Error checks and input validation routines should be incorporated whenever data is input and before the system accepts the data for processing. Error handling routines should be incorporated to warn the user that a fault has occurred and indicate from which source the error has arisen. This caveat also applies to operator input; an unacceptable input should be echoed to the user together with the range of acceptable responses. Care should be exercised when inputs are defaulted. The default response should result in activity rather than any form of positive action on the part of the system. Furthermore, the program should demand confirmation where a response or input condition will produce an irreversible outcome.

Testing

It is only possible to claim that a program has been validated after exhaustive testing in conjunction with the target hardware system. In many cases it may be possible to test individual code modules before they are linked into the final executable program. This may be instrumental in reducing debugging time at a later stage.

Testing the completed program requires simulating all conditions that can possibly arise and measuring the outcome in terms of the program's response. A common error is that of only presenting the system with a normal range of inputs. Comprehensive testing should also involve the simulation of each of the following:

- Unexpected or non-sensical responses from the operator or user.
- Failure of hardware components (including transducers, signal conditioning boards, disk drives etc.).
- Out-of-tolerance supplies (including complete power failure).

Documentation

Programmers are usually woefully lacking where program documentation is concerned. Documentation, which is essential to making the program understandable, takes various forms, the most obvious of which is the comments included in the lines of source code text.

Comments

Comments should explain the action of the source code within the program as a whole and, since the function of the operation code and operand will usually be obvious (or can be found by referring to the instruction set) there is no point in expanding on it. Comments should be reasonably brief (to save space in the source code file) but not so brief that they become cryptic. Also, there is no need to attempt to confine a comment to a single statement line. Comments can be quite effective if they read clearly and are continued over several statements to which they refer.

Headers

Headers are extended comments which are included at the start of a program module, macro definition, or subroutine. Headers should include all relevant information concerning the section of code in question and should follow a standard format. As a minimum, the following should be included:

- Name and purpose of the module or subroutine.
- Brief explanation of the action of the code (in terms of parameters passed, registers involved, etc.).
- Names of other modules, subroutines or macros on which the module depends and, where applicable, names of relevant macro libraries in which definitions are held.
- Entry requirements (in terms of register and/or buffer contents before the module is executed).
- Exit conditions (in terms of register contents, buffers and flags after the module has been executed).
- In the case of assembly language modules, a list of registers used during execution of the code (which may have their contents changed as a consequence).

When producing a program header, it is wise to include any information which may be required by another programmer who may subsequently need to debug or modify the code. Nothing should be taken for granted and all loose ends should be explained!

During the development phase, it is worth including a brief development history within the main program header, as shown in the following example:

```
' *
' *                 Program: DSM3.BAS                    *
' *                   Version: 0.1                       *
' *                Copyright USET 1989                   *
' *
'                  Development history
' 12/01/90 Creation date
' 13/01/90 Structure defined
' 14/01/90 New sub-programs added
' 15/01/90   ,,  ,,    ,,      ,,
' 16/01/90 64k block save and load added
```

```
'  18/01/90 View sample added
'  20/01/90 Block size increased to 256k
'  28/01/90 Mixed language interface added
'  28/01/90 Interrupt enable/disable added
'  29/01/90 Various flags added
'  29/01/90 Assembly language modules added
'  04/02/90 Block size modified to 128k
'  05/02/90 View sample removed
'  10/02/90 Multiple blocks added
'  15/02/90 Block parameter data file added
```

Names

Names used for variables, symbols and labels should be meaningful and any abbreviations used should be as obvious as possible. In the case of the names used for constants, where standard abbreviations are in common use (e.g. CR for Carriage Return), they should be adopted. In a large program, there may be a large number of labels and/or constants and it will be necessary to distinguish between them.

As an example of the use of names, comments, and headers consider the following examples which, while functionally identical, illustrate the extremes of programming style:

Case converter subroutine

```
;          CASE CONVERTER
;
con:       CMP    AL,61H       ;  Compare A with 61H.
           JE     exit         ;  Return if carry set.
           CMP    AL,7BH       ;  Compare A with 7BH.
           JNE    exit         ;  Return if carry reset.
           SUB    AL,20H       ;  Subtract 20H from A.
exit:      RET                 ;  Return.
```

Improved case converter subroutine (self-documenting)

```
;  LOWER TO UPPER CASE CHARACTER CONVERSION
;  PARAMETERS PASSED:
;  ENTRY:               AL=ASCII character (upper or lower case)
;  EXIT:                AL=ASCII character (upper case only)
;  REGISTERS:           AL, F
;
upcase:     CMP    AL,'a'         ;  Is it already upper case?
            JL     exit           ;  If so, do nothing.
            CMP    AL,'z'         ;  Or is it punctuation?
            JG     exit           ;  If so, do nothing.
            SUB    AL, 'a'-'A'    ;  Otherwise, change case.
exit:       RET
```

The second example shows how a program module can be made largely self-documenting by the inclusion of effective comments and a meaningful header. Note that the name of the routine has been changed so that it is easier to remember and is less likely to be confused with others. Finally, the code itself has been modified so that its action is much easier to understand.

Documentation is particularly important where software development is being carried out by several members of a team. Each development phase will rely on the documentation prepared in earlier stages, hence documentation should be considered an ongoing task and a folder should be prepared to contain the following items:

- A detailed program specification (including any notes relevant to the particular hardware configuration required).
- Flowcharts or descriptions of the program written in structured English.
- Lists of all definitions and variable names.
- Details of macro or sub-routine libraries used.
- Details of memory usage.
- A fully commented listing of the program (latest version).
- A diary giving the dates at which noteworthy modifications are made together with details of the changes incorporated and the name of the programmer responsible.
- A test specification for the program with descriptions and results of diagnostic checks performed.

Presentation

Finally, attention should be given to the way in which the program interacts with the user and the aim should be that of making the software as 'user friendly' as possible. Prompts and messages should always be meaningful and, whenever any doubt may exist, the user should be prompted with the range of acceptable values or valid responses. Chapters 6 and 7 contain further information on the general topic of making programs 'user friendly'.

5 Assembly language programming

This chapter aims to provide readers with an overview of assembly language programming techniques and explores the architecture and instruction set of the 8088/8086 microprocessor used in the PC and compatible equipment. Rather than provide a complete guide to 8088/8086 assembly language programming (which, in any event would require a complete book in its own right!) the aim has been that of providing readers with sufficient information to decide whether assembly language is appropriate for a particular application, to outline the advantages and disadvantages of assembly language programming, and to introduce techniques used in developing assembly language programs.

Readers wishing to develop their own assembly language programs will not only require complete documentation for the 8086 family of processors (including a comprehensive explanation of the microprocessor's instruction set) but will also require development software comprising, as a minimum, a macro assembler, a liner, and a debugger. Readers should not underestimate the investment required (in terms of both time and money) required to successfully follow this route.

Advantages of assembly language

Assembly language programs offer a number of advantages when compared with higher level alternatives. The principal advantages are that the executable code produced by an assembler (and linker) will:

- Invariably be more compact than an equivalent program written in a higher-level language.

- Invariably run faster than an equivalent program written in a higher-level language.
- Not require the services of a resident interpreter or a compiler run-time system.
- Be able to offer the programmer unprecedented control over the hardware in the system.

It is this last advantage, in particular, which makes assembly language a prime contender for use in bus-based process-control applications. No other programming language can hope to compete with assembly language where control of hardware is concerned. Indeed, an important requirement of high-level languages used in process-control applications is that they can be interfaced with machine code modules designed to cope with problems arising from limitations of the language where I/O control is concerned.

Disadvantages of assembly language

Unfortunately, when compared with higher-level languages, assembly language has a number of drawbacks, most notable of which are the following:

- Programs require considerably more development time (including writing, assembling, linking/loading and debugging) than their equivalent written in a high-level language.
- Programs are not readily transportable between microprocessors from different families. Different microprocessors have different internal architectures and, in particular, the provision of registers accessible to the programmer will vary from one microprocessor to another. Some microprocessors (e.g. 68000) offer a 'clean' internal architecture in which nearly every register can be described as 'general purpose'). Others (e.g. 6502) may have a limited number of internal registers largely dedicated to specific functions. Such obvious differences in architecture is reflected in corresponding differences in the type and function of the software instructions provided for the programmer.

  The situation is further compounded by the fact that microprocessor manufacturers frequently adopt different terminology to refer to the same thing. The variety of names used to describe the register used to indicate the outcome of the last ALU operation (and the internal status of the microprocessor) is a case in point. This is variously referred to as a Flag Register, Status Register, Condition Code Register, and Processor Status Word.

  In practice this means that the system designer is constrained to select one particular microprocessor type (or family) and develop code exclusively for this particular device. This, of course, is not a particular problem in the case of the PC and compatible equipment which are all based on the standard Intel 16-bit processor family. However, it should go without saying that assembly language programs written for the PC cannot, for example, be ported across to a Motorola 68000-based environment (e.g. Apple Macintosh, Commodore Amiga, or Atari ST).

- Unless liberally commented, the action of an assembly language program is not obvious from merely reading the source text. Programs written in high-level language are usually easy to comprehend and their structure is usually self-evident.
- The production of efficient assembly language programs requires a relatively high degree of proficiency on the part of the programmer. Such expertise can usually only be acquired as a result of practical experience aided by appropriate training.

Developing assembly language programs

The process of developing an assembly language program depends on a number of factors including the hardware configuration available for software development and the range of software tools available to the developer. As a minimum the task normally involves the following steps:

1. Analysing the problem and producing a specification for both hardware and software (see Chapter 4).
2. Developing the overall structure of the program, defining the individual elements and modules within it, and identifying those which already exist (or can be easily modified or extended) within the programmer's existing library.
3. Coding each new module required using assembly language mnemonics, entering the text using an editor, and saving each source code module to disk using an appropriate filename.
4. Assembling each source code module (using an assembler) to produce an intermediate relocatable object code file.
5. Linking modules (including those taken from the user's library) in order to produce an executable program.
6. Testing, debugging, and documenting the final program prior to evaluation and/or acceptance testing by the end-user (see Chapter 4).

In practice, the development process is largely iterative and there may also be some considerable overlap between phases. In order to ensure that the target specification is met (within the constraints of time and budget) an ongoing appraisal is necessary in order to maximize resources in the areas for which there is most need.

Software tools

The following items of utility software (software tools) are required in the development process:

- An ASCII text editor (e.g. Microsoft's M).
- A macro assembler (e.g. Microsoft's MASM).
- A linker (e.g. Microsoft's LINK).

In addition, three further software tools may be found to be invaluable. These are:

- A cross-referencing utility (e.g. Microsoft's CREF).
- A library manager (e.g. Microsoft's LIB).
- A utility which can help automate the program development cycle (e.g. Microsoft's MAKE).

(*Note:* MASM, LINK, CREF, LIB and MAKE are all supplied, together with the CodeView symbolic debugger as part of the Microsoft Macro Assembler package.)

For the benefit of the newcomer to assembly language programming, we shall briefly explain the function of each of these items of utility software in the production of assembly language programs.

Editors

Editors allow users to create and manipulate text files. Such files can be thought of as a sequence of keystrokes saved to disk. An assembly language source code file is simply a text file written using assembly language mnemonics and containing appropriate *assembler directives*.

The *Microsoft Editor* (M) is invoked using a command line of the form:

```
M <options><file list>
```

The options include that of allowing the user to load a previously save configuration file (TOOLS.INI). This file contains settings which will be used to initialize the editor and thus the user may easily customize the software to his or her own particular requirements. The file list is simply a list of files that will be loaded into the editor. The first file in the list will be the first to be edited. Then, when the user selects the exit option (F8), the next file in the list is loaded ready for editing.

The Microsoft editor is extremely powerful. It provides the usual 'cut and paste' and 'search and replace' facilities together with macros which can be invoked from a single keystroke. Furthermore, to reduce the overall edit/assemble cycle time, it is possible to assemble a program from within the editor, view and correct any errors that may have occurred, then re-assemble. Multiple source files can easily be handled and a split-screen windowing facility can be used to examine and edit different parts of the same file simultaneously.

Format of source code statements

When preparing source text using an editor, it is important to bear in mind the requirements of the assembler concerning the format of source code statements. In the case of most 8086 assemblers (and Microsoft's MASM in particular), each line of source code is divided into four fields, as shown in the example below:

Symbol	Operation	Argument	Comments
maxcount	DB	16	;initialize maximum count

The first entry in the line of code (i.e. the *symbol field*) is known as a *symbol*. The symbols used in a program are subject to certain constraints imposed by the assembler but are chosen by the individual programmer. *Labels* are a particular form of symbol which are referred to by one, or more, statements within a program.Labels are generally mark the entry point to the start of a particular section of code or the point at which a branch or loop is to be directed. During the assembly process, labels (wherever they appear in the program) are replaced by addresses.

Entries in the *operation field* may comprise an *operation code* (*opcode*), a *pseudo-operation code* (*pseudo-op*), an *expression*, or the name of a *macro*. Operation codes are those recognized by the microprocessor as part of its instruction set (e.g. MOV, ADD, JMP etc.) whereas pseudo-ops are *directives* which are recognized by the assembler and are used to control some aspect of the assembly process. Typical pseudo-ops are DB (define byte), DW (define word), ORG (origin/program start address), and INCLUDE. The last-named directive instructs the assembler to search a named macro library file and to expand macro definitions in terms of this library.

The *argument field* may contain constants or expressions (such as 0DH, 42, 64*32, 512/16, 'A', 'z'-'A') or the operands required by microprocessor operation codes (represented by numbers, characters, symbols are extended opcodes).

The *comment field* contains a line of text, added by the programmer, which is designed to clarify the action of the statement within the program as a whole (see Chapter 4).

In the example shown previously, the variable *maxcount* has been declared in the symbol column. The operation field contains a pseudo-op (*assembler directive*) which instructs the assembler to reserve a byte of storage and initialize its value to 16. Thereafter, any references to *maxcount* will take the value 16 (at least until the value is next modified by the program). The programmer has added a comment (following the obligatory semi-colon) which reminds him or her that *maxcount* is the symbol used to hold the current maximum value of the counter.

Not all source code lines involve entries in all four fields, as in the next example:

Symbol	*Operation*	*Argument*	*Comments*
	MOV	AL,maxcount;	get maximum count

Here, the symbol field is blank since the instruction does not form part of the start of a block of code. The operation, MOV, is an opcode which instructs the microprocessor to perform an operation which will move data from one location to another. The operand required by the instruction specifies the lower half of the 16-bit accumulator (AL) as the destination for the data and the contents of the variable storage location *maxcount* as the source of the data. The programmer has again added a brief comment to clarify the action of the line.

It should be noted that any line of source code starting with a semi-colon is ignored by the assembler and treated as a comment. This allows the programmer to include longer comments as well as program or module headers which provide lengthy information on the action of the statements which follow. Furthermore, since the four fields in each source code statement are each separated by 'white space' it is not essential that they are precisely aligned in columns. However, whereas the following lines of source text are perfectly legal:

```
start: MOV AH,05H;Printer output function code
MOV DL,0AH;First generate a
INT 21H;line feed
MOV DL,0DH;Next generate a
INT 21H;carriage return
MOV DL,41H;First character to print is A
MOV CL,3EH;Number of characters to print
MOV AH,05H;Set up the function code
prch: INT 21H;and print the character
INC DL;Get the next character
LOOP prch;and go round again
MOV AL,00H;Set up the return code
MOV AH,4CH;and the function code
INT 21H;for an exit to DOS
```

they would be much more readable had they been entered in the strict format shown below:

```
start:  MOV     AH,05H    ; Printer output function code
        MOV     DL,0AH    ; First generate a
        INT     21H       ; line feed
        MOV     DL,0DH    ; Next generate a
        INT     21H       ; carriage return
```

```
        MOV     DL,41H      ; Frist character to print is A
        MOV     CL,3EH      ; Number of characters to
                              print
        MOV     AH,05H      ; Set up the function code
prch:   INT     21H         ; and print the character
        INC     DL          ; Get and next character
        LOOP    prch        ; and go round again
        MOV     AL,00H      ; Set up the return code
        MOV     AH,4CH      ; and the function code
        INT     21H         ; for an exit to DOS
```

Macro assemblers

A macro facility allows the programmer to write blocks of often-used code and incorporate these in programs by referring to them by name. The blocks of code are each defined as a *macro*. Thereafter, the macro assembler expands the macro call by automatically assembling the block of instructions which it represents into the program. The macro call can also be used to pass parameters (e.g. symbols, constants or registers) to the assembler for use during the macro expansion.

As an example of the use of macros, the macro defined in the following code can be used to exchange the contents of two registers passed to the macro as parameters reg1 and reg2:

```
;       MACRO TO EXCHANGE 16-BIT REGISTER CONTENTS
;       PARAMETERS PASSED:    reg1, reg2
;       REGISTERS AFFECTED:   reg1, reg2
;
Swap    MACRO   reg1,reg2       ; Specify registers to
                                  swap
        PUSH    reg1            ; Stack reg1 first,
        PUSH    reg2            ; then reg2
        POP     reg1            ; reg1 receives reg2
        POP     reg2            ; reg2 receives reg1
        ENDM
```

The following line of code shows how the macro call is made:

```
        Swap    AX,CX           ; Call the macro
```

The macro assembler expands the call, replacing it with the code given in its definition. The code generated by the macro assembler (i.e. the *macro expansion*) will thus be:

```
PUSH    AX
PUSH    CX
POP     AX
POP     CX
```

A macro facility can be instrumental in making significant reductions in the size of source code modules. Furthermore, macros can be nested such that a macro definition can itself contain references to other macros which, in turn, can contain references to others. A notable disadvantage of using macros is that the resulting object code may contain a large number of identical sections of code and will also occupy more memory space than if an equivalent subroutine had been used. In practice, therefore, programmers should use macros with care since there may be occasions where subroutines would be more efficient even though they may not be quite so easy to implement.

As well as macros, most assemblers also support *conditional assembly*. This allows the programmer to specify conditions under which portions of the program are either assembled or not assembled. Conditional assembly allows the programmer to test for specific conditions (using statements such as IF . . . ELSE . . . ENDIF) and use the outcome to control the assembly process.

Assemblers generally make two passes through a source file. During the first pass, macro calls are expanded and a symbol table is generated. On the second pass, relocatable code is generated which can be saved in a disk file. Such files are, however, not directly executable and require the services of a linker in order to function as programs.

The Microsoft macro assembler (MASM) provides logical programming syntax which supports the segmented architecture of 8086, 8088, 80186, 80188, 80286 and 80386 microprocessors as well as the 8087, 80287, and 80387 mathematics coprocessors. The assembler produces relocatable object modules which are linked together using the Microsoft overlay linker, LINK.

MASM is invoked by a command line of the form:

```
M <options> <file list>
```

The options which may be selected include the generation of additional statistics, error information, and data which may be used by the Microsoft's CodeView debugger. The file list must contain the name of the assembly language source code file. This file name may be followed by the names of the object code file, the listing file, and the cross-reference file. Where these last three named files are not specified, MASM will prompt for them.

The default file extension for the objects code filename is OBJ while those for the source listing and cross-reference files are LST and CRF, respectively.

Linkers

The linker is used to combine one, or more, object code files into a single executable program file. The output file produced by the linker is not bound by specific memory addresses (i.e. it is relocatable) and the operating system is able to load and execute the file at any convenient address.

The linker must resolve address references between modules such that any module which directs program execution outside itself (by means of a CALL, an external symbol, or an 'include' directive) will be linked to the module which contains the corresponding code.

The Microsoft linker (LINK) is invoked by a command of the form:

```
LINK <options><file list>
```

The options which may be selected include the display of linker process information, the packing of executable files, and the listing of *public symbols*. The file list must contain the name of each of the object code files to be linked. These may be followed by the names of the executable program file, the map file, and the names of the library files. Where these last three named files are not specified, LINK will prompt for them.

The default file extension for the executable program file is EXE whilst those for the map and library files are MPA and LIB, respectively.

The screen output produced during a typical session with the Microsoft Macro Assembler (M, MASM and LINK) is shown in Figure 5.1. The assembly language source code file (ptest.asm) was produced by the Microsoft editor (M) and the code appears on p. 117.

Cross-reference utilities

Cross-reference utilities can be invaluable when debugging since they can greatly speed up the search for symbols within a source code file during a debugging session. A cross-reference utility can be used to produce a specially created listing of all of the symbols used in an assembly language program. The listing is invariably alphabetical and each symbol in the list is followed by one, or more, line numbers which indicate the lines in the source code file which contain a reference to the symbol.

The Microsoft cross-reference utility (CREF) is invoked by a command line of the form:

```
CREF <file list>
```

where the file list consists of the name of the cross-reference file generated by MASM followed by the name of the readable (ASCII format) cross-reference file.

Library managers

A library manager allows the programmer to gather a number of object code files (i.e. those with an OBJ extension) into a single library file

```
  (The source code file is entered using the Microsoft Editor)

C>m ptest.asm<RETURN>

  (When the editor is left, the source file PTEST.ASM is written
   to the hard disk. The Macro Assembler is then invoked...)

C>masm ptest<RETURN>
Microsoft (R) Macro Assembler Version 5.10
Copyright (C) Microsoft Corp 1981, 1988.  All rights reserved.

Object filename [ptest.OBJ]:<RETURN>
Source listing  [NUL.LST]:<RETURN>
Cross-reference [NUL.CRF]:<RETURN>

  50144 + 394061 Bytes symbol space free

       0 Warning Errors
       0 Severe  Errors

  (The Linker is now used to produce an executable program...)

C>link ptest<RETURN>

Microsoft (R) Overlay Linker  Version 3.64
Copyright (C) Microsoft Corp 1983-1988.  All rights reserved.

Run File [PTEST.EXE]:<RETURN>
List File [NUL.MAP]:<RETURN>
Libraries [.LIB]:<RETURN>
LINK : warning L4021: no stack segment

  (The executable program is then tested...)

C>ptest<RETURN>

ABCDEFGHIJKLMNOPQRSTUVWXYZ[\]^_'abcdefghijklmnopqrstuvwxyz{|}~
```

Figure 5.1 *Screen output produced during a typical session with the Microsoft Macro Assembler*

(having an LIB extension). This file will generally be used in the production of several different programs and the object code modules collected by the library manager may be special modules created by the programmer or modules taken from an existing library. An optional library list can usually also be created by the library manager.

The value of building a library is that the routines needed within a program can be very easily linked into an executable object code file. Routines taken from the library can be used to construct further libraries or combined, as necessary, into executable programs by the linker.

The Microsoft library manager (LIB) is invoked by a command of the form:

```
LIB <library name><file list>
```

where the file list contains the names of the object code modules (each preceded by a '+' and separated by a comma) which are to be added to the library. As an example, the command line:

```
LIB graphics +fill,+shape
```

will add the object code modules fill.obj and shape.obj to the *graphics* library.

Symbolic debuggers

A symbolic debugger is an item of utility software which facilitates interactive testing and debugging of programs. As a minimum, a debugger should provide the user with commands which can be used to:

- Examine and modify memory.
- Examine and modify CPU registers.
- Run a program (starting at a given address) with breakpoints at which execution may be halted to permit examination of the CPU registers.
- Single-step a program (starting at a given address) with a register dump at the completion of each instruction.
- Disassemble a block of memory into assembly language mnemonics.
- Relocate a given block of memory.
- Initialize a given block of memory with specified data.
- Load/save blocks of memory from/to disk.

The debugger provided with the MS-DOS operating system (DEBUG) can be used for simple debugging (see Chapter 3) however the Microsoft macro assembler provides a much enhanced debugger which is known as CodeView. This package is a powerful window-oriented software tool which allows the programmer to quickly locate logical errors in programs. The debugger can display source and object code simultaneously (indicating the line which is about to be executed), dynamically watch values (local or global), and switch screens to display program output. Use of the debugger is, to a large extent, intuitive and it greatly outperforms the DEBUG package supplied with DOS.

A full description of the facilities and use of CodeView is beyond the scope of this book. This package is, however, strongly recommended to all potential assembly language programmers as it can be instrumental in quickly and effectively dealing with the vast majority of bugs and defects in assembly language programs.

8086 assembly language

Before attempting to provide readers with an introduction to assembly language programming techniques, it is important that readers have an

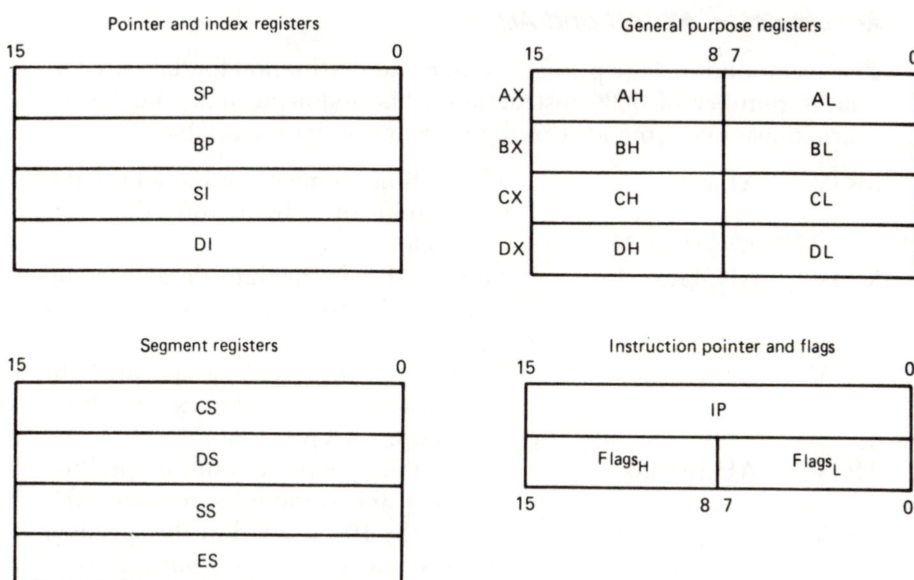

Figure 5.2 *8086 register model*

appreciation of the internal architecture and register model of the 8086 microprocessor.

The *register model* of the 8086 is shown in Figure 5.2. Of the fourteen 16-bit registers available, four may be described as general purpose and can be divided into separate 8-bit registers. As an example, the 16-bit extended accumulator (AX) can be divided into two 8-bit registers, AH and AL. The high-byte of a 16-bit word placed in AX is stored in AH whilst the low-byte is stored in AL. Instructions can be made to refer to various parts of the accumulator so that operations can be carried out on the word stored in AX, or the individual bytes stored in AH or AL.

The four segment registers are Code Segment (CS), Data Segment (DS), Stack Segment (SS), and Extra Segment (ES). By making appropriate changes to the contents of these registers, the programmer can dynamically change the allocation of work space.

As briefly mentioned in Chapter 1, the 8086 forms a 20-bit address from the contents of one of the segment registers (either CS, DS, SS or ES) and an offset taken from one (or more) of the other registers or from a memory reference within the program. The four segment registers (CS, DS, SS and ES) effectively allow the programmer to set up individual 64-kilobyte work space segments within the total 1-megabyte address range.

We shall now briefly consider each of the 8086 registers in turn:

Accumulator, AX (AH and AL)

The accumulator is the primary source and destination for data used in a large number of 8086 instructions. The following data movement instructions give some idea of the range of options available:

MOV	AL,data	Moves 8-bit immediate data into the least-significant byte of the accumulator, AL.
MOV	AH,data	Moves 8-bit immediate data into the most-significant byte of the accumulator, AH.
MOV	AX,register	Copies the contents of the specified 16-bit register to the 16-bit extended accumulator (AX).
MOV	AH,register	Copies the byte present in the specified 8-bit register to the 8-bit register, AH.
MOV	AX,[address]	Copies the 16-bit word at the specified address into the general-purpose base register, AX.

BX (BH and BL) register

The BX register is normally as a base register (address pointer). The following data movement instructions give some idea of the range of options that are available:

MOV	BX,data	Moves 16-bit immediate data into the general-purpose base register (BX).
MOV	BX,[address]	Copies the 16-bit word at the specified address into the general-purpose base register (BX).
MOV	BX,register	Copies the contents of the specified register to the base register (BX).
MOV	[BX],AL	Copies the contents of the AL register to the memory address specified by the BX register.

CX (CH and CL) register

The CX register is often employed as a *loop counter*. The 8086 LOOP instruction tests the contents of the CX register pair in order to determine whether the loop should be repeated, or not. This makes coding loops extremely simple as the following code fragment shows:

```
start:   MOV     CX,0C00H      ; Number of times round the
                                 loop
delay:   LOOP    delay         ; Count down
         RET                   ; Finished ?
```

The CX and CL registers are also used to implement repeated string moves, shifts and rotates. The following example shows how the contents of the accumulator can be rotated by the value placed in the CL register:

```
rote4:   MOV     CL,4          ; Number of bits to shift
         ROR     AX,CL         ; Rotate to the right
         RET
```

DX (DH and DL) register

The DX register is a general-purpose 16-bit register which can also be used as an extension of the AX register in 16-bit multiplication and division.

Stack pointer, SP

The SP register acts as a conventional Stack Pointer and points to the memory offset (relative to the paragraph address held in the Stack Segment register) of the current top of the stack. Adjustment of the SP register is automatic and programmers should avoid modifying the contents of this register if at all possible!

Base pointer, BP, destination index, DI, source index, SI

These registers are used in some of the more sophisticated of the 8086 addressing modes which permit the programmer to implement advanced data structures (such as two-dimensional arrays). All three registers are used to form addresses as shown in the following simple examples:

MOV	[BP+20],AX	Copies the word present in the AX register to an address offset by 20 bytes from the Base Pointer (BP).
MOV	[DI],space	Places 20H (previously defined by an equate of the form space EQU 20H) at the address pointed to by the Destination Index (DI).
MOV	SI,mess1	Move the start address of mess1 into the Source Index, SI. Thereafter, SI can be used with an offset to point to a particular character within the string, mess1.

Instruction pointer, IP

The Instruction Pointer is a 16-bit register which points to the address of the next instruction to be executed. The instruction Pointer is automatically updated by the CPU and the physical address of the instruction is found by adding the 16-bit value taken from the Instruction Pointer with the 16-bit value taken from the Code Segment register shifted four bits to the left (see Chapter 1).

Flag register, F

The 8086 has a 16-bit flag register which contains nine status bits which may be set (0) or reset (1) depending upon the internal state of the CPU. Flags keep their status (either set or reset) until an instruction is executed which has an effect on them. The 8086 flags are shown in Figure 5.3.

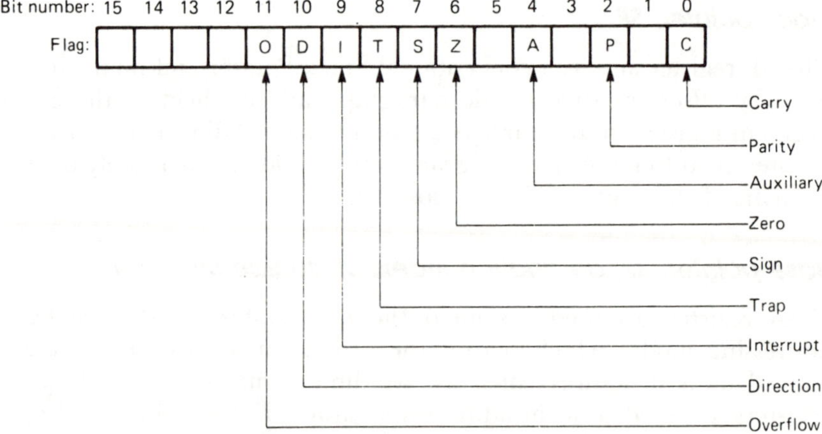

Figure 5.3 *8086 flags*

Segment Registers; CS, DS, SS, ES

We have already briefly mentioned the function of the four segment registers. Each register is associated with a separate workspace. The workspace defined by the Code Segment register will contain program instructions whilst the space defined by the Data and Extra Segments will generally contain data. In situations where RAM is limited (i.e. when only a small proportion of the total 1-megabyte address range is populated), there is no reason why the several segment registers should not have the same value (as in the case of a COM program). The code fragment:

```
MOV      AX,CS               ; Make Code and Data
MOV      DS,AX               ; Segments the same
```

can be used to make the Data Segment equal to the Code Segment (note that the instruction MOV DS,CS is not valid).

As a further example, the code fragment:

```
MOV      AX,vidram           ; Make Data Segment point
MOV      DS,AX               ; to Video RAM
```

can be used to make the data segment point to the start of a block of video RAM (vidram will previously have been the subject of an equate).

Interrupt handling

By comparison with earlier 8-bit microprocessors, the 8086 provides somewhat superior interrupt handling and uses a table of 256 4-byte pointers stored in the bottom 1 kilobytes of memory (addresses 0000H to 03FFH). Each of the locations in the Interrupt Pointer Table can be loaded with a pointer to a different interrupt service routine. Each pointer contains 2 bytes for loading into the Code Segment (CS) register and 2 bytes for loading into the Instruction Pointer (IP). This allows the programmer to place interrupt service routines in any appropriate place within the 1-megabyte physical address space.

Each of the 256 *interrupt pointers* is allocated a different type number. A Type 0 interrupt has its associated interrupt pointer in the lowest 4 bytes of memory (0000H to 0003H). A Type 1 interrupt will have its pointer located in the next 4 bytes of memory (0004H to 0007H), and so on.

The structure of the 8086 Interrupt Pointer Table is shown in Figure 5.4. Interrupt types 0 to 4 have dedicated functions while Types 5 to 31 are reserved. Hence there are 224 remaining locations in which interrupt pointers may be stored. The interrupting device places a byte on the data bus in response to an interrupt acknowledgement generated by the CPU. This byte gives the interrupt type and the 8086 loads its Code Segment and Instruction Pointer registers with the words stored at the appropriate locations in the Interrupt Pointer Table and then commences execution of the interrupt service routine.

The following code fragment shows how the Interrupt Pointer Table can be initialized to cope with three interrupt service routines:

```
; Initialize Interrupt Pointer Table
            MOV      AX,0          ; Point to start
            MOV      DS,AX         ; of memory.
            MOV      AX,CS         ; Get code segment.
```

```
; Type 32
            MOV     80H,dev1     ; Offset for dev1 ISR
            MOV     82H,AX       ; and segment address
; Type 33
            MOV     84H,dev2     ; Offset for dev2 ISR
            MOV     86H,AX       ; and segment address
; Type 255
            MOV     3FCH,dev3    ; Offset for dev3 ISR
            MOV     3FEH,AX      ; and segment address
```

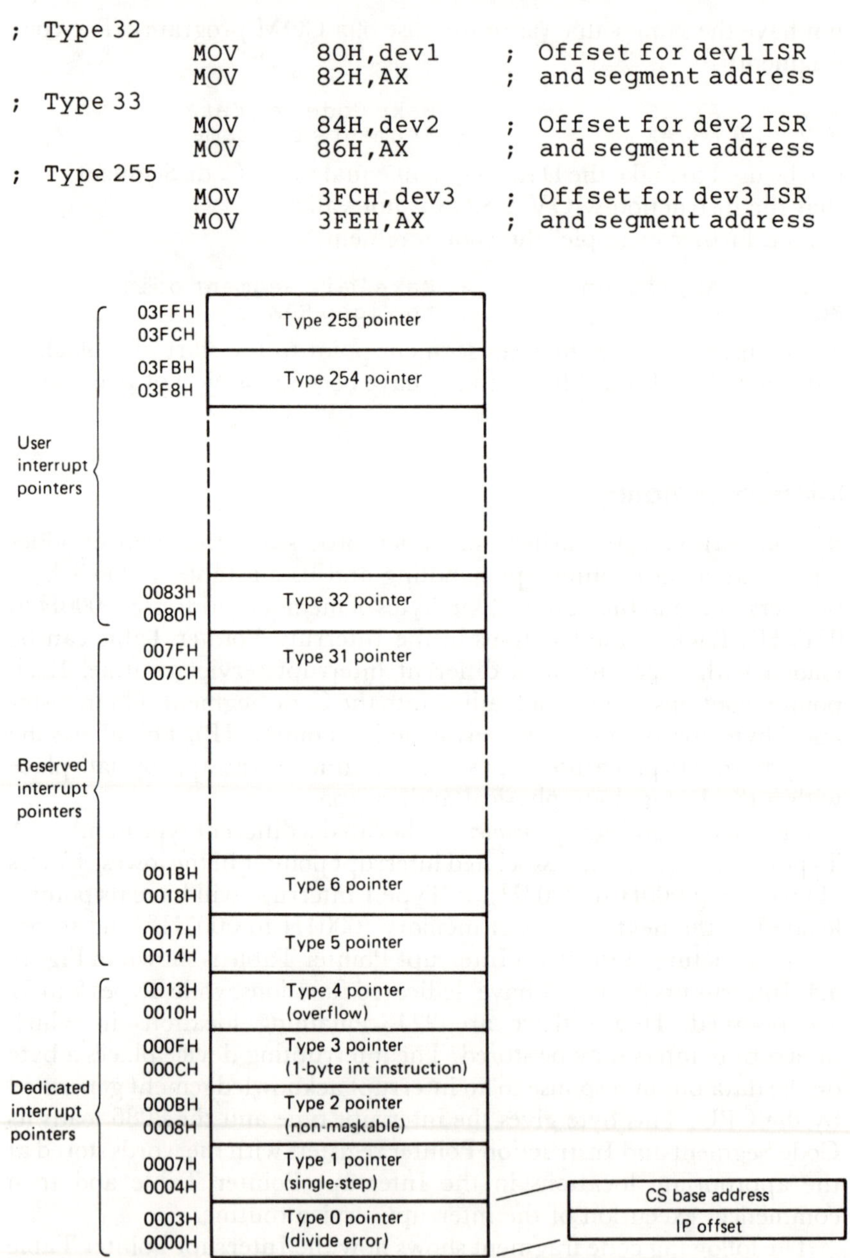

Figure 5.4 *8086 interrupt pointer table*

6 BASIC programming

Despite increasing competition from other languages (such as Pascal and C) BASIC remains extremely popular in the field of instrumentation and process control; the language is relatively easy to learn and programs can be quickly developed by those with little previous programming experience. Furthermore, modern implementations of the language put it on a par with many of its more powerful competitors. Gone are the days when BASIC programs were constrained to show a lack of structure by the absence of control structures such as DO . . . LOOP and WHILE . . . WEND. Furthermore, BASIC procedures, subprograms and user-defined functions all aid the programmer since they promote modularity and aid flexibility.

The availability of compilers adds a further dimension to the language since compiled BASIC programs can be indistinguishable from those written in other (supposedly superior) languages. Such programs are compact, execute at high speed, and are relatively 'tamper proof'. Such factors conspire to make modern structured and compiled BASICs worthy contenders for most applications within the field of instrumentation and control.

Since the majority of readers will have at least a passing acquaintance with the BASIC programming language, we shall deal only with topics which are directly relevant to the development of efficient programs for instrumentation and control applications. Readers with no previous knowledge are advised to consult one of the many tutorial books aimed at newcomers to BASIC programming (see Appendix E). There is no shortage of material to choose from and most texts will provide a more than adequate introduction to the subject.

Microsoft QuickBASIC

Microsoft QuickBASIC is an example of a powerful, yet easy to use, BASIC compiler. The BASIC programs which appear in this book were developed using QuickBASIC and Microsoft BASIC 6.0. Both of these excellent packages provide the user with a fully integrated development environment which allows program entry, editing, running, and debugging without having to leave the BASIC shell. Programs can be tested during development with minimal fuss and then free-standing executable programs (EXE files) can be produced when the user is reasonably confident that the program is robust and bug-free.

QuickBASIC also offers the user comprehensive context-sensitive on-line help. Using the resident BASIC text editor, syntax errors are reported immediately when the code is entered and debugging is aided by the availability of breakpoints and watchpoints which can be freely imbedded within the code.

Modular programming is encouraged and current modules are co-resident in memory during program development. Multiple editing windows allow the programmer to view the main code along with the code for a subprogram (procedure). The programmer can also exit to DOS, carry out a DOS operation (such as formatting a disk) and then return to the QuickBASIC environment at the point at which it was left.

Full graphics support is available for VGA, MCGA, EGA, CGA and Hercules standards. As a further bonus, QuickBASIC accepts IBM BASICA and Microsoft GWBASIC source files with minor changes (BASICA files must be saved in uncompressed format). QuickBASIC runs under MS-DOS Version 2.1, or later and BASIC 6.0 offers compatability with OS/2. Later versions of the Microsoft packages provide support for Microsoft's excellent CodeView debugger.

Developing BASIC programs

Since it is so easy to write and enter BASIC programs it is unfortunately all too easy to develop bad habits. Furthermore, the end result produced by an unstructured program may be indistinguishable from that produced by a program which is highly structured. The difference only becomes important when the time comes for extending, modifying or maintaining the program. With structured code this is a relatively simple matter. An unstructured program, on the other hand, may be a tangled nest of haphazard code and a major modification to the program may well result in the need for a complete rewrite. This can hardly be described as efficient!

There are a number of techniques which can assist in the production of efficient structured code. First and foremost, it is vitally important to get into the habit of being consistent in the layout of your programs and in the names used for variables. Failure to do this will make it extremely difficult to port sections of code from one program to another. This is a highly desirable feature which will save many hours of work. An efficient procedure for, say, accepting keyboard input and verifying that it is numeric, truncating it to integer and confirming that it is within a given range, can be useful in a huge variety of control applications. There is absolutely no reason why an efficient code module that performs such a function should not be included in every program that you write. Once written, you need never do it again!

Variable types

Wherever possible, integer numeric variables should be used in order to minimize storage space and increase processing speed. Floating point variables, which have considerable processing and storage overhead, should be avoided. Integer variables are normally recognized by a trailing %. Thus t represents a floating point numeric variable while t% represents an integer numeric variable and t$ represents a string variable.

Integer variables require 2 bytes for storage and values can be whole numbers (i.e. no decimal points) ranging from −32768 to +32767. Microsoft BASIC 6.0 and QuickBASIC also support long integers (which each occupy four bytes of storage) and both single and double precision floating point numbers (see Table 6.1).

String variables comprise a sequence of characters (letters, numbers and punctuation). QuickBASIC supports both fixed and variable length strings (the length of the former type must be declared). In either case, the maximum length permitted in QuickBASIC is 32767 characters.

Variable names

In order to aid readability, it has become fashionable to use relatively long names for variables. Happily, where a BASIC program will eventually be compiled, the overhead associated with long variable names applies only to the source files. It is thus permissible to use more meaningful variable names in such applications. Whether or not one is using a compiled BASIC, it is essential to maintain consistency with the choice of variable names.

Examples of acceptable variable names are:

chan%	channel number (integer)
col%	column number (integer)
date$	date (string – typical format mm/dd/yy)
day$	day of the week (string)
error$	error message (string)
file$	filename (string)

Table 6.1 QuickBASIC variable types

Class	*Type*	*Number of bytes for storage*	*Range of values*	*Suffix*	*Examples*	*Notes*
Numeric	Integer	2	−32768 to +32767	%	min%	Stored in 16-bit 2's complement format
Numeric	Long integer	4	−2147483648 to +2147483647	&	alt&	Stored in 32-bit 2's complement format
Numeric	Single precision floating point	4	−3.4E+38 to +3.4E+38 (approx.)	!	val!	Stored in IEEE format accurate to 7 decimal places
Numeric	Double precision floating point	8	−1.797E+308 to +1.797E+308 (approx.)	#	max#	Stored in IEEE format accurate to 15 or 16 digits
String	Character	Fixed	n/a	$	input$	Length must be declared (maximum 32767 characters)
String	Character	Variable	n/a	$	file$	Variable length (maximum 32767 characters)

in%	input port address (integer)
lin%	line number (integer)
lim%	limit value (integer number)
max%	maximum value (integer number)
min%	minimum value (integer number)
out%	output port address (integer)
prompt$	user prompt (string)
r$	general user response (string – usually a single character)
time$	time (string – typical format hh/mm/ss)

vel&	velocity (single precision floating point)
ver$	version number (string)
x%	*x*-axis displacement (integer)
y%	*y*-axis displacement (integer)

Subroutines

Subroutines can be instrumental in making very significant reductions in the size of BASIC programs and they should be employed whenever a section of code is to be executed more than once. Note, however, that if your version of BASIC supports the use of subprograms, procedures or user-defined functions, then these should normally be used instead! A typical example of the use of a subroutine might involve a delay routine which is required at various points in a program. Assuming that such a routine was to be used in a version of BASIC which employs line numbers and that the subroutine starts at line 10100, it might take the following form:

```
10100 REM Delay subroutine
10110 FOR c% = 0 TO 10000: NEXT c%
10120 RETURN
```

The subroutine may be called from several points within the main program as follows:

```
340 ..........
350 GOSUB 10100
360 ..........
```

etc.

```
440 ..........
450 GOSUB 10100
460 ..........
```

etc.

```
710 ..........
720 GOSUB 10100
730 ..........
```

In each case, program execution resumes at the line immediately following the GOSUB statement. Also note that, on exit from the subroutine, c% will have the value 10001.

We could make the delay subroutine even more flexible (allowing for variable length delays) by altering the upper limit of the loop using a variable which is set immediately prior to the subroutine call. The modified subroutine would then become:

```
10100 REM Delay subroutine
10110 FOR c% = 0 TO lim%: NEXT c%
10120 RETURN
```

As before, the routine may be called from several points in the main program as follows:

```
340 ..........
350 lim% = 10000: GOSUB 10100
360 ..........
```

etc.

```
440 ..........
450 lim% = 20000: GOSUB 10100
460 ..........
```

etc.

```
710 ..........
720 lim% = 15000: GOSUB 10100
730 ..........
```

On exit from the subroutine, the value of c% will have been modified to 1 greater than the value of lim% immediately prior to the subroutine call.

Since line numbers are not required when using Microsoft BASIC, a label must be used (instead of a line number) to mark the start of a subroutine, as shown below:

```
REM Start of main program
..........
GOSUB Delay
..........
```

etc.

```
..........
GOSUB Delay
..........
```

etc.

```
..........
GOSUB Delay
..........
```

etc.

```
..........
END
REM Delay subroutine
Delay:
FOR c% = 0 TO 10000: NEXT c%
RETURN
```

It is important to note that the label (Delay) is immediately followed by a colon and that the main body of program code must be terminated by an END statement in order to prevent execution of the subroutine when the end of the code has been reached. If this should ever happen, an error condition will result as the RETURN statement does not have a matching GOSUB.

Procedures

A user-defined procedure can be thought of as a named subroutine. The procedure is simply CALLed by name rather than by GOSUB followed by a label. This can be instrumental in not only making the resulting code more readable but it also ensures that the structure of the program can be easily understood. A further advantage of procedures is that parameters may be passed into procedures and values returned to the main program. Variables which are to be common with the main program may be declared at the start of the procedure using the SHARED statement (otherwise all variables internal to the procedure will be strictly local).

Procedures are defined using statement of the form SUB <name> and are terminated by END SUB. Procedures may also contain references to other procedures (i.e. procedures can be 'nested'). Procedure names should be chosen so they do not conflict with any variable names nor should they be BASIC reserved words.

The previous delay subroutine can be easily written as a procedure:

```
REM Delay procedure
SUB Delay(lim%)
FOR c% = 0 TO lim%: NEXT c%
END SUB
```

It should be noted that the Microsoft BASIC editor will recognize the SUB statement and will treat the procedure as a separate subprogram (with an automatic declaration inserted at the start of the main program code). Thereafter, the subprogram can be viewed and edited independently of the main program code.

The method of calling the delay procedure is more elegant than that used with the equivalent subroutine and takes the following format:

```
REM Start of main program
DECLARE SUB Delay(lim%)
..........
```

etc.

```
..........
CALL Delay(10000)
..........
```

etc.

```
..........
CALL Delay(20000)
..........
```

etc.

```
..........
CALL Delay(15000)
..........
```

The values within parentheses are parameters passed into the procedure as lim%. Such values are local to the procedure and external references to lim% will remain unchanged by the action of the procedure. It should also be noted that an overflow error will occur if values passed into the subprogram should ever exceed 32766.

Longer delays can be produced using floating point variables, as follows:

```
REM Delay procedure
SUB Delay(lim)
FOR c = 0 TO lim: NEXT c
END SUB
```

while the relevant code in the main program should run along the following lines:

```
REM Start of main program
DECLARE SUB Delay (lim)
..........
```

etc.

```
..........
CALL Delay(10000)
..........
```

etc.

```
..........
CALL Delay(20000)
..........
```

etc.

```
..........
CALL Delay(15000)
..........
```

Since we are using floating point variables, values passed into the subprogram can now exceed the 32766 limit imposed on integers (see Table 6.1 for details).

User-defined functions

User-defined functions are similar to user-defined procedures but return values (integer, float, or string) to the main program. As with user-defined procedures, functions are called by name (or FN name). An example of a user-defined function (FNConfirm%) appears later in this chapter.

Logical constructs

Modern BASICs provide us with a number of other useful constructs which can be instrumental in the production of efficient structured code.

As an example, a somewhat more elegant delay procedure can be produced using the WHILE . . . WEND construct. This routine uses a single variable rather than the two that were required in the FOR . . . NEXT construct used earlier.

```
REM Delay procedure
SUB Delay(lim%)
WHILE lim% > 0
  lim% = lim% – 1
WEND
END SUB
```

The condition in the WHILE stated is tested and, as long as it remains true (i.e. evaluates to non-zero), the code within the loop will be repeated. It should, perhaps, be stated that there is no particular advantage in using WHILE . . . WEND in this simple delay subroutine and a straightforward FOR . . . NEXT loop would, in practice, be perfectly adequate!

The DO . . . LOOP construct offers an even more powerful alternative to FOR . . . NEXT and WHILE . . . WEND. Several forms of DO . . . LOOP structure are available with tests for the loop condition at the start of the loop (DO WHILE . . . and DO UNTIL . . .

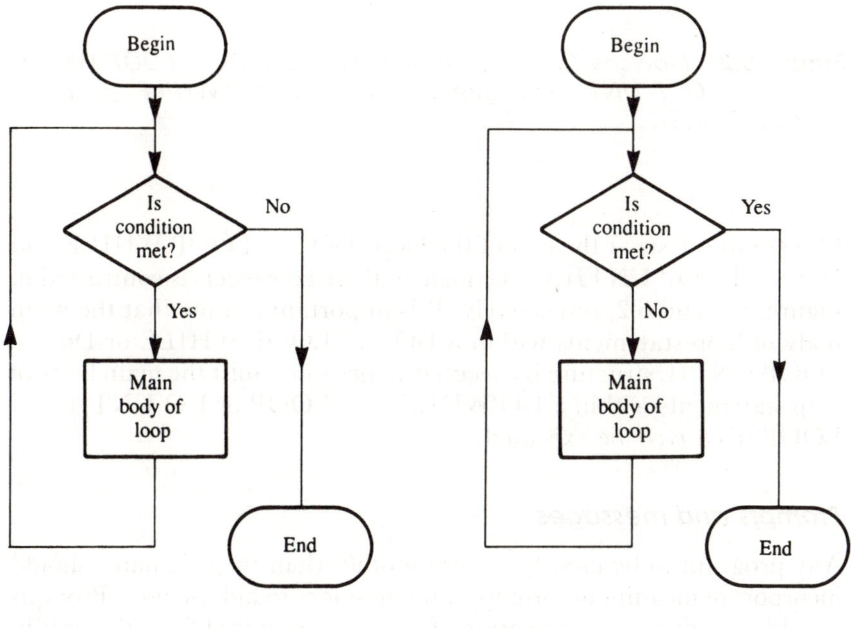

Figure 6.1 *Flowchart illustrating the logic of a DO WHILE . . . LOOP or DO UNTIL . . . LOOP structure. (a) DO WHILE . . . LOOP; (b) DO UNTIL . . . LOOP*

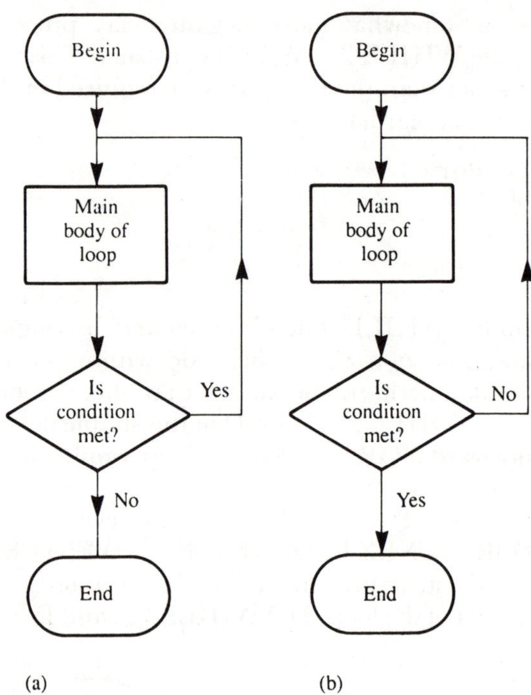

Figure 6.2 *Flowchart illustrating the logic of a DO . . . LOOP WHILE or DO . . . LOOP UNTIL structure. (a) DO . . . LOOP WHILE; (b) DO . . . LOOP UNTIL*

LOOP) and tests at the end of the loop (DO . . . LOOP WHILE and DO . . . LOOP UNTIL). The logic of these constructs is contrasted in Figures 6.1 and 6.2, respectively. It is important to note that the main body of loop statements within a DO . . . LOOP WHILE or Do . . . LOOP UNTIL structure is executed *at least once* whilst the main body of loop statements within a DO WHILE . . . LOOP or DO UNTIL . . . LOOP need *never* be executed.

Prompts and messages

Any program to be used by a person other than the originator should incorporate meaningful prompts and messages to aid the user. Prompts should also give some indication of the input required from the user in terms of the acceptable keystrokes, the length of an input string, and the need to include a RETURN keystroke. The following are examples of acceptable prompts:

```
Do you wish to quit ? (Y/N)
Press [SPACE] to continue . . .
Enter today's date (MM:DD:YY) followed by [RETURN] . . .
Enter filename (max. 8 characters) followed by [RETURN] . . .
```

Messages, unlike prompts, demand no immediate input from the user and should be included at any point in the program at which the user may require information concerning the state of the system. Messages should be written in plain English and should not assume any particular level of technical knowledge on the part of the user. The following are examples of acceptable messages:

```
Loading data file from disk . . . please wait!
Printer is not responding - please check paper.
Warning! Transducer on channel 4 is not responding.
```

Keyboard entry

Keyboard input from the user will be required in a variety of applications. Such input may take one of three basic forms summarized below:

1 Single keystrokes. Keystrokes may either be a letter, number, or punctuation and will generally not require the use of the RETURN or ENTER key.
2 Numerical inputs (comprising one or more keystrokes terminated by RETURN or ENTER). Each keystroke must be a number (or decimal point in the case of floats) and the input will normally be assigned to a numeric variable (either integer or floating point).
3 String inputs (comprising one or more keystrokes terminated by RETURN). Each keystroke may be a number, letter or punctuation. The string input by the user will normally be assigned to a string variable.

Single key inputs

Single key inputs will be required in a wide variety of applications. Such inputs can take various forms including menu selections or simple 'yes/no' confirmations. In either case, it is important to make the user aware of which keys are valid in each selection and, where the consequences of a user's input is irrevocable, a warning should be issued and further confirmation should be sought.

A simple typical 'yes/no' dialogue would take the following form:

```
INPUT "Are you sure (Y/N) "; r$
IF r$ = "Y" THEN . . . ELSE . . .
```

This piece of code has a number of shortcomings not the least of which is that it will accept any input from the user including a default (i.e. RETURN or ENTER used on its own). Other problems are listed below:

- The user may not realize that the input has to be terminated by ENTER or RETURN.
- A response of "N" is not distinguished from a default (or any input other than "Y").
- The routine does not allow a lower case input and the user may not realize that the SHIFT key has to be applied.
- If the user replies with "YES" or "yes", this would be equivalent to "N"!
- Finally, since we would probably want to use the routine at several points within the program, it should be coded as a procedure or user-defined function.

A much better solution to the problem would take the following form:

```
REM Confirm function
DEF FNConfirm%
r$ = ""
f% = -1
PRINT "Are you sure ? (Y/N)"
DO
 DO
  r$ = INKEY$
 LOOP WHILE r$=""
 IF r$ = "y" OR r$ = "Y" THEN f% = 1
 IF r$ = "n" OR r$ = "N" THEN f% = 0
LOOP WHILE f% <> 1 AND f% <> 0
FNConfirm% = f%
END DEF
REM Main program starts here
..........
..........
IF FNConfirm% = 1 THEN . . . ELSE . . .
..........
..........
```

The function returns a flag, f%, which is true (non-zero) if the user presses Y or y and is false (zero) if the user presses N or n. The function waits until a character is available from the keyboard (via the inner DO . . . LOOP) and then checks to ensure that it is one of four acceptable responses (i.e. upper and lower case Y and N). Any other keyboard input is invalid and the program continues to wait for further keyboard input until an acceptable value is returned (during this time the prompt message remains on the screen and does not scroll).

When an acceptable input is received, the function returns the appropriate flag in f%. It is important to note that INKEY$, unlike INPUT, does not require the use of the RETURN or ENTER key as a terminator and that a function definition must always precede the main body of code which calls it.

The problem could equally well have been solved by means of a procedure (rather than a user-defined function). In this case the code would have been as follows:

```
REM Main program starts here
DECLARE SUB Confirm(f%)
..........
..........
CALL Confirm(f%)
IF f% = 1 THEN . . . ELSE . . .
..........
..........
REM Confirm procedure
SUB Confirm (f%)
r$ = ""
f% = -1
PRINT "Are you sure ? (Y/N)"
DO
  DO
    r$ = INKEY$
  LOOP WHILE r$ = ""
  IF r$ = "y" OR r$ = "Y" THEN f% = 1
  IF r$ = "n" OR r$ = "N" THEN f% = 0
LOOP WHILE f% <> 1 AND f% <> 0
END SUB
```

Note that, as with all subprograms, the procedure is declared at the beginning of the main code and defined at the end.

Now, to take a more complex example, let's consider the case of a main menu selection. Suppose we are dealing with a control system which has four main functions (each of which is to be handled by a secondary menu) together with a function which closes down the system and exits from the program. The five main functions will be as follows:

1 Set parameters
2 Heater control
3 Pump control
4 Print report
5 Close down

The following code can be used for the main program loop:

```
DECLARE SUB Setparams()
DECLARE SUB Heatcontrol()
DECLARE SUB Pumpcontrol()
DECLARE SUB Printreport()
DECLARE SUB Closedown()
REM Main menu selection
WHILE 1
  CLS
  LOCATE 3, 36
  PRINT "MAIN MENU"
```

```
 LOCATE 6, 30
 PRINT "[1] Set parameters"
 LOCATE 8, 30
 PRINT "[2] Heater control"
 LOCATE 10, 30
 PRINT "[3] Pump control"
 LOCATE 12, 30
 PRINT "[4] Print report"
 LOCATE 14, 30
 PRINT "[5] Close down"
 LOCATE 16, 30
 PRINT "Option required (1-5)?"
 DO
   r$ = INKEY$
   k% = VAL(r$)
 LOOP UNTIL k% < 6 AND k %> 0
 IF k% = 1 THEN CALL Setparams
 IF k% = 2 THEN CALL Heatcontrol
 IF k% = 3 THEN CALL Pumpcontrol
 IF k% = 4 THEN CALL Printreport
 IF k% = 5 THEN CALL Closedown
WEND
```

Notice that the main program loop consists of an infinite WHILE . . . WEND loop. The single character string returned by INKEY$ is converted to an integer and then tested to see whether it is within range of the valid keyboard responses (note that depressing the 1 key returns a value in k% of 1, and so on). The DO . . . LOOP is only exited when a valid keystroke is detected. Having obtained a valid keystroke, the program checks the response to see which key was depressed using a series of IF . . . THEN statements so that the desired procedure can be CALLed.

If, for example, the user had pressed the 5 key, the result of the IF . . . THEN statement would have been found to be true (all others having been false) and program execution would be diverted to the procedure named Closedown. In this case, and since the result of the Closedown routine is irrevocable, the user should be given the option of returning to the main menu. Hence the Closedown procedure should take the following form:

```
REM Close down and exit
SUB Closedown
CLS
PRINT "You have selected the CLOSE DOWN option."
CALL Confirm(f%)
IF f% = 1 THEN END
END SUB
```

As before, the confirmation function returns a flag, f%, which is true (non-zero) if the user presses Y or y but is false (zero) if the user presses N or n. If the user decides not to continue with the Closedown procedure, END SUB ensures that the procedure is abandoned and execution resumes at the statement which follows the procedure CALL. The

WEND statement then diverts the program back to the beginning of the main menu selection routine.

QuickBASIC offers a more powerful logical construct which is particularly useful when making menu selections. The construct is based on SELECT CASE and eliminates the multiple use of IF . . . THEN. The equivalent SELECT CASE menu selection program is as follows:

```
DECLARE SUB Setparams()
DECLARE SUB Heatcontrol()
DECLARE SUB Pumpcontrol()
DECLARE SUB Printreport()
DECLARE SUB Closedown()
REM Main menu selection
WHILE 1
 CLS
 LOCATE 3, 36
 PRINT "MAIN MENU"
 LOCATE 6, 30
 PRINT "[1] Set parameters"
 LOCATE 8, 30
 PRINT "[2] Heater control"
 LOCATE 10, 30
 PRINT "[3] Pump control"
 LOCATE 12, 30
 PRINT "[4] Print report"
 LOCATE 14, 30
 PRINT "[5] Close down"
 LOCATE 16, 30
 PRINT "Option required (1-5)?"
 DO
  r$ = INKEY$
 LOOP UNTIL r$ <>""
 SELECT CASE r$
  CASE "1"
   CALL Setparams
  CASE "2"
   CALL Heatcontrol
  CASE "3"
   CALL Pumpcontrol
  CASE "4"
   CALL Printreport
  CASE "5"
   CALL Closedown
  CASE ELSE
   SOUND 60, 2
  END SELECT
WEND
```

Since we are no longer testing for a valid key entry (in the form of a figure in the range 1 to 5) we have included a 'catchall' in the form of the CASE ELSE statement. We have also made the program a little more 'user-friendly' by providing the user with an audible warning if an input keystroke is unacceptable.

Whilst on the topic of 'user-friendly' programs, it is perhaps worth mentioning that good use can be made of QuickBASIC's ability to trap key events. As an example, let us assume that the user is to be provided with an on-line help facility available from any point in the program when the F1 key is depressed. The following steps are required:

1 Code the subroutine (in this example we shall name it Help) along the lines described earlier in this Chapter.
2 Inform QuickBASIC that the Help subroutine is to be associated with the F1 key. The required statement is:

```
ON KEY(1) GOSUB Help
```

3 Enable trapping of the F1 key using the statement:

```
KEY(1) ON
```

4 If, at any time (e.g. during some critical process) it is subsequently necessary to disable F1 key trapping, simply include a statement of the form:

```
KEY(1) OFF
```

5 Finally, to temporarily inhibit F1 key trapping but, at the same time remembering whether or not the F1 key has been depressed (so that the event trap can later be executed when a subsequent KEY ON statement is encountered) the following statement can be used:

```
KEY(1) STOP
```

For readers who may wish to make further use of QuickBASIC's key event trapping facility, Table 6.2 gives the requisite numerical values associated with the other function and cursor keys within the KEY(n) statement.

Numerical inputs

The simple method of dealing with numerical input involves using a BASIC statement of the form:

```
INPUT "Value required"; n%
```

Sadly, this line of code will only work properly if the user realizes that a numeric value is required. Since BASIC cannot assign a letter to a numeric variable, the program will either crash or assign a value of zero if the user inadvertently presses a letter rather than a number. Furthermore, it would be useful to be able to impose a range of acceptable values on the user. The program should reject input values outside this range, warn the user that his input is invalid, and prompt again for further input. Again, such a routine would be ideally coded as a procedure. The procedure call could typically take the form:

```
prompt$ = "Temperature required"
CALL Numberin(prompt$, 60, 90, num%)
```

Table 6.2 Function and cursor key numbers for use in conjunction with the QuickBASIC KEY(n) statement

Key to be trapped	*Value of n*
F1	1
F2	2
F3	3
F4	4
F5	5
F6	6
F7	7
F8	8
F9	9
F10	10
F11	30
F12	31
Cursor up	11
Cursor down	12
Cursor right	13
Cursor down	14

while the procedure itself would be coded along the following lines:

```
REM General purpose integer numerical input
SUB Numberin (prompt$, min%, max%, num%)
DO
  PRINT prompt$;
  INPUT num$
  num% = VAL(num$)
  IF n% <= max% AND n% >= min% THEN EXIT SUB
  PRINT "Value outside permissible range!"
LOOP
END SUB
```

The procedure prints the prompt string (prompt$) and assigns the user's input to a string variable in order to avoid the program crashing if a letter is inadvertently pressed. The string is subsequently converted to an equivalent numeric variable using the VAL function. The resulting integer is then tested to see whether it lies within the acceptable range. If the integer is within range, the procedure is exited (via EXIT SUB) with num% containing a valid integer input. If the integer is not within range, the user is warned and prompted for further input. A similar routine can be produced for floating point input and, if desired, the prompt string can be included in the list of parameters to be passed into the function.

String inputs

The simple method of dealing with string input involves using a BASIC statement of the form:

```
INPUT "Filename"; n$
```

This line of code is fortunately not quite so prone to problems as its equivalent for numeric input. It is, however, worth considering what action we should take if the user should default the input (i.e. just presses RETURN or ENTER) or proceeds to input an unacceptably long string (the latter is an important consideration when dealing with filenames). Hence our general-purpose string input routine should allow for the substitution of a default string and should also truncate the user's input to a specified length. The procedure call might take the following form:

```
prompt$ = "Filename"
CALL Stringin(prompt$, 8, "MYSAMPLE", inputstr$)
```

while the procedure itself would be coded along the following lines:

```
REM General purpose string input
SUB Stringin (prompt$, length%, default$, inputstr$)
  PRINT prompt$; " ? ";
  LINE INPUT r$
  IF r$= "" THEN r$ = default$
  inputstr$ = LEFT$(r$, length%)
END SUB
```

As before, the procedure prints the prompt string (prompt$) and assigns the user's input to a string variable. The use of LINE INPUT (rather than just INPUT) ensures that the user can include punctuation. The user's response (r$) is then checked to determine whether it is a null string (i.e. the user has defaulted) and, if so, the specified default string is substituted. Lastly, the string is truncated to the specified length using the LEFT$ string function.

The following gives typical user entries and resulting values returned to the main program (in inputstr$) by the foregoing code when length% takes the value 8:

User input	*Value returned*
OLD_DATA	OLD_DATA
NEW_SAMPLE	NEW_SAMP
CONTROL_DATA	CONTROL_
(default)	MYSAMPLE

Data files

Finally, the ability to store data acquired by a control system is important where a detailed analysis of system performance is required. Data may be stored in one, or more, disk files in a disk-based system. Such files can readily be manipulated from BASIC.

The stages required for saving data in a disk file are as follows:

1 Open the file for output (using OPEN . . . FOR Output) and include a filename or complete file specification and an associated channel number which will be used to buffer operations on the disk file.
2 Send data to the file (using PRINT #).
3 Close the file (using CLOSE #).

As an example, let's assume that we have an integer array of 32 floating values, a(), to be stored in a disk file. If the data file is to be called "TEMP.DAT" and is to be stored on the disk in drive A, the following code can be used:

```
REM Save current data in disk file
SUB Save
SHARED a()
PRINT "Saving data on disk - please wait!"
OPEN "A:TEMP.DAT" FOR OUTPUT AS #1
FOR N% = 0 TO 31
 PRINT #1, a(n%)
NEXT n%
CLOSE #1
END SUB
```

The stages required for loading data from a disk file are as follows:

1 Open the file for input (using OPEN . . . FOR INPUT) and include a filename or complete file specification and an associated channel number for the buffer which will be used for subsequent operations on the file.
2 Retrieve data from the file (using INPUT #).
3 Close the file (using CLOSE #).

The following code can be used to retrieve the data stored by the previous example, loading it back into array a():

```
REM Load data from disk file
SUB Load
PRINT "Loading data from disk - please wait!"
OPENIN "A:TEMP.DAT" FOR INPUT AS #1
FOR n% = 0 TO 31
 INPUT #1, a(n%)
NEXT N%
CLOSE #1
END SUB
```

It is important to note that in both the Load and Save subprograms, the data array, a(), has been declared as SHARED between the procedure and the main code. The subprograms therefore have access to the data held in the array without the need for values to be passed in the form of a parameter list.

7 C programming

The C programming language was the brainchild of Dennis Ritchie and the language was originally implemented on a DEC PDP-11 running under the UNIX operating system. Despite its origins and close association with UNIX, C is now available in a variety of microcomputer implementations. These include the immensely popular Borland Turbo C and QuickC from Microsoft. Both of these packages will run happily on any PC or PC-compatible machine.

The C language is comparatively small but it employs a powerful range of control flow and data structures. It is, therefore, not surprising that it has become increasingly popular amongst programmers and software engineers. The language is well suited to the development of effective real-time applications and is ideally suited to the world of control and instrumentation (C is an excellent choice for small, tight, and fast applications).

The relatively small core of the language has been instrumental in ensuring a high degree of portability from one hardware configuration to another. C offers some significant advantages in the development of software for real-time applications. The language promotes the use of structure, is highly portable and it yields code which is relatively compact. Furthermore, when compiled, it can offer execution speeds which are far in excess of those which can be obtained with comparable interpreted languages.

To the newcomer, C source code can appear somewhat cryptic. Indeed, programmers experienced in other (less structured) languages may have difficulty when making the transition to C. Indeed, it is often said that it is easier to learn C if one has not had the misfortune of acquiring preconceptions developed as a result of familiarity with BASIC. Whilst this may be demonstrably true, the fact is that most of today's learners of C will already be proficient in one or more other languages and these will invariably include BASIC.

Those wishing to develop proficiency with C programming should not underestimate the amount of time required. As always, the best way to learn is to test out each new concept as it is introduced. Furthermore, it is best not to dwell on comparisons between C and other languages (such as BASIC). It is first necessary to understand something of the structure of C programs before progressing to such topics as data types, pointers, functions, and control structures. The rewards for perseverance are considerable!

The examples given in this chapter are written in Microsoft Quick C. This development package provides an integrated environment with full-screen editor, Microsoft Code View based debugger and a compiler which runs at 10000 lines per minute. The package is both source and object code compatible with Microsoft C version 5. Program development in Quick C is simple and straightforward and the process is automated to a large extent through the use of pull-down menus and dialogue boxes.

C programming techniques

There are numerous texts devoted to C programming. Hence, rather than devote space in this chapter to introducing readers to the basic

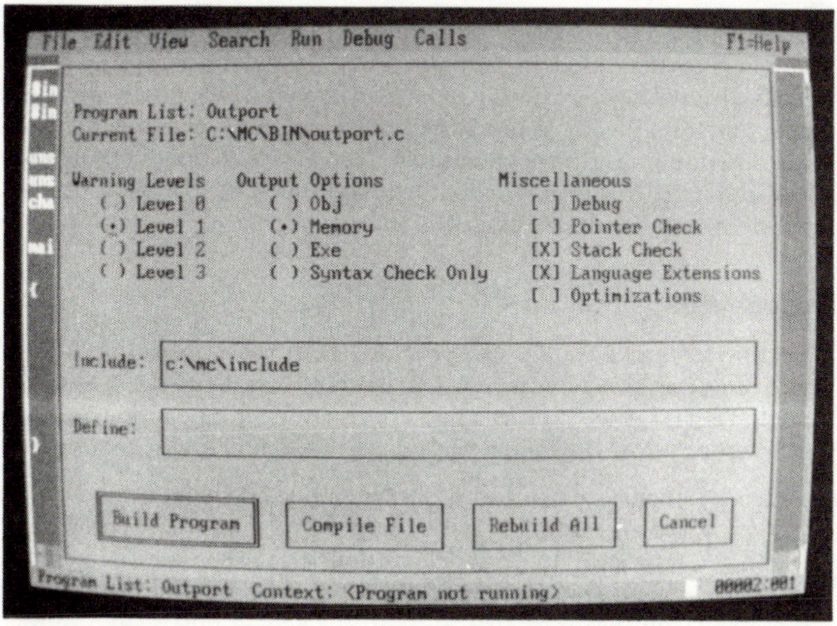

Plate 7.1 *Microsoft Quick C compiler options*

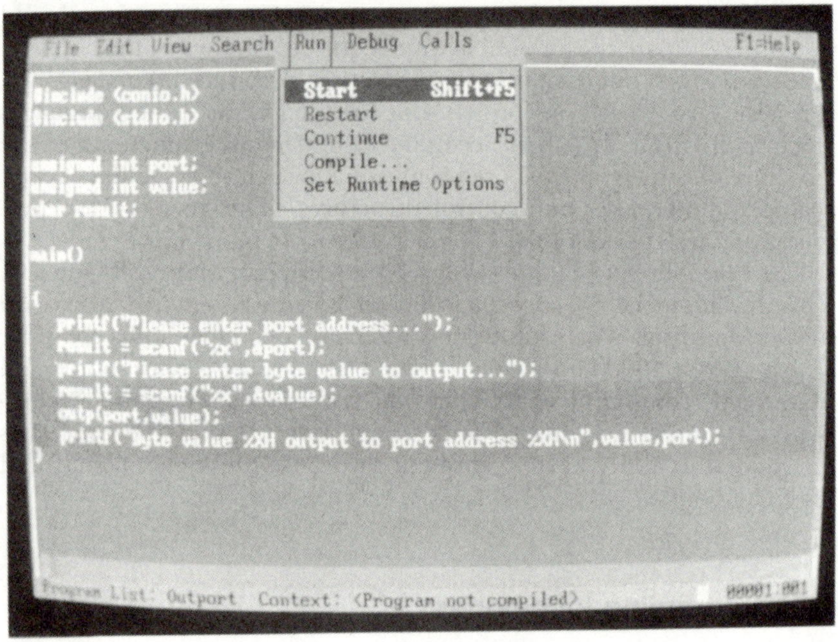

Plate 7.2 *Microsoft Quick C program development*

concepts associated with C programming, we shall adopt the same approach to that used in Chapter 6 by providing a tutorial aimed specifically at showing how C can be used in control applications. Topics have therefore been included which have particular relevance to instrumentation and process control. Newcomers to the language are advised to refer to one or more of the texts listed in Appendix E prior to, or concurrently with, reading this chapter.

Include files

A number of include files are provided with the C run-time library. These files contain macro and constant definitions, type definitions and function declarations. Include files are given the file extension, h. Some of the more useful include files are listed below:

bios.h — contains functions, declarations and structure definitions for the BIOS service routines (e.g. int86 and int86x).

conio.h — contains function declarations for the console and port I/O routines (e.g. cgets, cputs, getch, inp,inpw, outp and outpw).

ctype.h — defines macros and constants and declares global arrays used in character classification (e.g. isalnum, isalpha, islower, isupper, toascii, tolower, toupper, etc.).

dos.h contains macro definitions, function declarations, and type definitions for the MS-DOS interface.

io.h contains function declarations for file handling and low-level I/O functions such as open, close, read and write.

malloc.h contains function declarations for the memory allocation functions (e.g. malloc, calloc, free, etc.).

math.h contains function declarations for all floating point mathematics routines (e.g. abs, sin, cos, log, log10, exp, etc.).

stdio.h contains definitions of constant, macros, and types. Also contains function declarations for the stream I/O functions. The function definitions include fopen, fclose, fread, printf, and scanf. The constants defined within stdio.h include BUFSIZ (buffer size), EOF (end of file marker), and NULL.

stdlib.h contains function definitions which include abort, exit, and system.

string.h contains definitions for the string manipulation functions (e.g. strcpy, strlen, and strcat).

It is important to note that many run-time routines use macros, constants, and types which are defined in separate include files. These files must be specified within the source file (using the pre-processor directive #include), e.g.:

```
#include <stdio.h>
```

C functions

The fundamental building blocks of C programs are called functions. Once written, functions (like BASIC procedures) may be incorporated in a variety of programs whenever the need arises. The following function definition provides a delay:

```
delay()
{
  long x;
  for (x=1;x<200000;++x);
}
```

The delay function is called from a main program by a statement of the form:

```
delay();
```

A complete program to produce a delay would take the form:

```
/* delay1.c */
main()
```

```
{
   delay();
}
delay()
{
   long x;
   for (x=1;x<200000;++x);
}
```

It is important to note that no semi-colon follows the closing bracket of a function definition whereas, when the function is called, the program statement is terminated by a semi-colon. The main body of the function is enclosed between curly braces ({ and }). Since C is essentially a 'free-form' language (i.e. the compiler ignores white space within the source text) the programmer is able to adopt his or her own style of layout within the source text. The C functions and programs presented in this chapter will, however, follow the convention adopted by the author summarized below:

- Matching opening and closing braces ({ and }) are vertically aligned with one another.
- Statements within the body of a function are indented by three columns with respect to their opening and closing braces.
- Expressions (enclosed in brackets) used in conjunction with for and while statements are placed on the same line as the matching for or while.
- Blank lines are used to separate function definitions.
- The first function defined in a program is main().

Returning to the previous example, readers will probably have spotted a fundamental weakness in the simple delay function arising from the fact that it is only capable of providing a fixed delay. The function can be made more versatile by passing a parameter into it. The following modified delay function achieves this aim:

```
delay(limit)
long limit;
{
   long x;
   for (x=1;x<limit;++x);
}
```

The argument (contained in parentheses after the function name) is defined as an integer type before the function body. The function is then called using a statement of the form:

```
delay(200000);
```

Thereafter, the value 20000 is passed to the function and is used as the value for *limit*. An improved delay program would then take the form:

```
/* delay2.c */
main()
{
  delay(200000);
}
delay(limit)
long limit;
{
  long x;
  for (x=1;x<limit;++x);
}
```

Where more than one argument is to be passed to a function, they are simply listed and separated by commas. The data type for each argument must then be defined before the opening brace of the function body. A function definition dealing with port output, for example, might be declared with statements of the form:

```
out(port,byte)
int port, byte;
{
......
......
}
```

The corresponding function call would require a statement of the form:

```
out(255,128);
```

In this case, the value 255 would be passed into port whilst the value 128 would be passed into byte.

I/O functions

The following types of I/O function are available within C:

1 Stream I/O	In which a data file or data item is treated as a stream of individual characters. Examples of stream I/O functions include fopen, fgetc, fgets, and fclose.
2 Low-level I/O	Routines which do not perform buffering and formating but which, instead, directly invoke the I/O capabilities of the operating system. Examples of low-level I/O functions include open, close, read, and write.
3 Console and port I/O	An extension of stream I/O which permits reading and writing to a console/terminal or sending/receiving bytes of data via an I/O port. Examples of console and port I/O functions include getch, cgets, cputs, inp, and outp (the latter are used in many control applications).

Messages

Messages in C can be printed using statements of the form:

```
printf(message-string);
```

The standard C printf statement is, however, more versatile than its equivalent in BASIC as it allows a wide variety of formatting variations. Those available in Quick C include:

\b for backspace
\f for form feed
\n for new line
\t for tab

The following example prints the message "Warning!" immediately preceded and immediately followed by two blank lines:

```
printf("\n\nWarning!\n\n");
```

Variables can be included within the formatted print statement, as the following example shows:

```
printf("Tank number %d temperature %d\n",tankno,temp);
```

The current values of tankno and temp are printed as integer decimal numbers within the string. Thus if tankno and temp currently had the values 4 and 56, the resulting output generated would be:

```
Tank number 4 temperature 56
```

C allows a wide range of conversion characters to be included within formatted print strings. These usually include:

%c for single character
%d for signed decimal
%o for unsigned octal
%s for string
%u for unsigned decimal
%x for lower case hexadecimal
%X for upper case hexadecimal

The following example shows how conversion specifiers can be used to print the decimal, hexadecimal and octal value of the same number:

```
printf("Decimal %d, hexadecimal %X, octal %o", num,
num,num);
```

It is important to note that each conversion specifier must correspond to an argument within the list. The following program prints the hexadecimal and octal equivalents of the decimal number, 191:

```
/* bconv1.c */
main()
```

```
{
 int num;
 num=191;
 printf("Decimal %d, Hex. %X, octal %o", num, num,num);
}
```

Loops

Loops can be easily implemented in C programs. The following program prints the ASCII character set and uses the %d, %X, and %c conversion specifiers to provide the decimal, hexadecimal and ASCII representation of the loop index (byte):

```
/* bconv2.c */
main()
{
 int byte;
 for (byte=32;byte<128;++byte)
 {
  printf("Decimal  %d,  hex.,  %X,  ASCII  %c\n",  byte,
byte,byte);
 }
}
```

The loop is executed for byte values in the range 32 to 127 and byte is increased by 1 on at the end of each pass round the loop.

A somewhat better formatted output can be achieved by printing column headings before the loop is entered and including field width specifiers within the format string. The following shows how:

```
/* bconv3.c */
main()
{
 int byte;
 printf("Decimal Hex. ASCII\n");
 for (byte=32;byte<128;++byte)
 {
  printf("Decimal %3d, %2X, %c\n",byte,byte,byte);
 {
{
```

Loops can also be nested to any required depth. The following program provides a simple example based on the use of while rather than for:

```
/* loop1.c */
main()
{
 int s;
 s=0;
 while (s<4)
 {
  ++s;
  printf("Outer loop count = %d\n",s);
  inner();
 }
}
```

```
inner()
{
 int t;
 t=0;
 while (t<4)
 {
 ++t;
 printf(" Inner loop count = %d\n",t);
 }
{
```

The outer loop is executed four times (with *s* taking the values 0 to 3 in the expression following while). The inner loop is executed four times (with *t* taking the values 0 to 3 in the expression following while) on each pass through the outer loop.

Inputs and prompts

A single character can be returned from the standard input (usually the terminal) by means of the getchar() function. The following routine shows how a single character can be returned from the keyboard:

```
/* getin1.c */
main()
{
 char c;
 c=inchar("Enter option required ...");
 printf("\n\nOption selected = %c\n",c);
}
int inchar(prompt)
char *prompt;
{
 printf("\n%s",prompt);
 return(getchar());
}
```

Here we have defined a function, inchar, which returns an integer to main. This is automatically converted to a character and assigned to c. It is important to note that the return key is used to terminate user input and, where the user provides more than one input character before pressing the return key, only the first character is returned by getchar().

Where a multiple (rather than single) character string is required, the scanf() function can be used, as shown in the code fragment below:

```
char code[16];
getcode()
{
 printf("Enter operator code ...");
 scanf("%s",code);
}
```

The following example shows how a string of characters can be accepted from the user and then printed on the screen:

```
/* getin2.c */
main()
{
  char code[16]
  printf("Enter operator code ...");
  scanf("%s",code);
  printf("\n\nCode entered: %s",code);
}
```

The scanf() function allows a similar set of conversion characters to that available for use within printf(). It is important to note that scanf() terminates input when a return, space or tab character is detected. Furthermore, the array must be sufficiently large to accommodate the longest string likely to be input. Since the string is automatically terminated by a null character, the array must be dimensioned so that its number of elements is one greater than the maximum string length.

Multiple arguments may be included within scanf(). It is important to note that, unlike printf(), the arguments to scanf() are pointers (not variables themselves). This point regularly causes confusion. Finally, since scanf() involves considerable overhead, simpler functions may be preferred where the space for code is strictly limited.

The following example illustrates the combined use of getchar(), printf(), and scanf() in a simple decimal to hexadecimal conversion utility:

```
/* hexdec.c */
#include <stdio.h>
char number[16];
main()
{
  int num, c;
  num=1;
  printf("\n\nDECIMAL TO HEXADECIMAL CONVERSION\n\n");
  while (num!=0);
  {
    printf("\nEnter decimal number (max. 65535) or 0 to
    quit");
    scanf("%s",number);
    num=atoi(number);
    printf("\nDecimal %u = %X hexadecimal",num,num);
  }
}
```

The expression following while evaluates true if the current value of num is non-zero. In such cases, the code following while is executed and the ASCII character string is converted to an integer by means of the atoi() function. If the user responds with 0 (or with a non-numeric character string) and expression evaluates false, the code following while is not executed and the program terminates.

Menu selection

It is often necessary to provide users with a choice of several options at some point in a control program. Fortunately, C offers the switch . . . case statement which is ready made for this particular purpose. Complex menu selections can be very easily implemented using switch . . . case statement. The following example shows how:

```
/* menu1.c */
#include <stdio.h>
#define FOREVER 1
main()
{
 char c;
 while (FOREVER)
 {
  menu();
  c=getchar();
  switch(c)
  {
   case '1':
    init();
    break;
   case '2':
    pump();
    break;
   case '3':
    mix();
    break;
   case '4':
    deliver();
    break;
   case '5':
    exit();
   default:
    printf("Invalid input!\n");
    printf("Please enter a number in the range [1] to
    [5]\n\n");
  }
  c=getchar();
 }
}
scroll(lines)
int nolines;
{
 int x;
 for (x=0;x<lines;++x)
 {
  printf("\n");
 }
}
menu()
{
 scroll(6);
 printf(" MAIN MENU\n\n");
 printf("[1] Initialise the system\n");
 printf("[2] Pump control\n");
```

```
  printf(" [3] Mixer control\n");
  printf(" [4] Delivery control\n");
  printf(" [5] Close down and exit\n");
  printf("Enter option required ... ");
}
init()
{
  scroll(4);
  printf("INITIALISING SYSTEM\n");
  ......
  ......
  ......
}
pump()
{
  scroll(4);
  printf("PUMP CONTROL\n");
  ......
  ......
  ......
}
mix()
{
  scroll(4);
  printf("MIXER CONTROL\n");
  ......
  ......
  ......
}
deliver()
{
  scroll(6);
  printf("DELIVERY CONTROL\n");
  ......
  ......
  ......
}
```

The foregoing program, which should largely be self-explanatory, may readily be adapted to users own requirements and clearly shows how C encourages structured programming style.

Passing arguments into main

A useful facility available within C running under MS-DOS is that of passing arguments into programs from which the command line input by the user when the program is first loaded. The main() function allows two arguments, argc and argv. When main is called, argc is the number of elements in argv, and argv is an array of pointers to the strings which appear in the command line.

The following demonstration illustrates the method of passing parameters:

```
/* argdemo.c */
#include <stdio.h>
main(argc,argv)
char *argv[];
int argc;
{
  int i;
  printf("argc: %d\n", argc);
  for (i=0;i<argc;i++)
    printf ("argv[%d]: %s\n", i, argv[i]);
}
```

After compilation into an executable program the routine is invoked from the MS-DOS in the following manner:

```
ARGDEMO ONE TWO THREE
```

The parameters to be passed are, in this case, the strings; one, two, three. The program generates the following output:

```
argc: 4
argv[0]: C\MC\BIN\ARGDEMO.EXE
argv[1]: one
argv[2]: two
argv[3]: three
```

The total number of parameters passed is given in argc. In this case, four parameters have been passed (including the current directory and program name which appears as argv[0]). The three strings (one, two, and three) appear as argv[1], argv[2], and argv[3].

The following program shows how the technique of parameter passing can be used to display the contents of a named file.

```
/* sprint.c */
#include <stdio.h>
#define EOF 26
FILE *stream;
main(argc,argv)
char *argv[];
int argc;
{
  int c;
  stream = fopen(argv[1],"rb");
  while (c!=EOF)
  {
   c=getc(stream);
   printf("%c",c);
  }
  fclose(stream);
}
```

Assuming that the compiled program is named SPRINT.EXE, the command line entered after the operating system prompt, would take the form:

```
SPRINT filename
```

To print a file called HELLO.DOC, the command would be:

```
SPRINT HELLO.DOC
```

More than one argument can be passed into main. The following is a simple utility program for renaming files which passes two arguments into main:

```
/* rname.c */
#include <io.h>
#include <stdio.h>
main(argc,argv)
char *argv[];
int argc;
{
  int result;
  result = rename(argv[1],argv[2]);
  if (result !=0)
   perror("\nUnable to rename file!\n");
  else
   print("\nRename successful!\n");
}
```

Assuming that the program is called RNAME, the command line entered after the operating system prompt, would take the form:

```
RNAME oldname newname
```

To change the name of a file called HELLO.DOC to GOODBYE.DOC the command would be:

```
RNAME HELLO.DOC GOODBYE.DOC
```

Disk files

File handling is quite straightforward in C. Files must be opened before use using a statement of the form:

```
stream = fopen (filename, mode);
```

The filename can be a file specification or the name of a logical device. The mode can be "r" for read, "w" for write, and "u" for update. If the file cannot be opened (e.g. it is not present on the disk) fopen() returns 0 otherwise fopen returns the stream number to be used in conjunction with subsequent read or write operations.

After use, files must be closed using statement of the form:

```
fclose(stream);
```

where stream is the channel number returned by a previous fopen statement.

As a further example of file handling in C, the following program converts the case of an ASCII file to upper case:

```
/* ucase.c */
#include <stdio.h>
```

```
FILE *fp, *fq;
int pc;
main(argc,argv)
int argc;
char *argv[];
{
 printf("CONVERTING FILE TO UPPER CASE\n");
 fp=fopen(argv[1],"r");
 if (fp==0)
   {
    printf("Unable to open input file: %s \n",argv[1]);
    exit();
   }
 fq=fopen(argv[2],"w");
 if (fq==0)

    printf("Unable  to  create  output  file:  %s\n")
 ,argv[2]);
    exit();
   }
 printf("\nInput file: %s \n",argv[1]);
 printf("\nOutput file: %s \n\n", argv[2]);
 while ((pc=getc(fp))!=EOF)
   {
    pc=toupper(pc);
    putc(pc,fq);
    putchar(pc);
   }
 fclose(fq);
 fclose(fp);
 exit();
}
```

Making DOS calls

Quick C allows the programmer to execute DOS interrupts by means of the intdos(), int86(), and int86x(), functions. The intdos() function is based on INT 21H and its usage is as follows:

```
intdos(inregs, outregs);
```

Before executing the instruction, the function copies the contents of inregs to the corresponding registers. After executing the instruction, the register values are copied to outregs. Both the inregs and outregs arguments are unions of type REGS (defined in the dos.h header file).

```
/* lprint.c */
#include <stdio.h>
#include <dos.h>
main()
{
 lprint("Hello world\13");
}
lprint(str)
char *str;
{
```

```
 union REGS regs;
 regs.h.ah=0x05;
 while (*str)
   {
     regs.h.dl=*str;
     intdos(&regs,&regs);
     str++;
   }
}
```

The function code (5H) is passed via register AH while the pointer to the character is passed in DL. The intdos() function is contained in a loop which is repeated until all characters of the string have been printed.

MS-DOS interrupts other than INT 21H can be made using the int86() and int86x() functions. The int86() function requires three arguments, as follows:

```
int86(intno, inregs, outregs);
```

The first of the three arguments is the requisite interrupt number. Before executing the interrupt, the function copies the contents of inregs to the corresponding registers. When the interrupt returns, the function copies the current register values to outregs. Note that inregs and outregs are again unions of type REGS (defined in the dos.h header file).

The int86x() function is similar to int86() but, to cater for a 'large memory model', the function also allows for setting of the segment register, DS and ES. The function requires three arguments and its syntax is as follows:

```
int86(intno, inregs, outregs, segregs);
```

One common use of int86() is that of providing better control over screen output than can be obtained with the printf() function. The final example contains four function definitions which show just how useful int86() can be:

`cursor(r,c)`	Positions the cursor at the given row and column arguments.
`cls()`	Simply clears the screen (no arguments are required).
`prstr(str)`	Prints a string argument at the current cursor position.
`setmode(mode)`	Sets the video display mode (see p. 30 for values of the mode argument).

The complete example program is as follows:

```
/* scnfunc.c */
#include <stdio.h>
#include <dos.h>
#define VIDEO 0x10
main()
```

```
{
  char password[16];
  setmode(3);           /* 80 x 25 text mode */
  cls();                /* clear the screen */
  cursor(12,25);        /* position cursor to centre text */
  prtstr("Enter password: ");
  scanf("%s",password);
  cls();                /* clear the screen */
}
/* Position cursor function */
cursor(r,c)
int r,c;
{
  union REGS regs;
  regs.h.ah=0x02;
  regs.h.bh=0;   /* Page 0 */
  regs.h.dh=r;
  regs.h.dl=c;
  int86(VIDEO,&regs,&regs);
}
/* Clear screen function */
cls()
{
  union REGS regs;
  regs.h.ah=0x06;
  regs.h.al=0;
  regs.h.bh=7;
  regs.h.ch=0;
  regs.h.cl=0;
  regs.h.dh=25;
  regs.h.dl=80;
  int86(VIDEO,&regs,&regs);
}
/* Print string function */
prtstr(str)
char *str;
{
  union REGS regs;
  regs.h.ah=0x0E;
  regs.h.bh=0;   /* Page 0 */
  while (*str)
  {
    regs.h.al=*str;
    int86(VIDEO,&regs,&regs);
    str++;
  }
}
/* Set video mode */
setmode(mode)
int mode;
{
  union REGS regs;
  regs.h.ah=0x00;
  regs.h.al=mode;
  int86(VIDEO,&regs,&regs);
}
```

8 The IEEE-488 bus

The IEEE-488 bus (also known as the *Hewlett Packard Instrument Bus* and the *General Purpose Instrument Bus*) provides a means of interconnecting a microcomputer controller with a vast range of test and measuring instruments. The bus is ideally suited to the implementation of *automatic test equipment (ATE)* and it has become increasingly popular in the last decade with a myriad of applications which range from routine production test to the solution of highly complex and specialized measurement problems.

In the past, IEEE-488 facilities have tended to be available within only on the more expensive test equipment. The necessary interface is, however, becoming increasingly commonplace in medium- and low-priced instruments. This trend reflects not only an increased demand from the test equipment user but also the availability of low-cost dedicated IEEE-488 controller chips.

Nowadays, most items of modern electronic test equipment (such as digital voltmeters and signal generators) are either fitted with the necessary IEEE-488 interface as standard or can be upgraded with optional IEEE-488 interface cards. This provision allows them to be connected to a microcomputer controller via the IEEE-488 bus such that the controller can be used to both supervise their operation and process the data which they collect.

Automated measurement is important in many applications, not just within the production test environment. Advantages of IEEE-488 based measurement systems incorporating PC-based controllers include:

1 Elimination of repetitive manual operation (freeing the test technician for more demanding tasks).
2 Equipment settings are highly repeatable (thus ensuring consistency of measurement).
3 Increased measurement throughput (measurement rates are typi-

cally between 10 and 100 times faster than those which can be achieved by conventional manual methods).

4 Reduction of errors caused by maladjustment or incorrect readings.
5 Consistency of measurement (important in applications where many identical measurements are made).
6 Added functionality (stored data may be analysed and processed in a variety of ways).
7 Reduction in skill level of operators (despite the complexity of equipment, user-friendly software can guide operators through the process of connection and adjustment prior to making a measurement).

This chapter introduces the IEEE-488 standard and describes a typical IEEE-488 interface for an IBM PC or compatible microcomputer.

IEEE-488 devices

The IEEE-488 standard provides for the following categories of device: Listeners, talkers, talkers and listeners, and controllers.

Listeners

Listeners can receive data and control signals from other devices connected to the bus but are not capable of generating data. An obvious example of a listener is a signal generator.

Talkers

Talkers are only capable of placing data on the bus and cannot receive data. Typical examples of talkers are magnetic tape, magnetic stripe, and bar code readers. Note that, whilst only one talker can be active (i.e. presenting data to the bus) at a given time, it is possible for a number of listeners to be simultaneously active (i.e. receiving and/or processing the data).

Talkers and listeners

The function of a talker and listener can be combined in a single instrument. Such instruments can both send data to and receive from the bus. A digital multimeter is a typical example of a talker and listener. Data is sent to it in order to change ranges and returned to the bus in the form of digitized readings of voltage, current, and resistance.

Controllers

Controllers are used to supervize the flow of data on the bus and provide processing facilities. The controller within an IEEE-488 system is invariably a microcomputer and, whilst some manufacturers provide dedicated microprocessor based IEEE-488 controllers, this function is often provided by means of a PC or PC-compatible.

IEEE-488 bus signals

The IEEE-488 bus uses eight multi-purpose bi-directional parallel data lines. These are used to transfer data, addresses, commands and status bytes. In addition, five bus management and three handshake lines are provided.

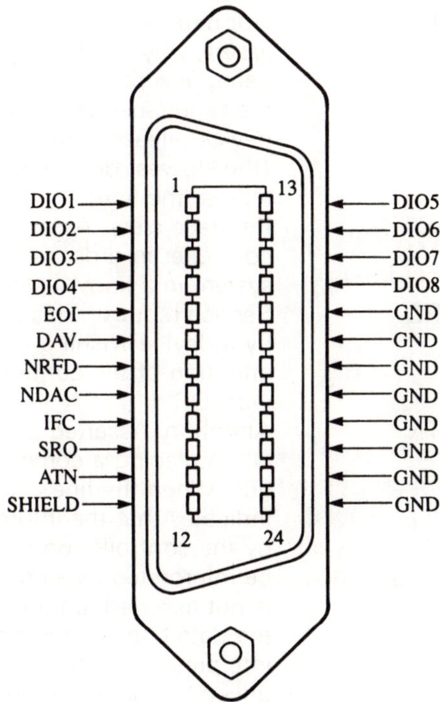

Figure 8.1 *IEEE-488 bus connector*

The connector used for the IEEE-488 bus is invariably a 24-pin type (as shown in Figure 8.1) having the following pin assignment:

Pin number	*Abbreviation*	*Function*
1	DI1	Data line 1
2	DIO2	Data line 2
3	DIO3	Data line 3
4	DIO4	Data line 4
5	EOI	End or identify. This signal is generated by a talker to indicate the last byte of data in a multi-byte data transfer. EOI is also issued by the active controller to perform a parallel poll by simultaneously asserting EOI and ATN.
6	DAV	Data valid. Thus signal is asserted by a talker to indicate that valid data has been placed on the bus.
7	NRFD	Not ready for data. This signal is asserted by a listener to indicate that it is not yet ready to accept data.
8	NDAC	Not data accepted. This signal is asserted by a listener whilst data is being accepted. When several devices are simultaneously listening, each device releases this line at its own rate (the slowest device will be the last to release the line).
9	IFC	Interface clear. Asserted by the controller in order to initialize the system in a known state.
10	SRQ	Service request. This signal is asserted by a device wishing to gain the attention of the controller. This line is wire-OR'd.
11	ATN	Attention. Asserted by the controller when placing a command on to the bus. When the line is asserted this indicates that the information placed by the controller on the data lines is to be interpreted as a command. When it is not asserted, information placed on the data lines by the controller must be interpreted as data. ATN is always driven by the active controller.
12	SHIELD	Shield.
13	DIO5	Data line 5.
14	DIO6	Data line 6.
15	DIO7	Data line 7.
16	DIO8	Data line 8.
17	REN	Remote enable. This line is used to

Pin number	Abbreviation	Function
		enable or disable bus control (thus permitting an instrument to be controlled from its own front panel rather than from the bus).
18–24	GND	Ground/common signal return.

Notes:
1 Handshake signals (DAV, NRFD and NDAC) employ active low open-collector outputs which may be used in a wired-OR configuration.
2 All remaining signals are fully TTL compatible and are active low.

Commands

Bus commands are signalled by taking the ATN line low. Commands are then placed on the bus by the controller and directed to individual devices by placing a unique address on the lower five data bus lines. Alternatively, universal commands may be issued to all of the participating devices (see p. 193).

Handshaking

The IEEE-488 bus uses three handshake lines (DAV, NRFD, and NDAC). The handshake protocol adopted ensures that reliable data transfer occurs at a rate determined by the *slowest* listener.

A talker wishing to place data on the bus first ensures that NDAC is in a released state. This indicates that all of the listeners have accepted the previous data byte. The talker than places the byte on the bus and waits until NRFD is released. This indicates that all of the addressed listeners are ready to accept the data. Finally, the talker asserts DAV to indicate that the data on the bus is valid. Figure 8.2 illustrates this sequence of events.

Service requests

The service request (SRQ) line is asserted whenever a device wishes to attract the attention of the active controller. SRQ essentially behaves as a shared interrupt line since all devices have common access to it. In order to determine which device has generated a service request, it is necessary for the controller to carry out a poll of the devices present. The polling process may be carried out either serially or in parallel.

In the case of *serial polling*, each device will respond to the controller by placing a status byte on the bus. DIO7 will be set if the device in question is requesting service, otherwise this data bit will be reset. The active

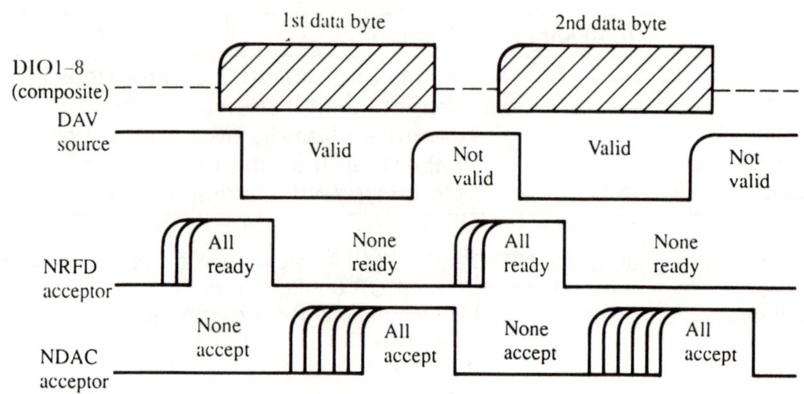

Figure 8.2 *IEEE-488 handshake sequence*

controller continues to poll each device present in order to determine which one has generated the service request. The remaining bits within the status byte are used to indicate the status of a device and, once the controller has located the device which requires service, it is a fairly simple matter to determine its status and instigate the appropriate action.

In the case of *parallel polling*, each device asserts an individual data line. The controller can thus very quickly determine which device requires attention. The controller cannot, however, at the same time ascertain the status of the device which has generated the service request. In some cases it will, therefore, be necessary to carry out a subsequent serial poll of the same device in order to determine its status.

Multiline commands

The controller sends multiline commands over the bus as data bytes with ATN asserted. Multiline commands are divided into five groups, as in the table opposite. Figure 8.3 summarizes the IEEE-488 command codes.

Bus configurations

Since the physical distance between devices is usually quite small (less than 20 metres), data rates may be relatively fast. Data rates of between 50 kilobyte/second and 250 kilobyte/second are typical however, to cater for variations in speed of response; the slowest listener governs the speed at which data transfer takes place. Figure 8.4 shows a typical

Command group	*Abbreviation*	*Function*	*Command byte*
Addressed command	ACG	Used to select bus functions affecting listeners (e.g. GTL which restores local front-panel control of an instrument)	00–0F
Universal command	UCG	Used to select bus functions which apply to all devices (e.g. SPE which instructs all devices to output their serial poll status byte when they become the active talker)	10–1F
Listen address	LAG	Sets a specified device to listen	20–3E
	UNL	Sets all devices to unlisten status	3F
Talk address	TAG	Sets a specified device to talk	40–5E
	UNL	Sets all devices to untalk status	5F
Secondary command	SCG	Used to specify a device sub-address or sub-function (also used in a parallel poll configure sequence)	60–7F

IEEE-488 bus arrangement in which a microcomputer is used as the controlling device.

IEEE-488 software

In order to make use of an IEEE-488 bus interface, it is necessary to have a DOS resident driver to simplify the task of interfacing with control software. The requisite driver is invariably supplied with the interface hardware (i.e. the IEEE-488 expansion card). The driver is installed as part of the normal system initialization and configuration routine and,

ACG — Addressed Command Group
UCG — Universal Command Group
LAG — Listen Address Group
TAG — Talk Address Group
SCG — Secondary Command Group

ACG	UCG	LAG	LAG	TAG	TAG	SCG	SCG
$00 0 NUL	$10 16 DLE	$20 32 SP 00	$30 48 0 16	$40 64 @ 00	$50 80 P 16	$60 96 SCG	$70 112 p SCG
$01 1 SOH GTL	$11 17 DC1 LLO	$21 33 ! 01	$31 49 1 17	$41 65 A 01	$51 81 Q 17	$61 97 a SCG	$71 113 q SCG
$02 2 STX	$12 18 DC2	$22 34 " 02	$32 50 2 18	$42 66 B 02	$52 82 R 18	$62 98 b SCG	$72 114 r SCG
$03 3 ETX	$13 19 DC3	$23 35 # 03	$33 51 3 19	$43 67 C 03	$53 83 S 19	$63 99 c SCG	$73 115 s SCG
$04 4 EOT SDC	$14 20 DC4 DCL	$24 36 $ 04	$34 52 4 20	$44 68 D 04	$54 84 T 20	$64 100 d SCG	$74 116 t SCG
$05 5 ENQ PPC	$15 21 NAK PPU	$25 37 % 05	$35 53 5 21	$45 69 E 05	$55 85 U 21	$65 101 e SCG	$75 117 u SCG
$06 6 ACK	$16 22 SYN	$26 38 & 06	$36 54 6 22	$46 70 F 06	$56 86 V 22	$66 102 f SCG	$76 118 v SCG
$07 7 BEL	$17 23 ETB	$27 39 ' 07	$37 55 7 23	$47 71 G 07	$57 87 W 23	$67 103 g SCG	$77 119 w SCG
$08 8 BS GET	$18 24 CAN SPE	$28 40 (08	$38 56 8 24	$48 72 H 08	$58 88 X 24	$68 104 h SCG	$78 120 x SCG
$09 9 HT TCT	$19 25 EM SPD	$29 41) 09	$39 57 9 25	$49 73 I 09	$59 89 Y 25	$69 105 i SCG	$79 121 y SCG
$0A 10 LF	$1A 26 SUB	$2A 42 * 10	$3A 58 : 26	$4A 74 J 10	$5A 90 Z 26	$6A 106 j SCG	$7A 122 z SCG
$0B 11 VT	$1B 27 ESC	$2B 43 + 11	$3B 59 ; 27	$4B 75 K 11	$5B 91 [27	$6B 107 k SCG	$7B 123 { SCG
$0C 12 FF	$1C 28 FS	$2C 44 , 12	$3C 60 < 28	$4C 76 L 12	$5C 92 \ 28	$6C 108 l SCG	$7C 124 | SCG
$0D 13 CR	$1D 29 GS	$2D 45 - 13	$3D 61 = 29	$4D 77 M 13	$5D 93] 29	$6D 109 m SCG	$7D 125 } SCG
$0E 14 SO	$1E 30 RS	$2E 46 . 14	$3E 62 > 30	$4E 78 N 14	$5E 94 ^ 30	$6E 110 n SCG	$7E 126 ~ SCG
$0F 15 SI	$1F 31 US	$2F 47 / 15	$3F 63 ? UNL	$4F 79 O 15	$5F 95 _ UNT	$6F 111 o SCG	$7F 127 DEL SCG

Example:

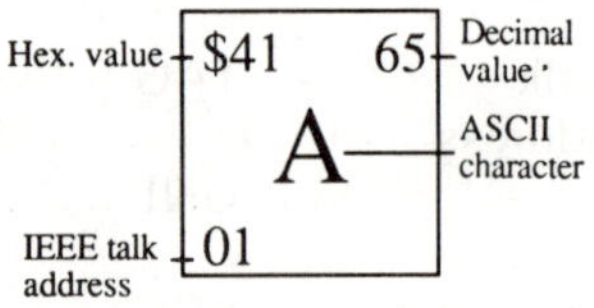

Hex. value — $41; Decimal value — 65; ASCII character — A; IEEE talk address — 01

Figure 8.3 *IEEE-488 command codes*

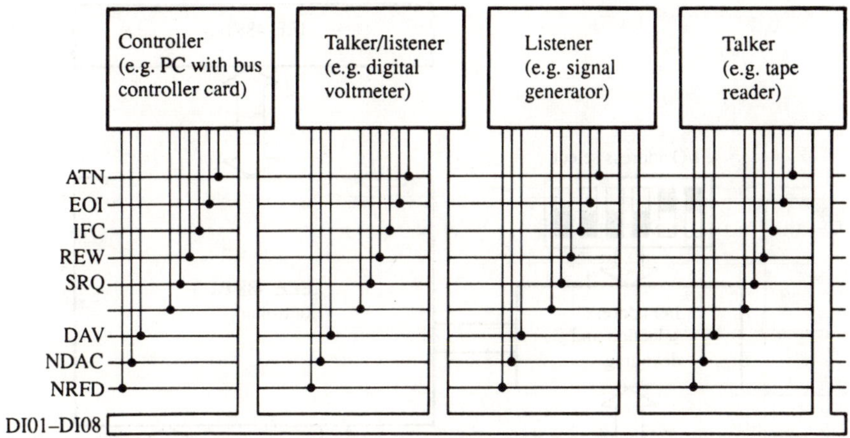

Figure 8.4 *Typical IEEE-488 bus configuration*

thereafter, will provide a software interface to applications packages or bespoke software written in a variety of languages (e.g. BASIC, Pascal and C). The user and/or programmer is then able to access the facilities offered by the IEEE-488 bus using high-level IEEE-488 commands such as REMOTE, LOCAL, ENTER, OUTPUT, etc.

MetraByte MBC-488

MetraByte's MBC-488 IEEE-488 is an example of a popular IEEE-488 interface which is designed to plug directly into a standard 8-bit PC expansion bus connector. A standard IEEE-488 connector is fitted to the rear metal plate on the card.

A complete software driver is supplied with the board (on disk). This driver handles initialization and protocol conversions required to provide access to all of the functions described in the 1978 IEEE-488 specification.

The board is based on the NEC 7210 IEEE-488 controller chip. This device is fully compatible with the 1978 IEEE-488 standard. The board handles all of the system timing for talking, listening and controlling the bus. The controller may be the PC itself or any of the other 14 devices which may be present on the bus. A maximum number of two MBC-488 cards may be fitted to any one PC. The simplified block schematic of the MBC-488 is shown in Figure 8.5.

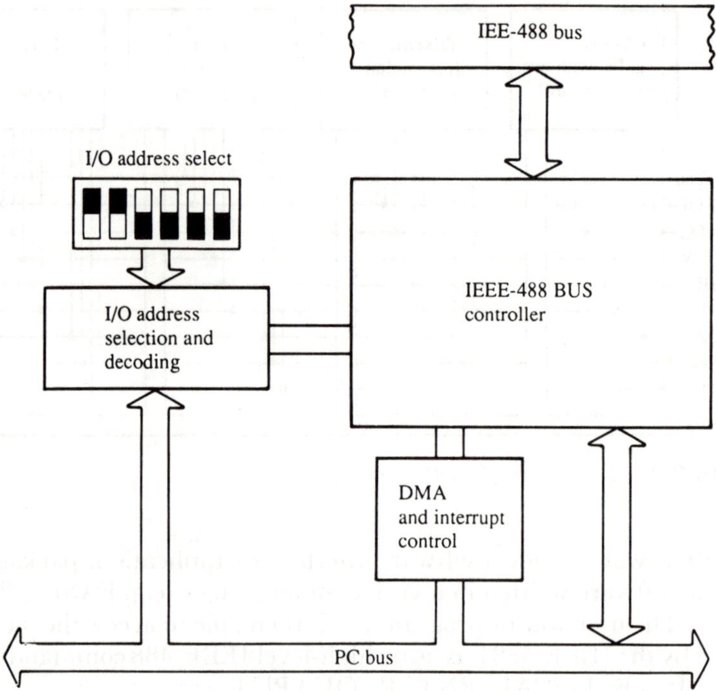

Figure 8.5 *Simplified block diagram of the MBC-488*

MBC-488 commands

The MBC-488 may be controlled from a high-level language using the following commands:

Command	*Function*
ABORT	Terminate the current selected device and command. If no device is given, the bus is cleared and set to the state given in the last CONFIG command. *Example:* `ABORT 1` terminates device 1.
CLEAR	Clear or reset the selected devices or all devices. If no device is given, the bus is cleared and set to the state given in the last CONFIG command. *Example:* `CLEAR 10` resets device 10.
CONFIG	Configures the bus to a given set of requirements.

Command	Function
	The bus will remain in the configured state until it is reconfigured. *Example:* `CONFIG TALK=2 LISTEN=1,3,4` configures device number 2 as a talker and devices 1, 3, and 4 as listeners.
ENTER	Enters bus data from the selected device into a specified string array (the array must have been previously dimensioned. A flag (FLAG%) will contain any error codes returned. *Example:* `ENTER 10[$, 0, 15]` enters data from address 10, array elements 0 to 18.
EOI	Sends a data byte to the selected device with EOI asserted. The bus must have been programmed to talk before the command is executed. The variable contains the data to be transferred. *Example:* `EOI 12[$]` issues an EOI with the last byte of the string to listener 12.
LOCAL	Sets the selected device(s) to the local state. If no device is specified then all devices on the bus are set to local. *Example:* `LOCAL 10, 11` sets devices 10 and 11 to local state.
LOCKOUT	Locks out (on a local basis) the specified device(s). The devices cannot be set to local except by the bus controller. *Example:* `LOCKOUT 9, 10` performs a local lockout on devices 9 and 10.
OUTPUT	Outputs a string to the selected listener(s). If no listener is specified in the command then all listeners will receive the specified string. *Example:* `OUTPUT 9, 11 [$E]` outputs the specified string using even parity.
PASCTL	Passes control of the bus to the specified device. Thereafter, the issuing PC controller will adopt the role of talker/listener.

Command	*Function*
	Example: `PASCTL 5` passes control of the bus to device 5 (which must be a bus controller).
PPCONF	Sets the parallel polling configuration for the specified device. *Example:* `PPCONF 12` selects parallel polling for device 12.
PPUNCF	Resets the parallel polling configuration for the specified device. *Example:* `PPUNCF 12` de-selects parallel polling for device 12.
PARPOL	Reads the status byte for the devices which have been set for parallel polling. *Example:* `PARPOL` reads status byte from a parallel polled device.
REMOTE	Selects remote operation for the specified device(s). *Example:* `REMOTE 9, 10, 11` selects remote operation for devices 9, 10 and 11.
REQUEST	Requests service from an active bus controller (used only when the PC is itself not the current bus controller). *Example:* `REQUEST` requests service from the current bus controller.
STATUS	Reads a (serial polled) status byte from the selected device. *Example:* `STATUS 8` reads the status byte (serial polled) from device 8.
SYSCON	Configures the system for a particular user configuration. The command initialises a number of system variables including: MAD the address of the PC system controller CIC the controller board in charge (more than one IEEE-488 bus controller board may be fitted to a PC) NOB the number of IEEE-488 bus controller boards fitted (1 or 2)

Command	Function
	BA0 the base I/O address for the first bus controller board (i.e. board 1) BA1 the base I/O address for the second bus controller board (i.e. board 2) *Example:* `SYSCON MAD=3, CIC=1, NOB=1, BA0=&H300` configures the system as follows: PC bus controller address =3 Controller board in charge =1 Number of board fitted =1 Base address of the controller board = 300 hex.
TIMEOUT	Sets the timeout duration when transferring data to and from devices. An integer number (e.g. VAR%) in the range 0 to 65000 is used to specify the time. For a standard IBM-PC/XT the time (in seconds) is equivalent to 3.5*VAR% whilst for an IBM-PC/AT the time is approximately 1.5*VAR%.
TRIGGER	Sends a trigger message to the selected device (or group of devices). *Example:* `TRIGGER 9, 10` triggers devices 9 and 10.

Note: If a command is issued by a device which is not the current controller then an error condition will exist.

Programming the MBC-488

Programming the MBC-488 IEEE-488 controller board is relatively straightforward as all control information is passed to the DOS resident software driver in the form of an ASCII encoded string. The command string is followed by three further parameters:

1 The variable to be used for output or input (either an integer number or a string).
2 A flag (integer number) which contains the status of the data transaction (e.g. an error or transfer message code).
3 The address of the interface board (either 0 or 1 or the physical I/O base address).

A command is executed by means of a CALL to the relevant DOS interrupt. The syntax of an interpreted BASIC (BASIC-A or GWBASIC) statement would thus be:

```
CALL IEEE(CMD$, VAR$, FLAG%, BRD%)
```

where IEEE is the DOS interrupt number, CMD$ is the ASCII command string, VAR$ is the variable to be passed (where numeric data is to be passed, VAR$ is replaced by VAR%), FLAG% is the status or error code, and BRD% is the board number (0 or 1).

As an example, the following code configures a system and then received data from device 10, printing the value received on the screen:

```
100 REM System configuration
110 DEF SEG=&H2000
120 BLOAD "GPIBBASI.BIN",0
130 IEEE=0
140 FLAG%=0
150 BRD%=&H300
160 CMD$="SYSCON MAD=3, CIC=1, NOB=1, BA0=768"
170 CALL IEEE(CMD$, A$, FLAG%, BRD%)
180 PRINT "System configuration status: ";HEX$(FLAG%)
200 REM Get string data from device 10
210 B$=SPACE$(18)
220 CMD$="REMOTE 10"
230 CALL IEEE(CMD$, B$, FLAG%, BRD%)
240 PRINT "Remote device 10 return flag: ";HEX$(FLAG%)
250 CMD$="ENTER 10[$, 0, 17]"
260 CALL IEEE(CMD$, B$, FLAG%, BRD%)
270 PRINT "Enter from device 10 return flag: ";HEX$(FLAG%)
280 PRINT "Data received from device 10:     ";B$
290 END
```

Line 110 defines the start address of a block of RAM into which the low level interrupt code is loaded from the binary file GPIBBASI.BIN (line 120). The IEEE interrupt number (0) is allocated in line 130 while the message/status code is initialized in line 140. Line 150 selects the base I/O address for the MBC-488 board (300 hex.) and the system configuration command string is defined in line 160 (note that the PC bus controller is given address 3 and a single IEEE-488 bus interface board is present).

The status flag (returned after configuring the system by means of the CALL made in line 170) is displayed on the screen in hexadecimal format (line 180). An empty string (B$) is initialized in line 210 (this will later receive the data returned from device 10). Device 10 is selected as the remote device in lines 220 and 230 whilst line 240 prints the returned status flag for this operation. Data is then read from device 10 (lines 250 and 260) and, finally, the status code and returned data are displayed in lines 270 and 280.

In most cases, it will not be necessary to display returned status codes. However, it is usually necessary to check these codes in order to ascertain whether a particular bus transaction has been successful and that no errors have occurred. Furthermore, a more modern BASIC (e.g. Microsoft QuickBASIC) will allow erstwhile programmers to develop a more structured approach to controlling the IEEE-488 interface with command definitions, error checks, and CALLs consigned to subprograms (see Chapter 6).

Troubleshooting the IEEE-488 bus

The IEEE-488 bus is generally well-tempered and easy use. Despite this, occasions do arise when the would-be system integrator is confounded by recalcitrant hardware and software which just will not behave as planned. Fortunately, fault finding on the IEEE-488 bus is usually very much simpler than when performing a similar task on an asynchronous serially-based system (e.g. an RS-422 based network). There are two main reasons for this; firstly, the IEEE-488 bus standard is open to much less variation in implementation and secondly, all signals use standard TTL voltage levels. This latter fact permits the use of conventional digital instruments (such as logic probes and pulsers). Furthermore, the controlling software often contains its own diagnostic routines and will warn the user if, for example, an external device is not responding to commands placed on the bus.

Where necessary, simple routines can be generated to exercize the bus (reading and displaying status codes for each device and transaction). It should be a relatively easy matter to isolate the fault by this means. Alternatively, remote instruments can be checked by interfacing in a different (perhaps simpler) bus configuration and checking that they perform correctly.

Finally, before delving into hardware, it is always worth checking the configuration of the software and the assignment of addresses to the various devices employed within the system at an early stage. If it is necessary to check the state of the various control signal lines (including EOI, SRQ, NRFD, NDAC etc.), a common-or-garden logic probe can be used to check for activity (remember that lines are active low).

9 Interfacing

This chapter aims to introduce readers to the general principles of interfacing sensors and transducers to PC bus I/O cards. We shall describe a variety of common sensors and transducers and, for those who do not wish to make use of 'off-the-shelf' signal conditioning modules, details of the circuitry necessary to interface such devices to several commonly available I/O cards has been provided. Before embarking on this task, it is perhaps worth mentioning some of the more important characteristics and limitations of conventional digital and analogue digital I/O ports.

Characteristics of digital I/O ports

The digital I/O ports provided by most PC expansion cards are invariably byte wide (i.e. each port comprises eight individual I/O lines). Such ports are usually implemented with the aid of one, or more, programmable parallel I/O devices (e.g. the 8255 described on p. 30).

Where expansion card parallel I/O devices are connected directly to the outside world via a rear panel-mounted I/O connector, care should be taken to ensure that no output line is excessively loaded nor that any input level exceeds the manufacturer's recommended limits.

As far as outputs are concerned, the Port B lines of a programmable parallel I/O device are usually able to source sufficient current to permit the direct connection of the base of a high current gain (preferably Darlington) NPN transistor. To minimize loading on the remaining I/O lines it will generally be necessary to employ the services of one, or more, octal TTL buffers. In any event, it is important to note that, when sourcing appreciable current, the high-level output voltage present on a port line may fall to below 1.5 V. This will be acceptable when driving a conventional or Darlington transistor but represents an illegal voltage level as far as TTL devices are concerned.

Some digital I/O expansion cards incorporate buffers between the parallel I/O device and the rear panel-mounted expansion I/O connector. Others make use of octal tri-state buffer/transceivers (e.g. 74LS245) rather than a VLSI parallel I/O device. Such devices can often source and sink as much as 15 mA and 24 mA, respectively.

Where a much higher output current capacity is required, external circuitry will generally be required in order to boost the output current. Alternatively (and provided that switching speed is unimportant) an interface card fitted with medium/high relays may be used. Such a card may also be employed when a high degree of isolation is required between an output load and a PC-based controller.

An expansion card which uses programmable I/O devices (rather than conventional buffers and latches) will require software configuration. A typical configuration routine for an interface based on two 8255 PPI devices (providing 48 digital I/O lines in six groups of eight lines) would involve initializing Ports A, B and C of both devices as either inputs or outputs, as required. This is carried out by simply writing appropriate control words to the *control register* of each device.

Having configured the I/O port, it is then relatively easy matter to send data to it or read data from it. Each port will appear as a unique address within the PC I/O map and data can be read from or written to the port using appropriate IN and OUT statements (or equivalent). Where the digital I/O lines within a port group have individual functions, appropriate *bit masks* can be included in the software so that only the state of the line in question is affected during execution of an OUT command.

Characteristics of analogue I/O ports

PC bus expansion cards for analogue I/O generally provide up to 16 analogue input lines and several analogue output lines. Analogue I/O ports are often based on one or more of the following devices:

Device	*Resolution*	*Function*	*Package*	*Notes*
AD573JN	10-bit	ADC	20-pin	
AD574	12-bit	ADC	28-pin	
AD667JN	12-bit	DAC	28-pin	
AD7226KN	8-bit	DAC	20-pin	4 channel
AD7528JN	8-bit	DAC	20-pin	2 channel
AD7528JN	8-bit	DAC	20-pin	2 channel
AD7542KN	12-bit	DAC	16-pin	
AD7545KN	12-bit	DAC	20-pin	
AD7547JN	12-bit	DAC	24-pin	2 channel

Device	Resolution	Function	Package	Notes
AD7569JN	8-bit	ADC/DAC	24-pin	I/O port
AD7579KN	12-bit	ADC	24-pin	
AD7580JN	10-bit	ADC	24-pin	
AD7681JN	8-bit	ADC	28-pin	8 channel
AD7672KN	12-bit	ADC	24-pin	high speed
AD7846AD	16-bit	DAC	28-pin	
AD7870JN	12-bit	ADC	24-pin	high speed
DAC08CP	8-bit	DAC	16-pin	
DAC0800LCN	8-bit	DAC	16-pin	
ZN425E	8-bit	DAC	16-pin	
ZN427E	8-bit	ADC	18-pin	
ZN428E	8-bit	DAC	16-pin	
ZN435E	8-bit	DAC	18-pin	
ZN439E	8-bit	ADC	22-pin	
ZN448E	8-bit	ADC	18-pin	
ZN502E	10-bit	ADC	28-pin	

Analogue inputs generally exhibit a high resistance (50 kΩ or more) and operational amplifier buffers are usually fitted to provide voltage gain adjustment and additional buffering between the analogue input and the input of the ADC chip.

Analogue outputs are usually available at a relatively low output impedance (100 Ω or less) and are invariably buffered from the DAC by means of operational amplifier stages. Typical output voltages produced by an analogue output port utilizing an 8-bit DAC range from 0 V to 5.1 V (20 mV/bit) when configured for *unipolar operation* or −5.1 V to +5.1 V (40 mV/bit) when *bipolar operation* is selected.

The procedure for reading values returned by an analogue input port will vary depending upon the type of ADC used. A typical sequence of operations for use with a multi-channel analogue input card with 8-bit resolution based on the ZN448E ADC would take the following form:

1 Select the desired input channel and start conversion. Send the appropriate byte to the status latch in order to select the required channel and input multiplexer. Conversion starts automatically when data is written to the status latch address.

2 Either:

   (i) Wait 10 μs (this is just greater than the 'worst-case' conversion time) using an appropriate software delay.

   or

   (ii) Continuously poll the ADC to sense the state of the end-of-conversion (EOC) line. This signal appears as a single bit in the status byte and, when low, it indicates that conversion is complete and valid data is available from the ADC.

3 Read the data. Having ensured that conversion is complete, the valid data byte can be read from the appropriate ADC address.

The byte read from the port will take a value between 00H and FFH. If the ADC has been configured for unipolar operation, a value of 00H will correspond to 0 V while a value of FFH will correspond to full-scale positive input (typically 5.1 V). When bipolar operation is used, a data byte of 00H will indicate the most negative voltage (typically −5.1 V) whilst FFH will indicate the most positive voltage (typically +5.1 V).

It is important to note that the values returned by conventional successive approximation ADCs will not be accurate unless the input voltage has remained substantially constant during the conversion process. Furthermore, where some variation is inevitable, several samples should be taken and averaged.

Analogue output ports are generally much easier to use than their analogue input counterparts. It is usually merely sufficient to output a byte to the appropriate port address. In most cases, analogue output ports will be configured for unipolar operation and, in the case of an 8-bit DAC, a byte value of 00H will result in an output of 0 V whilst a byte value of FFH will result in a full-scale positive output (typically 5.1 V).

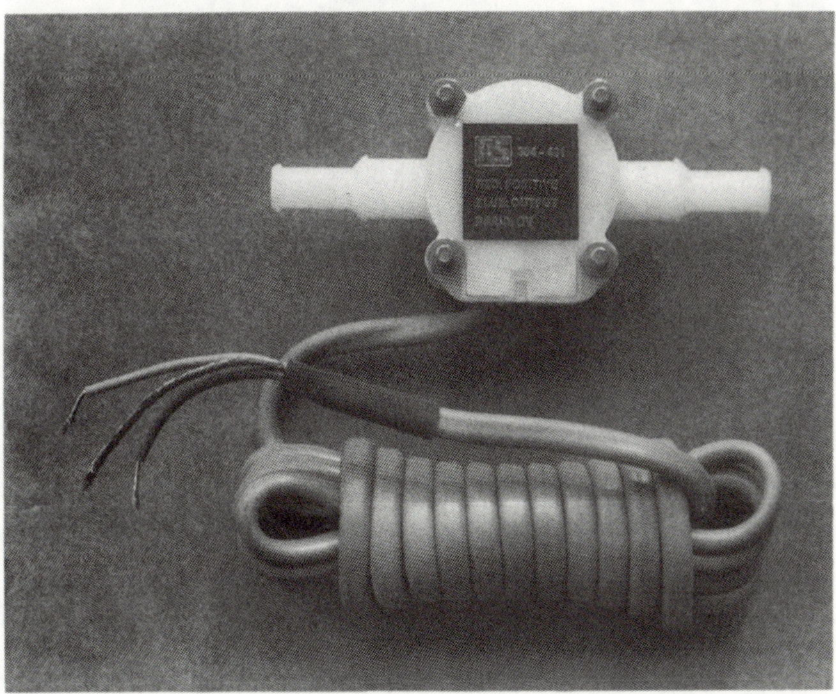

Plate 9.1 *Liquid flow sensor (digital output)*

Sensors

Sensors provide a means of inputting information to a process control system. This information relates to external physical conditions such as temperature, position and pressure. The data returned from the sensors together with control inputs from the operator (where appropriate) will subsequently be used to determine the behaviour of the system.

Any practical industrial process control system will involve the use of a number of devices for sensing a variety of physical parameters. The choice of sensor will be governed by a number of factors including accuracy, resolution, cost, and physical size. The following table covers the range of sensors and inputs most commonly encountered in industrial process control systems. The list is not exhaustive and details of other types of sensor can be found in most texts devoted to measurement, instrumentation and control systems (see Appendix E).

Plate 9.2 *Linear position sensor (analogue output)*

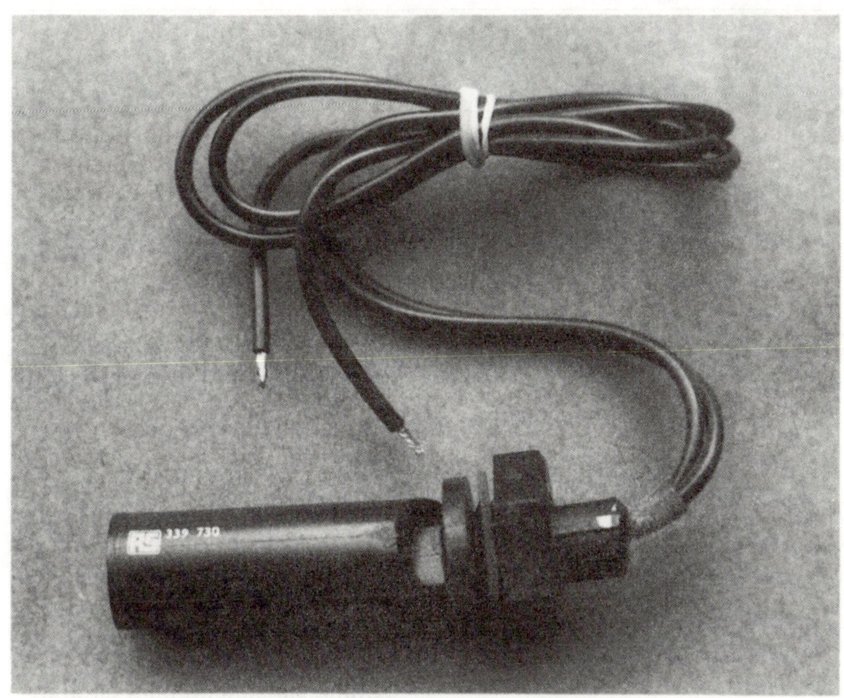

Plate 9.3 *Liquid level float switch*

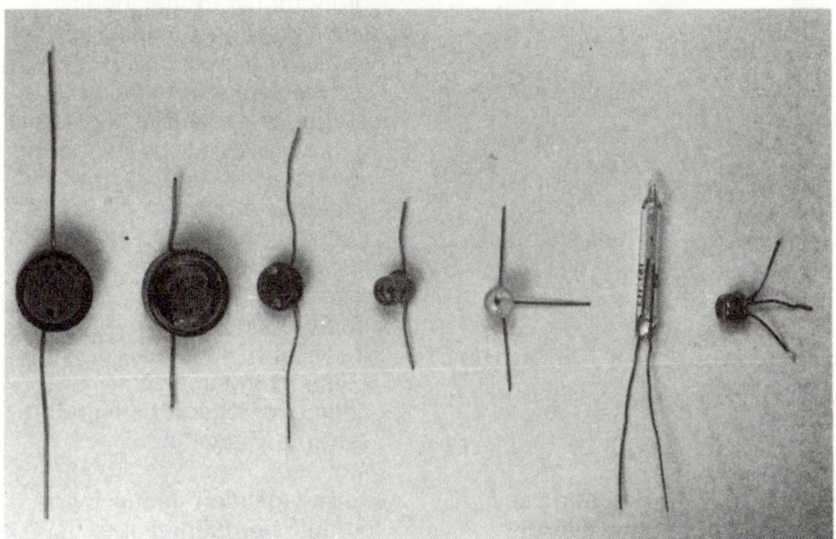

Plate 9.4 *Various optical and temperature sensors*

Physical parameter or input	*Type of sensor*	*Notes*
Angular position	Resistive rotary position sensor*	Rotary track potentiometer with linear law produces analogue voltage proportional to angular position. Limited angular range. Analogue input port required.
	Shaft encoder*	Encoded disk interposed between optical transmitter and receiver (infra-red LED and photodiode or photo-transistor). Usually requires signal conditioning based on operational amplifiers. Digital input port required.
Angular velocity (rotational)	Toothed rotor tachometer	Magnetic pick-up responds to the movement of a toothed ferrous disk. May require signal conditioning (typically an operational amplifier and Schmitt input logic gate). Some sensors contain circuitry to provide TTL-compatible outputs. The pulse repetition frequency of the output is proportional to the angular velocity. Digital input port required.
	Tachogenerator	d.c. generator with linear output characteristic. Analogue output voltage proportional to shaft speed. Requires an analogue input port.
	Shaft encoder*	Encoded disk interposed between optical transmitter and receiver (infra-red LED and photodiode or photo-transistor). Requires signal conditioning and some additional logic for direction sensing. Digital input port required.
Flow	Rotating vane flow sensor*	Pulse repetition frequency of output is proportional to flow rate. A counter/timer chip can be used to minimize software

Physical parameter or input	*Type of sensor*	*Notes*
		required. Digital input port required.
Light level	Photocell	Voltage-generating device. The analogue output voltage produced is proportional to light level. Analogue input port required.
	Light-dependent resistor	Usually connected as part of a potential divider or bridge. An analogue output voltage results from a change of resistance within the sensing element. Analogue input port required.
	Photodiode	Two-terminal device connected as a current source. An analogue output voltage is developed across a series resistor of appropriate value. Analogue input port required.
	Phototransistor	Three-terminal device connected as a current source. An analogue output voltage is developed across a series resistor of appropriate value. Analogue input port required.
Linear position	Resistive linear position sensor*	Linear track potentiometer with linear law produces analogue voltage proportional to linear position. Limited linear range. Analogue input port required.
	Linear variable differential transformer (LVDT)	Miniature transformer with split secondary windings and moving core attached to a plunger. Requires a.c. excitation and phase-sensitive detector. Analogue input port required.
Linear velocity	Magnetic sensor*	Magnetic pick-up responds to movement of a toothed ferrous track. May require signal conditioning (typically using an operational amplifier and Schmitt input logic gates) but some sensors contain circuitry

Physical parameter or input	*Type of sensor*	*Notes*
		to provide TTL-compatible outputs. The pulse repetition frequency of the output is proportional to the linear velocity. Digital input port required.
Liquid level	Float switch*	Simple switch element which operates when a particular level is detected. Digital input port required.
	Capacitive proximity switch*	Switching device which operates when a particular level is detected. Ineffective with some liquids. Digital input port required.
	Diffuse scan proximity switch*	Switching device which operates when a particular level is detected. Ineffective with some liquids. Digital input port required.
Operator	Switch or push-button*	Suitable for providing simple on/off control. Available in various styles including keyswitches and foot operated types. Digital input port required.
	DIL switch*	Only suitable for infrequent operation (e.g. setting parameters when re-configuring a system). Digital input port required.
	Keypad*	More cost-effective than using a large number of push-button switches when several options are required. Also suitable for numeric data entry. Keypads fitted with encoders require digital input ports. Unencoded keypads are conventionally configured in the form of a 4 × 4 array and will require at least one digital I/O port.
	Keyboard	Provides the ultimate in data entry (including generation of

Physical parameter or input	*Type of sensor*	*Notes*
		the full set of ASCII characters). Encoded keyboards are generally easier to use with digital I/O cards than unencoded types which are more suitable for 'memory mapped' I/O. Digital input port required.
	Joystick	Available in digital and analogue forms. The former type is generally based on four microswitches (two for each axis) whilst the latter is based on conventional rotary potentiometers. Either form is suitable for providing accurate position control. Analogue or digital input ports required, as appropriate.
Pressure	Microswitch pressure sensor	Microswitch fitted with actuator mechanism and range setting springs. Suitable for high-pressure applications. Digital input port required.
	Differential pressure/vacuum switch	Microswitch with actuator driven by a diaphragm. May be used to sense differential pressure. Alternatively, one chamber may be evacuated and the sensed pressure applied to the remaining port. Digital input port required.
	Piezo-resistive pressure sensor	Pressure exerted on diaphragm causes changes of resistance in attached piezo-resistive transducers. Transducers are usually arranged in the form of a four active element bridge which produces an analogue output voltage. Analogue input port required.
Proximity	Microswitch*	Microswitch fitted with actuator. Requires physical contact and small operating

Physical parameter or input	*Type of sensor*	*Notes*
		force. Digital input port required.
	Reed switch*	Reed switch and permanent magnet actuator. Only effective over short distances. Digital input port required.
	Inductive proximity switch*	Target object modifies magnetic field generated by the sensor. Only suitable for metals (non-ferrous metals with reduced sensitivity). Digital input port required.
	Capacitive proximity switch*	Target object modifies electric field generated by the sensor. Suitable for metals, plastics, wood, and certain powders and liquids. Digital input port required.
	Optical proximity switch*	Available in diffuse and through scan types. Diffuse scan types require reflective targets. Both types employ optical transmitters and receivers (usually infra-red emitting LEDs and photo-diodes or photo-transistors). Digital input port required.
Strain	Resistive strain gauge	Foil type resistive element with polyester backing for attachment to body under stress. Normally connected in full bridge configuration with temperature-compensating gauges to provide an analogue output voltage. Analogue input port required.
	Semiconductor strain gauge	Piezo-resistive elements provide greater outputs than comparable resistive foil types. More prone to temperature changes and also inherently non-linear. Analogue input port required.
Temperature	Thermocouple*	Output appears as a small

Physical parameter or input	*Type of sensor*	*Notes*
		e.m.f. generated by a sensing junction. Requires compensated connecting cables and specialized interface. Analogue input port required.
	Thermistor	Usually connected as part of a potential divider or bridge. An analogue output voltage results from resistance changes within the sensing element. Analogue input port required.
	Semiconductor temperature sensor*	Two-terminal device connected as a current source. An analogue output voltage is developed across a series resistor of appropriate value. Analogue input port required.
Weight	Load cell	Usually comprises four strain gauges attached to a metal frame. This assembly is then loaded and the analogue output voltage produced is proportional to the weight of the load. Requires analogue input port.
Vibration	Electromagnetic vibration sensor	Permanent magnet seismic mass suspended by springs within a cylindrical coil. The frequency and amplitude of the analogue output voltage are respectively proportional to the frequency and amplitude of vibration. Analogue input port required.

*See later in this chapter.

Interfacing switches and sensors

Sensors can be divided into two main groups according to whether they are *active* (generating) or *passive*. Another, arguably more important distinction in the case of PC-based process control systems, is whether they provide digital or analogue outputs. In the former case, one or more digital I/O boards will be required whereas, in the latter case one or more analogue input ports must be provided.

We shall deal first with techniques of interfacing switches and sensors which provide *digital outputs* (such as switches and promixity detectors) before examining methods used for interfacing sensors which provide *analogue outputs*. It should be noted that the majority of sensors (of either type) will require some form of signal conditioning circuitry in order to make their outputs acceptable to conventional PC expansion cards.

Sensors with digital outputs

Sensors which provide digital (rather than analogue) outputs can generally be quite easily interfaced with conventional PC bus expansion cards. However, since the signals generated by such sensors are seldom TTL compatible, it is usually necessary to include additional circuitry between the sensor and input port.

Switches

Switches can be readily interfaced to expansion cards in order to provide manual inputs to the system. Simple toggle and push-button switches are generally available with *normally open* (NO), *normally closed* (NC), or *changeover* contacts. In the latter case, the switch may be configured as either an NO or an NC type, depending upon the connections used.

Toggle, lever, rocker, rotary, slide, and push-button types are all commonly available in a variety of styles. Illuminated switches and key switches are also available for special applications. The choice of switch type will obviously depend upon the application and operational environment.

An NO switch or push-button may be interfaced to a digital I/O card using nothing more than a single pull-up resistor as shown in Figure 9.1.

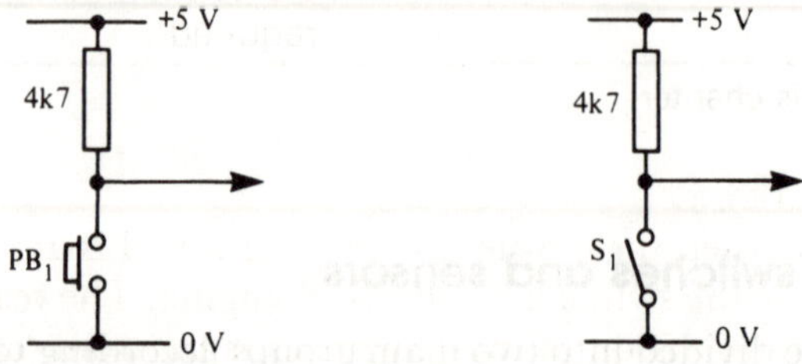

Figure 9.1 *Interfacing a normally open switch or push-button to a digital input port*

The relevant bit of the input port will then return 0 when the switch contacts are closed (i.e. when the switch is operated or where the push-button is depressed). When the switch is inactive, the relevant port bit will return 1.

Unfortunately, this simple method of interfacing has a limitation when the state of a switch is regularly changing during program execution. However, a typical application which is unaffected by this problem is that of using one or more PCB mounted switches (e.g. a DIL switch package) to configure a system in one of a number of different modes. In such cases, the switches would be set once only and the software would read the state of the switches and use the values returned to configure the system upon reset. Thereafter, the state of the switches would then only be changed in order to modify the operational parameters of the system (e.g. when adding additional I/O facilities). A typical DIL switch input interface to a digital input port is shown in Figure 9.2.

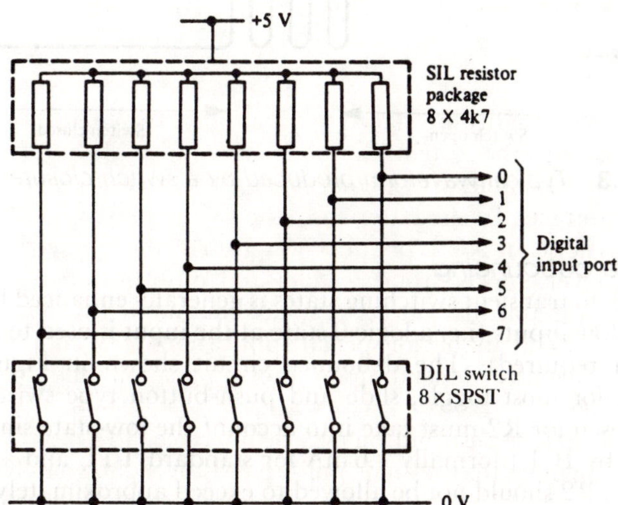

Figure 9.2 *Interfacing a DIL switch input to a digital input port*

Switch debouncing

As mentioned earlier, the simple circuit of Figure 9.1 is unsuitable for use when the state of the switch is regularly changing. The reason for this is that the switching action of most switches is far from 'clean' (i.e. the switch contacts make and break several times whenever the switch is operated). This may not be a problem when the state of a switch remains static during program execution but it can give rise to serious problems when dealing with, for example, an operator switch bank or keypad.

The contact 'bounce' which occurs when a switch is operated results in rapid making and breaking of the switch until it settles into its new state. Figure 9.3 shows the waveform generated by the simple switch input circuit of Figure 9.1 as the contacts close. The spurious states can cause problems if the switch is sensed during the period in which the switch contacts are in motion, and hence steps must be taken to minimize the effects of bounce. This may be achieved by means of additional hardware in the form of a 'debounce' circuit or by including appropriate software delays (of typically 4–20 ms) so that spurious switching states are ignored. We shall discuss these two techniques separately.

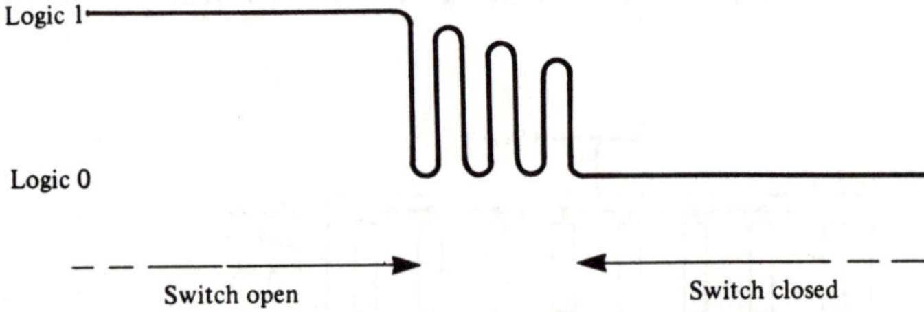

Figure 9.3 *Typical waveform produced by a switch closure*

Hardware debouncing

Immunity to transient switching states is generally enhanced by the use of active-low inputs (i.e. a logic 0 state at the input is used to assert the condition required). The debounce circuit shown in Figure 9.4 is adequate for most toggle, slide and push-button type switches. The value chosen for R2 must take into account the low-state sink current required by IC1 (normally 1.6 mA for standard TTL and 400 μA for LS-TTL). R2 should not be allowed to exceed approximately 470 Ω in order to maintain a valid logic 0 input state. The values quoted generate an approximate 1-ms delay (during which the switch contacts will have settled into their final state). It should be noted that, on power-up, this circuit generates a logic 1 level for approximately 1 ms before the output reverts to a logic 0 in the inactive state. The circuit obeys the following state table:

Switch condition	*Logic output*
closed	1
open	0

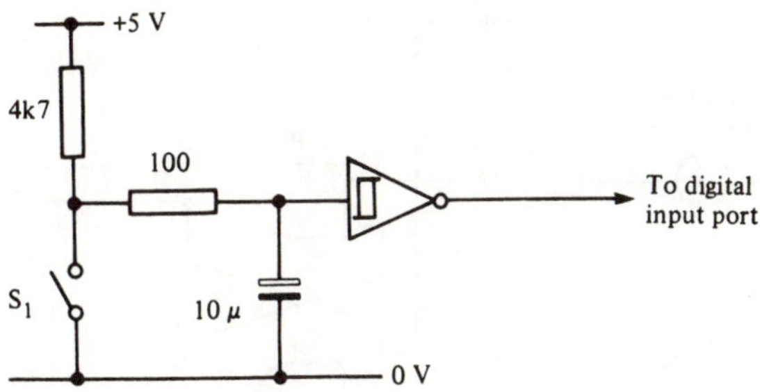

Figure 9.4 *Simple debounce circuit*

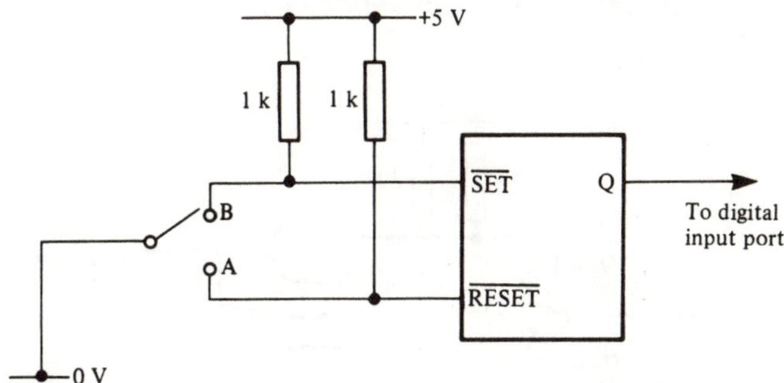

Figure 9.5 *Debounce circuit based on an RS bi-stable*

An alternative, but somewhat more complex, switch de-bouncing arrangement is shown in Figure 9.5. Here a single-pole double-throw (SPDT) changeover switch is employed. This arrangement has the advantage of providing complementary outputs and it obeys the following state table:

Switch condition	*Logic output*	
	Q	$\overline{Q}$
position A	0	1
position B	1	0

Rather than use an i.c. RS bistable in the configuration of Figure 9.5 it is often expedient to make use of 'spare' two-input NAND or NOR gates arranged to form bistables using the circuits shown in Figures 9.6(a) and 9.6(b), respectively. Figure 9.7 shows a rather neat extension of this theme in the form of a touch-operated switch. This arrangement is based

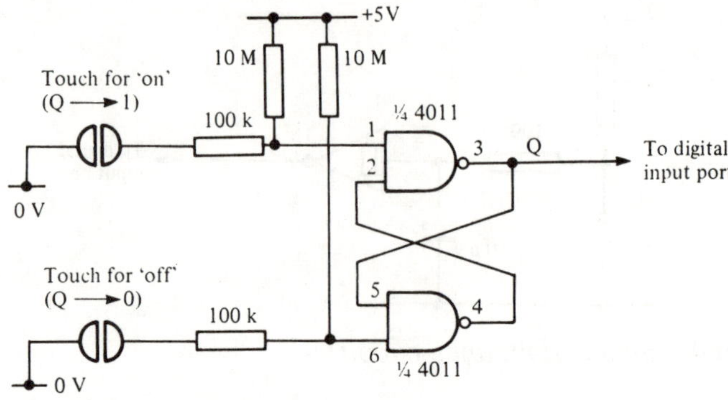

Figure 9.6 *Alternative switch debounce circuits: (a) based on NAND gates; (b) based on NOR gates*

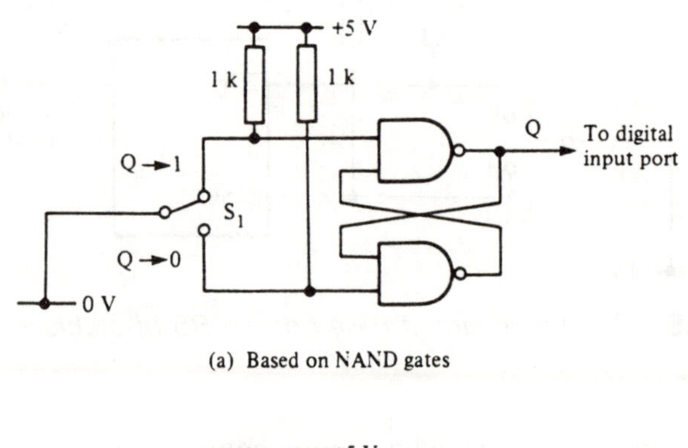

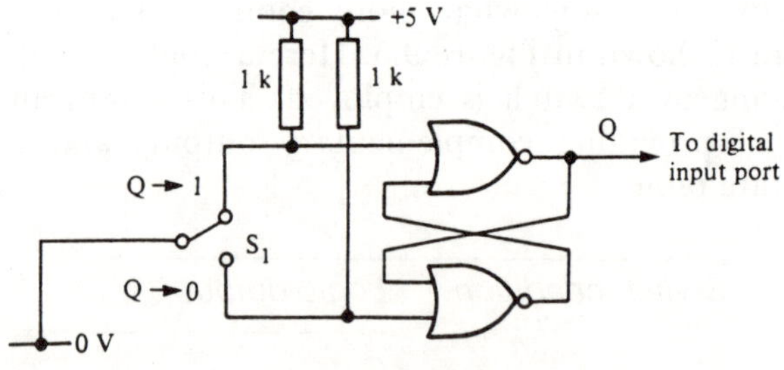

(b) Based on NOR gates

Figure 9.7 *Touch-operated switch*

on a 4011 CMOS quad two-input NAND gate (thought only two gates of the package are actually used in this particular configuration).

Finally, it is sometimes necessary to generate a latching action from an NO push-button switch. Figure 9.8 shows an arrangement in which a 74LS73 JK bistable is clocked from the output of a debounded switch.

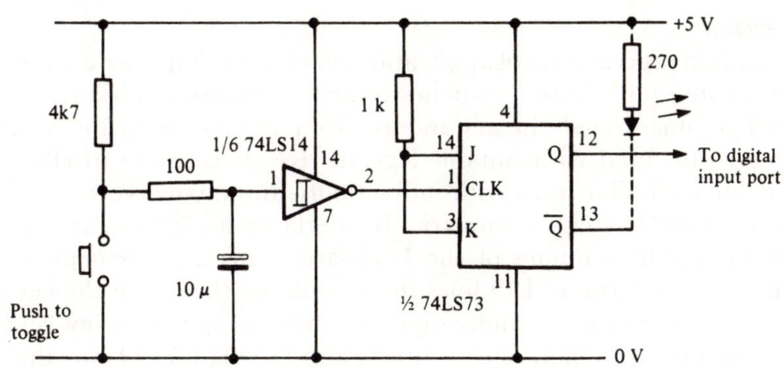

Figure 9.8 *Latching action switch*

Pressing the switch causes the bistable to change state. The bistable then remains in that state until the switch is depressed a second time. If desired, the complementary outputs provided by the bistable may be used to good effect by allowing the $\overline{Q}$ output to drive an LED. This will become illuminated whenever the Q output is high.

Software debouncing

Software debouncing involves the execution of a delay routine whenever the state of a switch is read. The state of the switch at the start of the delay routine is compared with that at the end. If the same value is returned in both cases, the last value returned is assumed to represent the state of the switch. If the value has changed, the switch is read again. The period of the delay routine is chosen so that it is just greater than the maximum period of contact bounce expected (typically 4–10 ms).

A typical software debounce routine is given below:

```
readsw:   CALL  switch     ; Read the switch and
          MOV   BL,AL      ; store the value.
          CALL  swdelay    ; Debounce and
          CALL  switch     ; read it again.
          CMP   AL,BL      ; Has it changed ?
          JNE   NZ,readsw  ; Yes, so try again.
          RET              ; No, so return with bit set
                           ; in AL
switch:   IN    AL,porta   ; Get value from Port A and
          AND   AL,mask    ; check bit concerned.
          RET              ; Go back . . .
swdelay:  PUSH  AX         ; preserve the set bit and
          MOV   CX,0800H   ; delay for a while.
sloop:    LOOP  sloop
          POP   AX
          RET
```

Keypads

Keypads in process control applications vary from simple arrangements of dedicated push-button switches to arrangements of 16-keys (either coded or unencoded) in a standard 4 × 4 matrix. Keycaps may be engraved or fitted with suitable legends. Keypads sealed to IP65 are available as similar units with individually illuminated keys.

Unencoded keypads are invariably interfaced using row and column lines to enable scanning of the keyboard. This arrangement is less demanding in terms of I/O lines than would be the case if the keypad contacts were treated as individual switches. A typical 16-key keypad arranged on a 4 × 4 matrix would make use of 12 digital I/O lines though it is possible to use just eight lines of a single port by alternately configuring the port for input and output. A representative arrangement is shown in Figure 9.9.

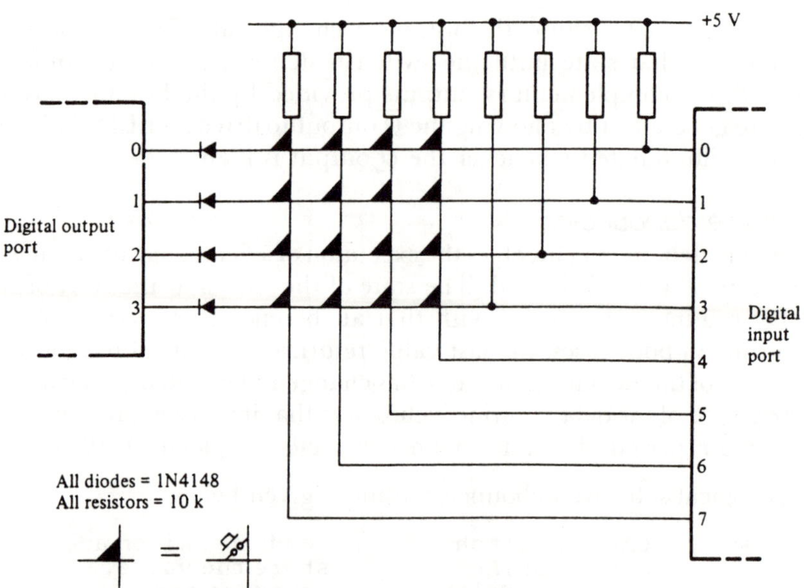

Figure 9.9 *Typical 4 × 4 matrix keypad interface*

Unencoded keypads are generally preferred in high-volume applications where the cost of interfacing hardware has to be balanced at the expense of the extra overhead required by the software involved with scanning the keyboard, detecting and decoding a keypress. In low-volume applications, and where software overheads have to be minimized, the use of a fully encoded keyboard is much to be preferred.

Encoded keypads employ dedicated encoder chips such as the

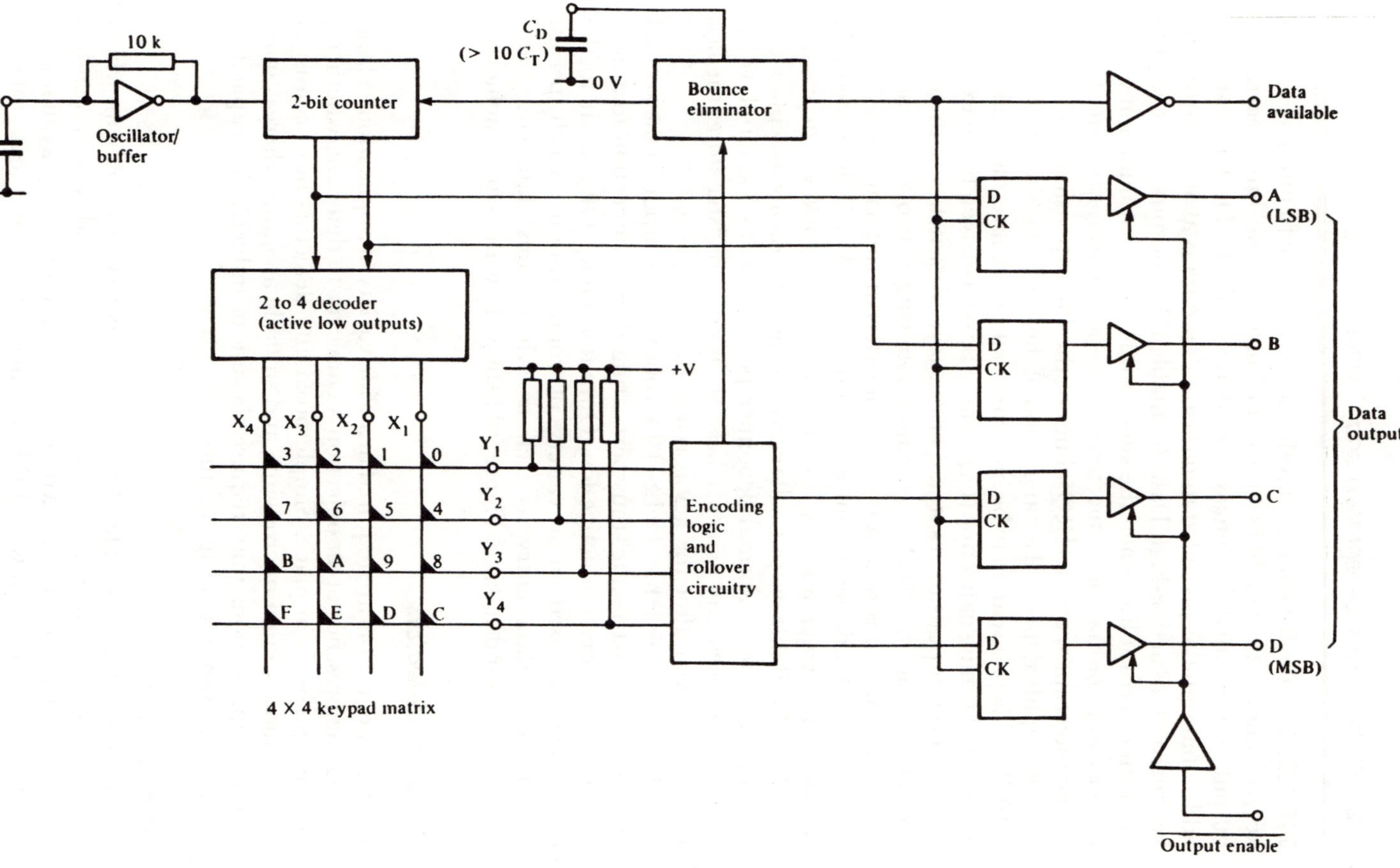

Figure 9.10 *Simplified internal arrangement of the 74C922 keypad encoder*

74C922. This device contains all the necessary logic to interface a 4 × 4 keypad matrix to four lines of a data bus or digital input port. The output is presented in binary coded decimal (BCD) form and an additional signal is provided to indicate that data is available from the keyboard. This active-high Data Available (DA) output can be used to drive an interrupt line when the keyboard is used in conjunction with a bus processor or may be connected via an open-collector inverter to one of the interrupt request (IRQ) lines of the PC expansion bus.

The simplified internal arrangement of the 74C922 is shown in Figure 9.10. The keypad scan may be implemented by the internal clock using an external timing capacitor (C_T) or may be over-driven by an external clock. On-chip pull-up resistors permit keypad switches with contact resistance of up to 50 kΩ. Internal debouncing is provided, the time constant of which is determined by an external capacitor (C_D).

The Data Available output goes high when a key is depressed and returns to low when a key is released even if another key is depressed. The Data Available output will return to high to indicate acceptance of the new key after a normal debounce period; this two key rollover is provided between any two switches. An internal register stores the last key pressed even after the key is released.

It should be noted that the LS-TTL compatible outputs of the keypad encoder chip are tri-state, thus permitting direct connection to a data bus. Furthermore, the active-low Output Enable ($\overline{OE}$) input to the device can be used in a variety of configurations which permit asynchronous data entry as well as synchronous data entry and synchronous handshaking. Figure 9.11 shows how this can be achieved.

Proximity detectors

Proximity detectors are required in a wide variety of applications – from sensing the presence of an object on a conveyor to detecting whether a machine guard is in place. Simple proximity detectors need consist of nothing more than a microswitch and suitable actuator whereas more complex applications may require the use of inductive or capacitive sensors, or even optical techniques.

Microswitches

A microswitch is a simple electromechanical switch element which requires minimal operating and release force and which exhibits minimal differential travel. Microswitches are normally available in single-pole double-throw (SPDT) configurations and can thus be configured as either normally open (NO) or normally closed (NC).

The principal disadvantage of the humble microswitch is that it not only requires physical contact with the object sensed but also requires a

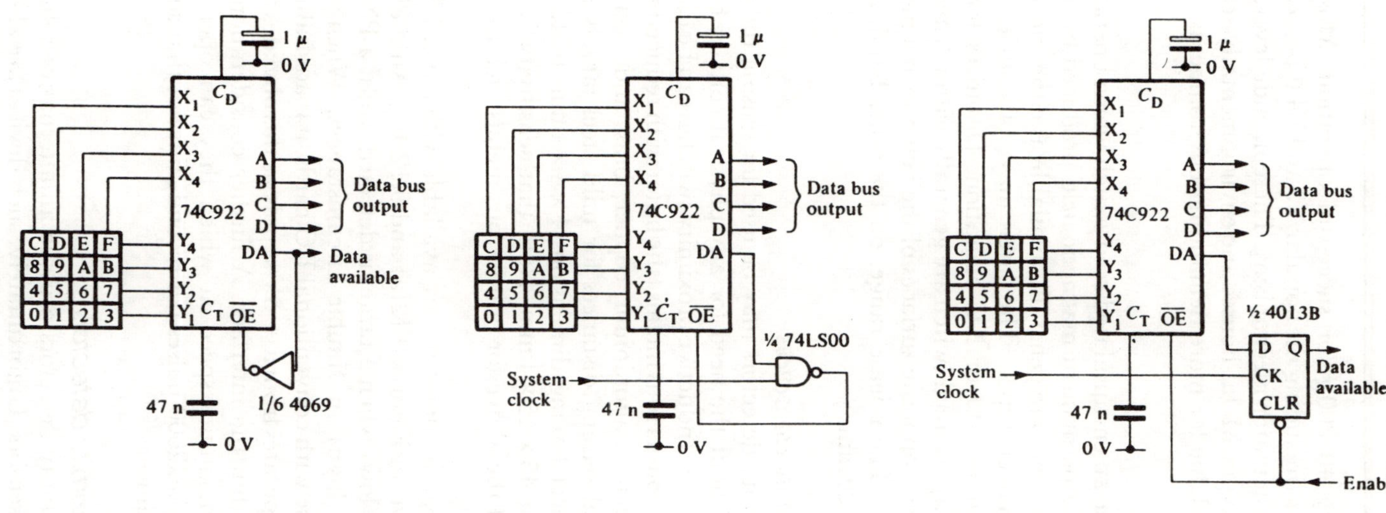

Figure 9.11 *Modes of operation for the 74C922: (a) asynchronous data entry; (b) synchronous data entry; (c) synchronous handshaking*

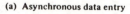

force of typically 40–200 g for successful operation. Most common microswitch types (including the popular V3 and V4 types) can be fitted with a variety of actuator mechanisms. These include lever, roller and standard button types. Metal-housed and environmentally-sealed microswitches are available for more demanding environments.

Reed switches

Reed switches use an encapsulated reed switch which operates when in the proximity of a permanent magnetic field produced by an actuator magnet. Reed switches are generally available as either normally open (NO) or changeover types. The later may, of course, be readily configured for either NO or NC operation. Distances for successful operation (pull-in) of a reed switch are generally within 8 mm to 15 mm (measured between opposite surfaces of the actuator magnet and reed switch assembly). The release range, on the other hand, is generally between 10 and 20 mm.

Inductive proximity detectors

Inductive proximity detectors may be used for sensing the presence of metal objects without the need for any physical contact between the object and the sensor. Inductive proximity switches can be used to detect both ferrous and non-ferrous metals (the latter with reduced sensitivity). Hence metals such as aluminium, copper, brass and steel can all be detected. Typical sensing distances for mild steel targets range from 1 mm for an object having dimensions 4 × 4 × 1 mm to 15 mm for an object measuring 45 × 45 × 1 mm. Note that sensitivity is reduced to typically 35% of the above for non-ferrous metals such as aluminium, brass and copper.

Inductive proximity detectors are available with either NPN or PNP outputs (as shown in Figures 9.12(a) and 9.12(b)). An NPN type will return a logic 0 (low) when a target is detected whilst a PNP type will return logic 1 (high) in similar circumstances. When selecting a transducer for use with conventional I/O cards, it is advisable to choose a device which operates from a + 5V supply as this obviates the need for level shifting within the interface. A further consideration with such devices is the maximum speed at which they can operate. This is typically 2 kHz (i.e. 2000 pulses per second) but not that some devices are very much slower.

Capacitive proximity detectors

Capacitive proximity detectors provide an alternative solution to the use of inductive sensors. Unfortunately, such devices are also limited in their speed of response (typically 250 Hz maximum) and often require supply voltages in excess of the conventional +5 V associated with TTL

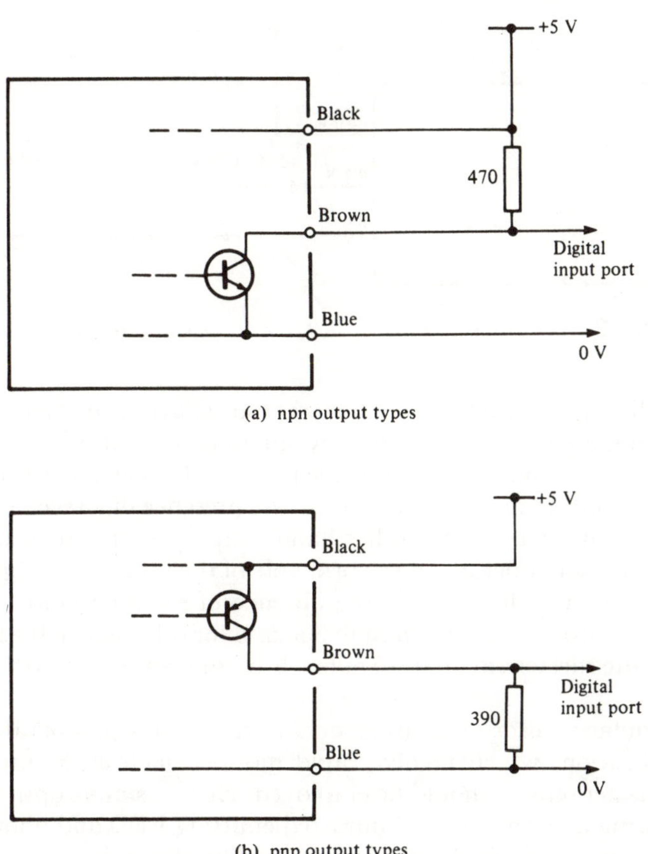

Figure 9.12 *Interfacing inductive proximity sensors: (a) NPN output types; (b) PNP output types*

signals. Capacitive proximity sensors will, however, detect the presence of materials such as cardboard, wood and plastics, as well as certain powders and liquids. Typical sensing distances range from 20 mm for metals to 4 mm for cardboard. As with inductive proximity sensors, the sensitivity of the detector is proportional to target size. A typical interface circuit for a d.c.-powered capacitive proximity detector is shown in Figure 9.13. This circuit provides a logic 0 (low) when a target is detected.

Optical proximity detectors

Optical proximity detectors generally offer increased sensing ranges in comparison with both capacitive and inductive types. Optical proximity sensors are available in two basic forms; *diffuse scan* and *through scan*

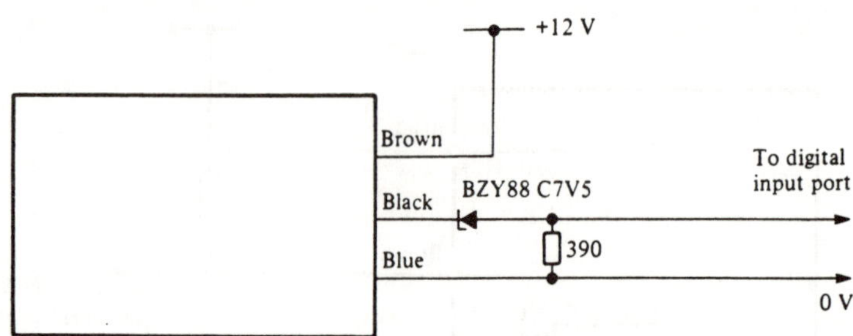

Figure 9.13 *Interface circuit for a typical capacitive proximity sensor*

types. The former types rely on the target surface returning a proportion of the modulated light emitted by an optical transmitter which is mounted in the same enclosure as the receiver. In such an arrangement, a reflective target may be detected by the presence of a received signal. Through scan types, on the other hand, employ a separate transmitter and receiver and operate on the principle of the interrupted light beam (i.e. the target is detected by the absence of received light). Typical ranges vary from about 100 mm to 300 mm for diffuse scan sensors with plane white surfaces to up to 15 m for through scan sensors with opaque targets.

Proprietary sensor units are generally rather slow in operation and, for applications which involve rapid motion (such as counting shaft speeds) faster sensors should be employed. Here, a simple optical sensor (comprising an unmodulated infra-red emitting LED and photodiode) may be employed. Such devices are readily available in a variety of packages including miniature diffuse scan types and slotted through scan units. Figure 9.14 shows the circuitry required to interface such a device to a typical digital input port.

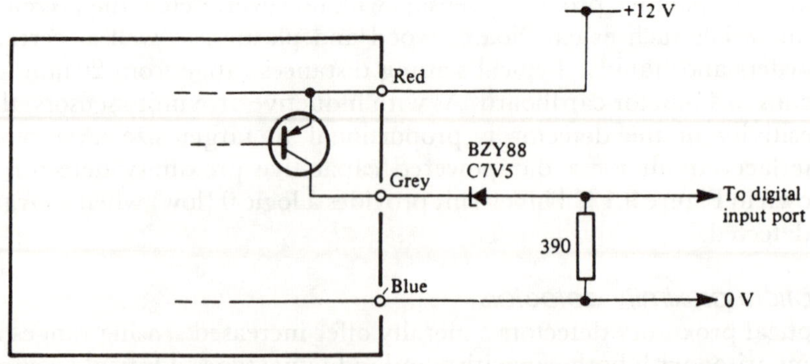

Figure 9.14 *Interface circuit for an optical proximity sensor*

Position transducers

Position transducers can be used to provide an accurate indication of the position of an object and are available in a variety of forms (including linear and rotary types). Linear position sensors use linear law potentiometer elements (of typically 5 kΩ) and offer strokes of typically 10 mm or 100 mm. Rotary position sensors are also available. These provide indications over typically 105° and use linear law potentiometer elements similar to those found in conventional rotary potentiometer controls. A typical value for the resistive element is again 5 kΩ.

The output of linear and rotary position sensors is usually made available as an analogue voltage and a typical arrangement is shown in Figure 9.15. Note that the analogue input port should have a high impedance (say 500 kΩ or more) in order to avoid non-linearity caused by loading of the sensing potentiometer.

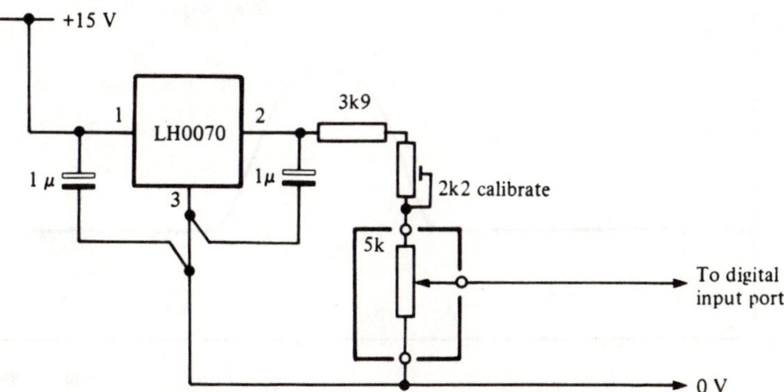

Figure 9.15 *Interface circuit for a resistive position transducer (either linear or rotary type)*

Shaft encoders

Shaft encoders can be used for sensing both rotary position and shaft speed. A typical shaft encoder produces 100 pulses per revolution and can thus provide a resolution of better than 1°. Such a device generally produces two phase-shifted outputs (to enable detection of direction of rotation) plus a third synchronizing pulse output (one pulse per revolution).

Shaft encoders are generally supplied in kit form comprising an encoder module, slotted disc and hub. The encoder module usually contains three infra-red emitting LEDs and three matching photo-detectors. The slotted disc is bonded to the hub ring which is, in turn,

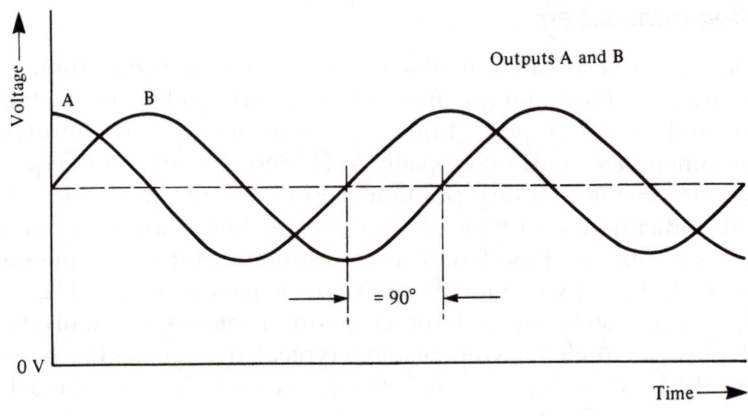

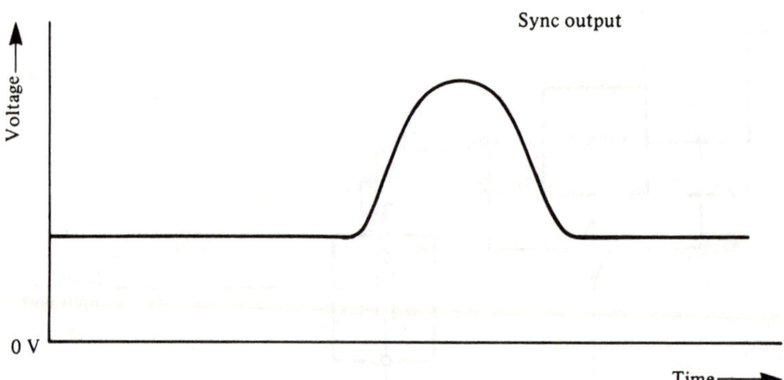

Figure 9.16 *Output waveforms produced by a typical shaft encoder*

fitted to the rotating shaft. The encoder module is then mounted so that the disc is interposed between the LEDs and photo-detectors.

The outputs of the encoder module are sinusoidal (as shown in Figure 9.16) and these must be converted to TTL compatible input pulses in order to interface with a standard digital input port. For simple speed-sensing applications, a typical input stage based on an operational comparator and low-pass filter is shown in Figure 9.17.

Unfortunately, the simple circuit of figure 9.17 is ineffective at very low frequencies and for stationary position indication. In such cases, the circuit shown in Figure 9.18 may be employed. Here, the potentiometer (RV1) must be adjusted so that the potential at the inverting input of the comparator is equal to that present at the non-inverting input. In this condition, the comparator produces a near 50% duty cycle.

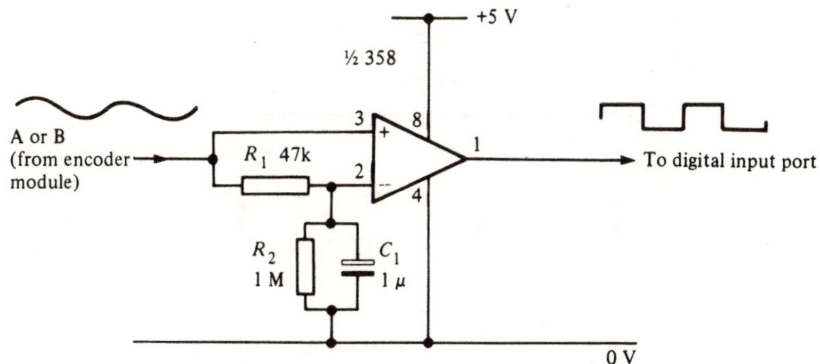

Figure 9.17 *Shaft encoder signal conditioning for measurement of rotational speed*

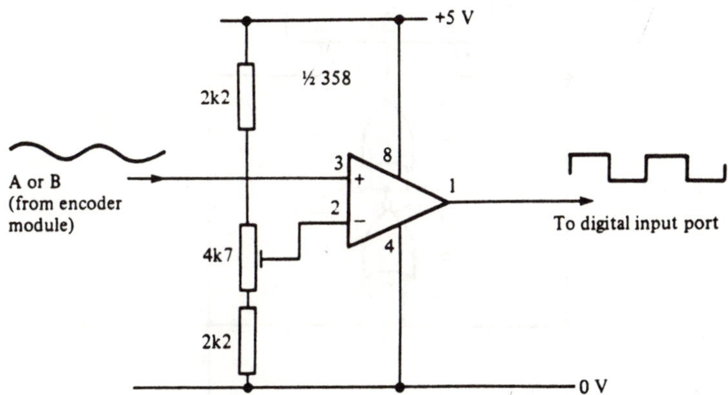

Figure 9.18 *Shaft encoder signal conditioning for low-speed applications and position sensing*

A further refinement is that of providing an output which indicates the sense of rotation (i.e. clockwise or anticlockwise). This may be achieved with the aid of some additional logic and a single JK bi-stable element as shown in Figure 9.19. The Q output of the bi-stable goes high (logic 1) for clockwise rotation and low (logic 0) for anticlockwise rotation. Figure 9.20 shows typical waveforms for the logic shown in Figure 9.19.

Fluid sensors

A number of specialized sensors are available for use with fluids. These sensors include float switches (both horizontal and vertical types) and flow sensors. These latter devices incorporate a rotating vane and are

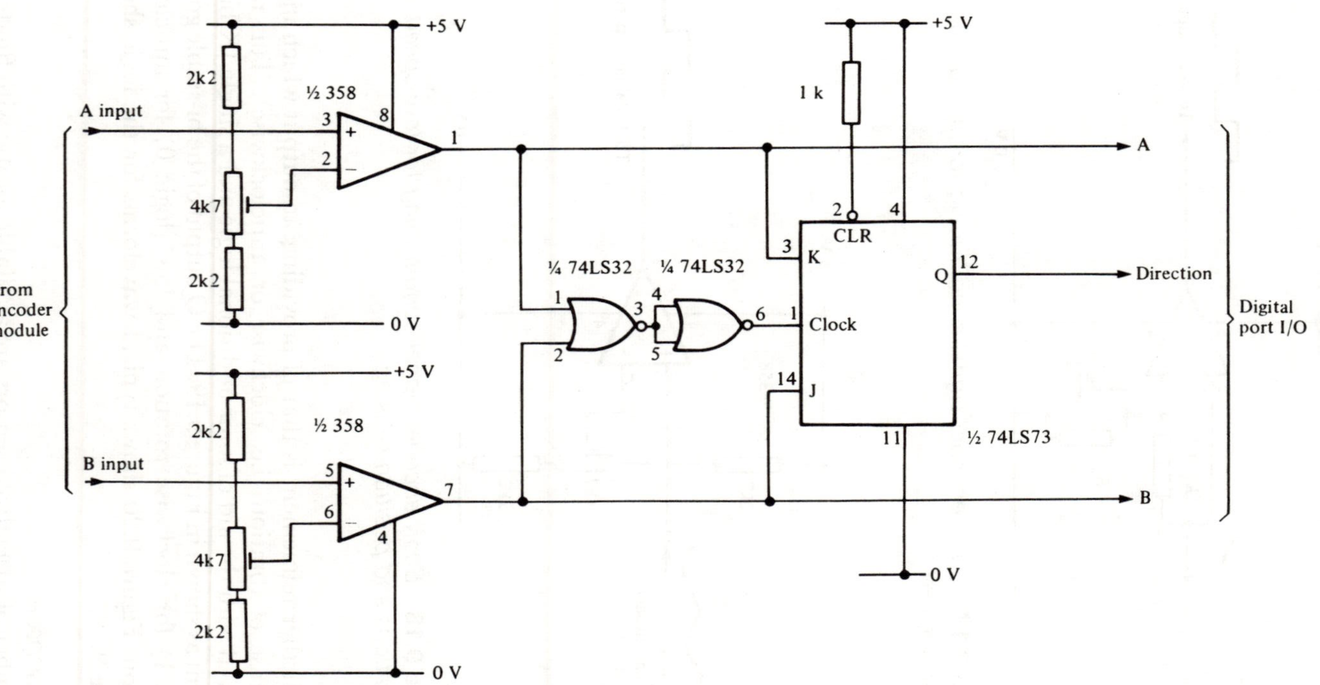

Figure 9.19 *Additional logic required to provide direction sensing*

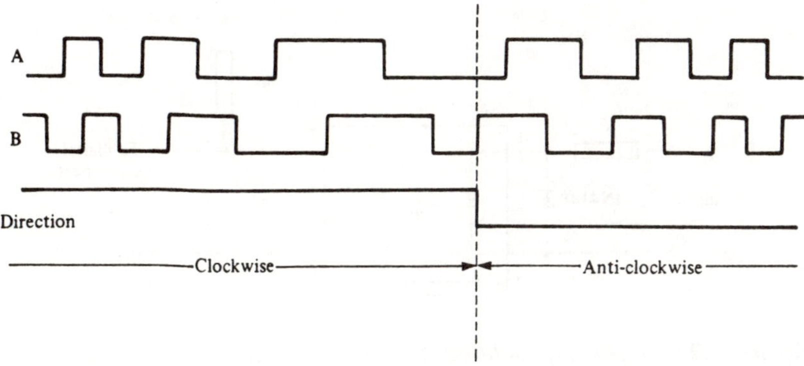

Figure 9.20 *Typical waveforms produced by the circuit of Figure 9.19*

suitable for use with flow rates ranging from 3 litre/hour to 500 litre/hour. Typical outputs range from 24 Hz at 10 litre/hour to 52 Hz at 20 litre/hour.

Optically isolated inputs

In a number of applications, it may be necessary to provide a high degree of electrical isolation between the source of a digital signal and its eventual connection to a digital input port. Such isolation can be achieved with the aid of an opto-isolator. These units comprise an optically coupled infra-red emitting LED and photodetector encapsulated in DIL package. The photodetector may take various forms including a *photodiode*, *phototransistor* and *photoDarlington*. Typical isolation voltages provided by such devices range from 500 V to 3 kV and switching rates may be up to 300 kHz, or so. High-voltage opto-isolators are available which will work reliably at voltages of up to 10 kV.

A typical single-channel optically isolated input arrangement is depicted in Figure 9.21. The external diode protects the infra-red emitting LED from inadvertent reversal of the input polarity and the value of the series resistor should be selected from the following table:

Input voltage range (*V*)	*Series resistor, R* (Ω)
3–4	330
4–5	560
5–6	680
6–8	1000
8–11	1500
11–15	2200

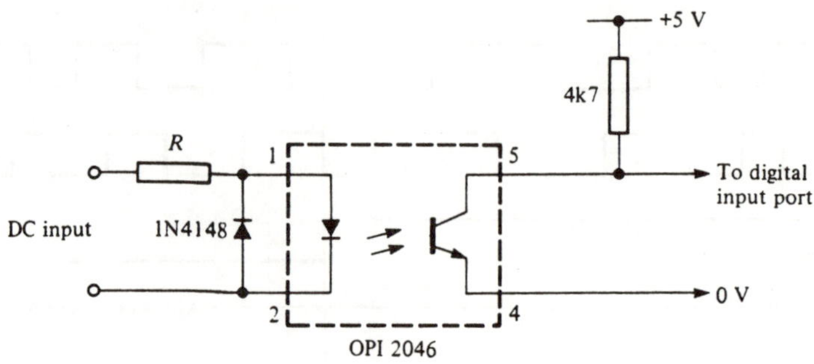

Figure 9.21 *Optically isolated digital input*

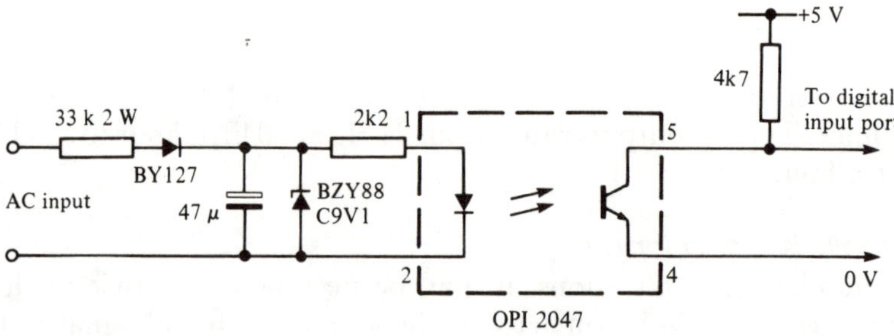

Figure 9.22 *Modification of Figure 9.21 to permit a.c. input*

The optically-isolated input stage can be extended for monitoring a.c. voltages as shown in Figure 9.22. This arrangement is suitable for a.c. inputs of up to 240 V 50 Hz and may be used to sense the presence or absence of a mains supply.

Sensors with analogue outputs

Having dealt with a number of common sensors which provide digital outputs, we shall now turn our attention to a range of transducers which provide analogue outputs. These outputs may manifest themselves as changes in e.m.f., resistance, or current and, in the latter cases it will usually be necessary to incorporate additional signal conditioning circuitry so that an analogue input voltage can be provided for use with a standard PC analogue I/O card.

Semiconductor temperature sensors

Semiconductor temperature sensors are ideal for a wide range of temperature-sensing applications. The popular 590 kH semiconductor temperature sensor, for example, produces an output current which is

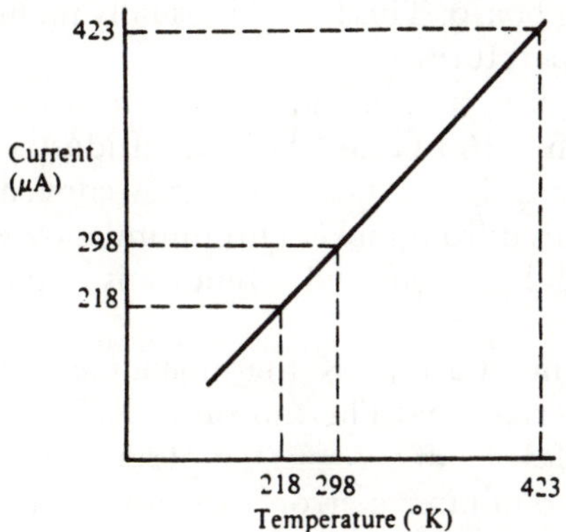

Figure 9.23 *Characteristic of the 590 kH semiconductor temperature sensor*

proportional to absolute temperature and which increases at the rate of 1 μA/K. The characteristic of the device is illustrated in Figure 9.23.

The 590 kH is laser trimmed to produce a current of 298.2 μA (±2.5 μA) at a temperature of 298.2°C (i.e. 25°C). A typical interface between the 590 kH and an analogue port is shown in Figure 9.24.

Thermocouples

Thermocouples comprise a junction of dissimilar metals which generate an e.m.f. proportional to the temperature differential which exists between the measuring junction and a reference junction. Since the measuring junction is usually at a greater temperature than that of the reference junction, it is sometimes referred to as the *hot junction*. Furthermore, the reference junction (i.e. the *cold junction*) is often omitted in which case the sensing junction is simply terminated at the

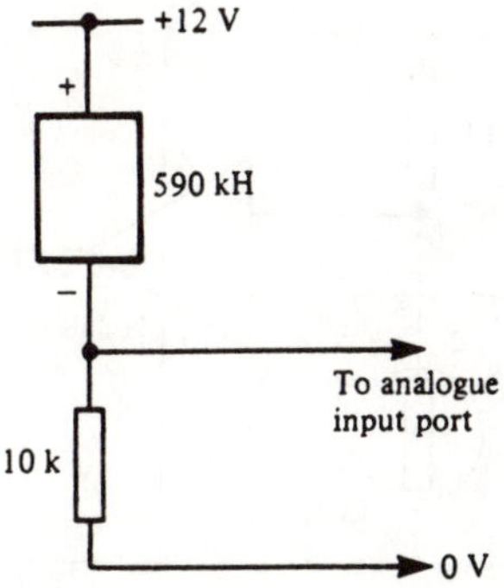

Figure 9.24 *Typical input interface for the 590 kH semiconductor temperature sensor*

signal conditioning board. This board is usually maintained at, or near, normal room temperatures.

Thermocouples are suitable for use over a very wide range of temperatures (from −100°C to +1100°C). Industry standard 'type K' thermocouples comprise a positive arm (conventionally coloured brown) manufactured from nickel/chromium alloy whilst the negative arm (conventionally coloured blue) is manufactured from nickel/aluminium.

The characteristic of a type K thermocouple is defined in BS 4937 Part 4 of 1973 (International Thermocouple Reference Tables) and this standard gives tables of e.m.f. versus temperature over the range 0°C to 1100°C. In order to minimize errors, it is usually necessary to connect thermocouples to appropriate signal conditioning using compensated cables and matching connectors. Such cables and connectors are available from a variety of suppliers and are usually specified for use with type K thermocouples.

Where thermocouples are to be used as sensors in conjunction with PC-based instrumentation systems, proprietary signal conditioning cards are available. These cards incorporate cable terminators and provide cold junction compensation as well as low-pass filtering to reduce the effects of 50 Hz noise induced in the thermocouple cables. The signal conditioning boards are then used in conjunction with one, or more, multi-channel analogue input ports.

Threshold detection with analogue output transducers

Analogue sensors are sometimes used in situations where it is only necessary to respond to a pre-determined threshold value. In effect, a two-state digital output is required. In such cases a simple one-bit analogue-to-digital converter based on a comparator can be used. Such

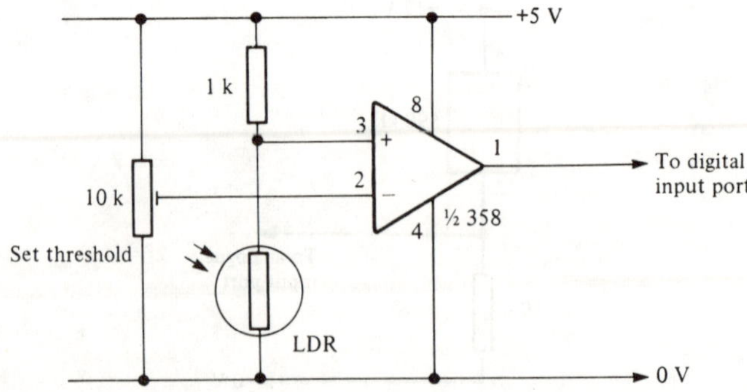

Figure 9.25 *Light-level threshold detector based on a light-dependent resistor (LDR)*

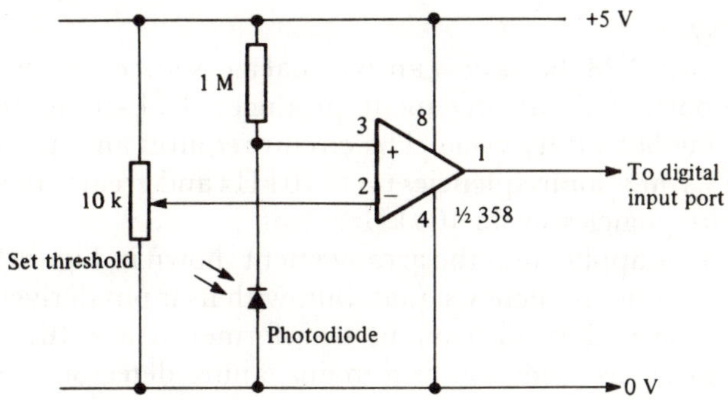

Figure 9.26 *Light-level threshold detector based on a photodiode*

an arrangement is, of course, very much simpler and more cost-effective than making use of a conventional analogue input port!

Simple threshold detectors for light level and temperature are shown in Figures 9.25 to 9.27. These circuits produce TTL-compatible outputs suitable for direct connection to a digital input port.

Figure 9.25 shows a light level threshold detector based on a comparator and light-dependent resistor (LDR). This arrangement generates a logic 0 input whenever the light level exceeds the threshold setting, and vice versa. Figure 9.26 shows how light level can be sensed using a photodiode. This circuit behaves in thc same manner as the LDR equivalent but it is important to be aware that circuit achieves peak sensitivity in the near infra-red region.

Figure 9.27 shows how temperature thresholds can be sensed using the 590 kH sensor described earlier. This arrangement generates a logic 0 input whenever the temperature level exceeds the threshold setting, and vice versa.

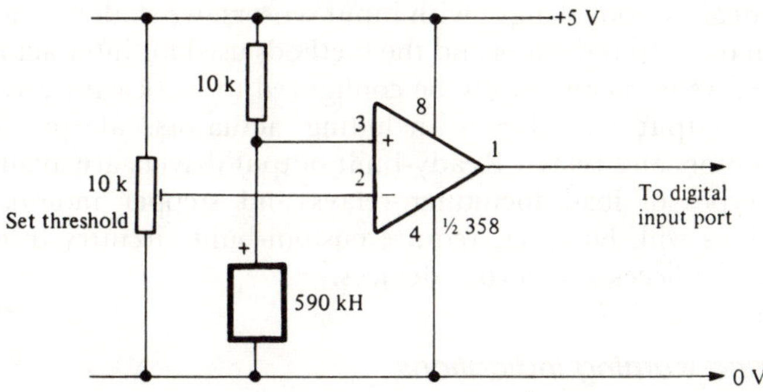

Figure 9.27 *Temperature threshold detector based on a 590 kH semiconductor temperature sensor*

AC sensing

Finally, Figure 9.28 shows how an external a.c. source can be coupled to an input port. This arrangement produces TTL-compatible input pulses having 50% duty cycle. The circuit requires an input of greater than 10 mV r.m.s. for frequencies up to 10 kHz and greater than 100 mV r.m.s. for frequencies up to 100 kHz.

The obvious application the arrangement shown in Figure 9.28 is the detection of audio frequency signals but, with its input derived from the low voltage secondary of a mains transformer (via a 10:1 potential divider), it can also function as a mains failure detector.

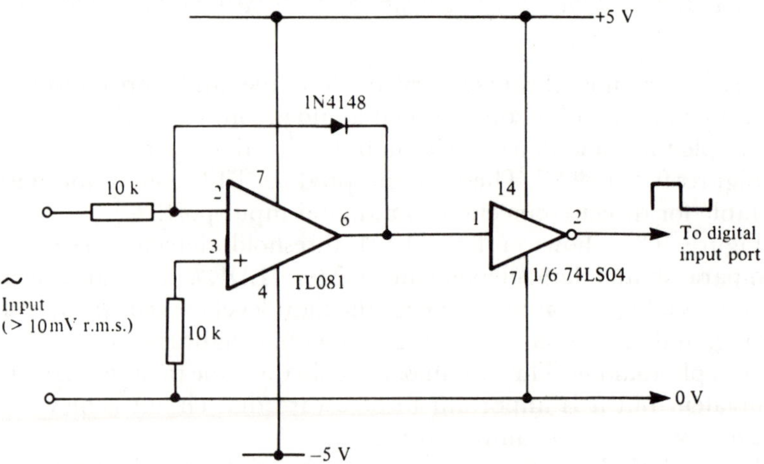

Figure 9.28 *Interface circuit to permit a.c. sensing*

Output devices

Having dealt at some length with input sensors, we shall now focus our attention on output devices and the methods used for interfacing them. PC-based systems can readily be configured to work with a variety of different output transducers including actuators, alarms, heaters, lamps, motors and relays. Ready-built output drivers are available for several types of load including relays and stepper motors. Many applications will, however, require custom-built circuitry in order to interface the necessary output devices.

Status and warning indications

Indicators based on light-emitting diodes (LEDs) are inherently more reliable than small filament lamps and also consume considerably less power. They are thus ideal for providing visual status and warning displays. LEDs are available in a variety of styles and colours and 'high brightness' types can be employed where high-intensity displays are required.

A typical red LED requires a current of around 10 mA to provide a reasonably bright display and such a device may be directly driven from a buffered digital output port. Different connections are used depending upon whether the LED is to be illuminated for a logic 0 or logic 1 state. Several possibilities are shown in Figure 9.29.

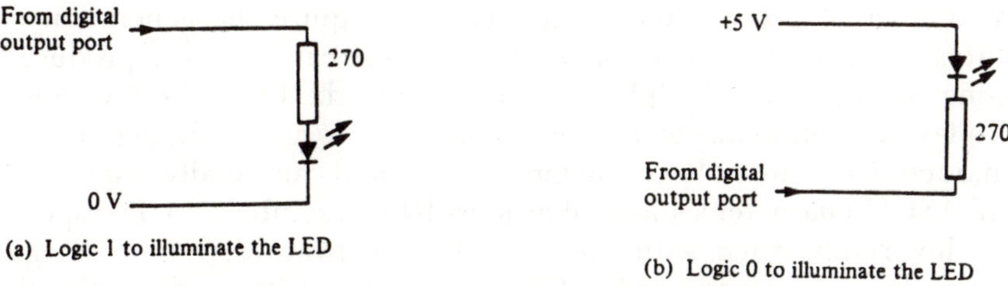

Figure 9.29 *Driving an LED from a buffered digital I/O port: (a) logic 1 to illuminate the LED; (b) logic 0 to illuminate the LED*

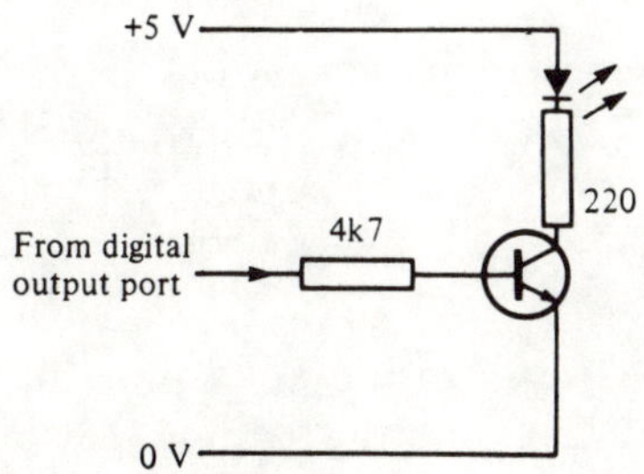

Figure 9.30 *Using an auxiliary transistor to drive an LED*

Where a buffered output port is not available, an auxiliary transistor may be employed as shown in Figure 9.30. The LED will operate when the output from a PC expansion card is taken to logic 1 and the operating current should be approximately 15 mA (thereby providing a brighter display than the arrangements previously described). The value of R2 will be dependent upon the supply voltage and should be selected from the table shown below:

Supply voltage (V)	*R2 (Ω)*
3–4	100
4–5	150
5–8	220
8–12	470
10–15	820

Driving LCD displays

A number of process-control applications require the generation of status messages and operator prompts. These can be easily produced using a conventional alphanumeric dot-matrix LCD display. Such displays are commonly available in a variety of formats ranging from 16 characters × 1 line to 40 characters × 4 lines and can usually display the full ASCII character set as well as user-defined symbols. LCD displays are invariably fitted with the necessary hardware drivers and logic (sometimes in the form of a CMOS microprocessor) to interface directly with a digital I/O port.

Plate 9.5 *Various types of relay (including conventional electro-mechanical, solid-state and encapsulated reed types)*

Driving medium- and high-current loads

Due to the limited output current and voltage capability of most standard digital I/O expansion cards, external circuitry will normally be required to drive anything other than the most modest of loads. Figure 9.31 shows some typical arrangements for operating various types of medium- and high-current load. Figure 9.31 (a) shows how an NPN transistor can be used to operate a low-power relay. Where the relay requires an appreciable operating current (say, 150 mA, or more) a plastic encapsulated Darlington power transistor should be used as shown in Figure 9.31 (b). Alternatively, a VMOS power FET may be preferred, as shown in Figure 9.31 (c). Such devices offer very low values of 'on' resistance coupled with a very high 'off' resistance. Furthermore, unlike conventional bipolar transistors, a power FET will impose a negligible load on an I/O port. Figure 9.31 (d) shows a filament lamp driver based on a plastic Darlington power transistor. This circuit will drive lamps rated at up to 24 V, 500 mA.

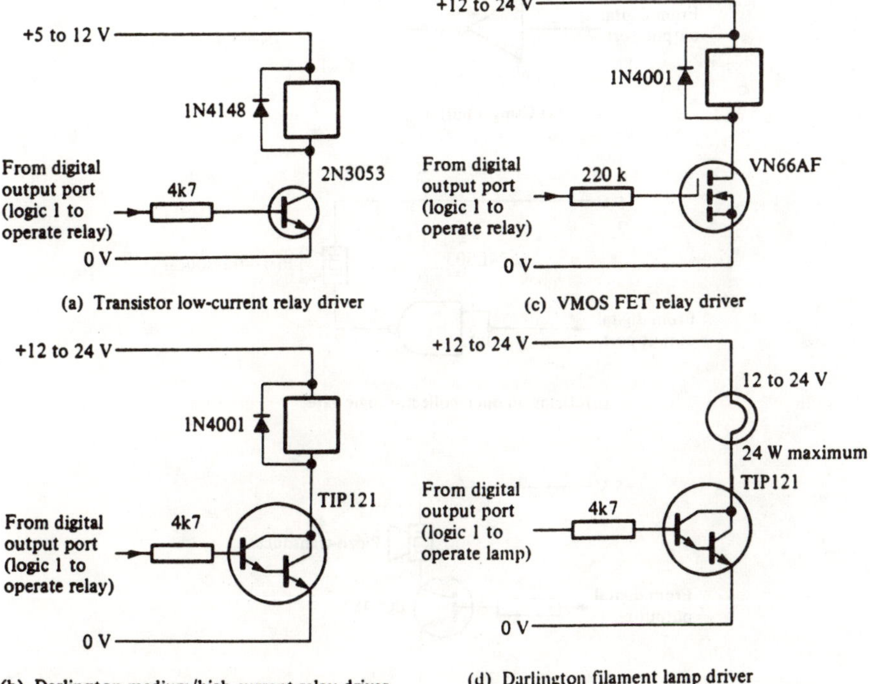

Figure 9.31 *Typical medium- and high-current driver circuits: (a) transistor low-current relay driver; (b) Darlington medium/high current relay driver; (c) VMOS FET relay driver; (d) Darlington filament lamp driver*

Audible outputs

Where simple audible warnings are required, miniature piezo-electric transducers may be used. Such devices operate at low voltages (typically in the range 3–15 V) and can be interfaced with the aid of a buffer, open-collector logic gate, or transistor. Figures 9.32(a)–(c) show typical interface circuits which produce an audible output when the port output line is at logic 1.

Where a pulsed rather than continuous audible alarm is required, a circuit of the type shown in Figure 9.33 can be employed. This circuit is based on a standard 555 timer operating in astable mode and operates at approximately 1 Hz. A logic 1 from the port output enables the 555 and activates the pulsed audio output.

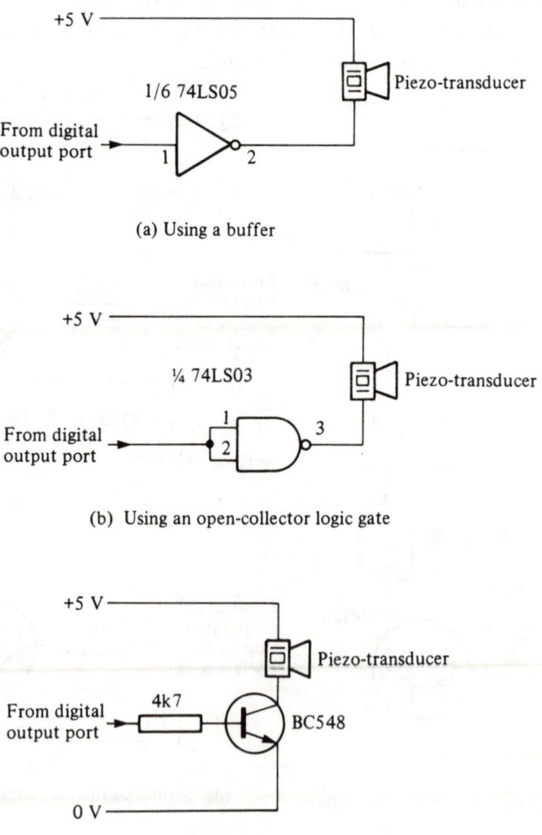

Figure 9.32 *Audible output driver: (a) using a buffer; (b) using an open-collector logic gate; (c) using a transistor*

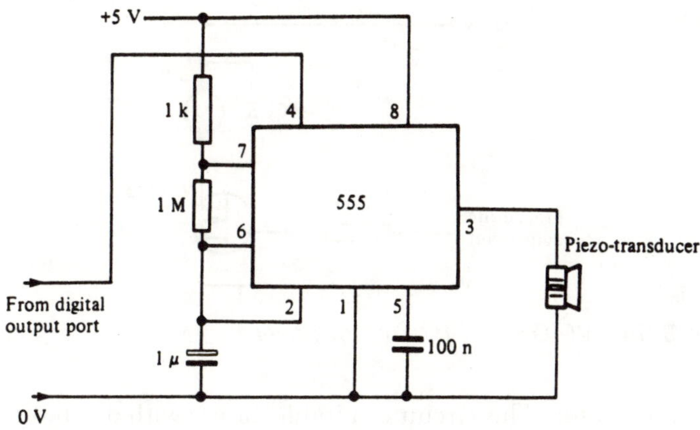

Figure 9.33 *Pulsed audible alarm based on a 555 astable*

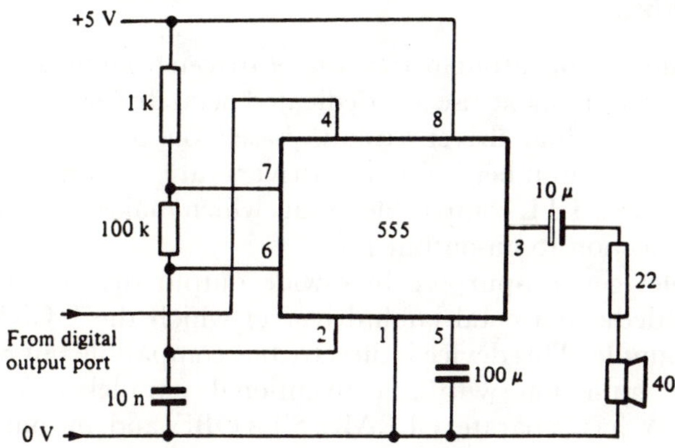

Figure 9.34 *Audible alarm with output to a moving coil loudspeaker*

Finally, the circuit shown in Figure 9.34 can be used where a conventional moving-coil loudspeaker is to be used in preference to a piezo-electric transducer. This circuit is again based on the 555 timer and provides a continuous output at approximately 1 kHz whenever the port output is at logic 1.

DC motors

Circuit arrangements used for driving d.c. motors generally follow the same lines as those described earlier for use with relays. As an example, the circuit shown in Figure 9.35 uses a VMOS FET to drive a low-

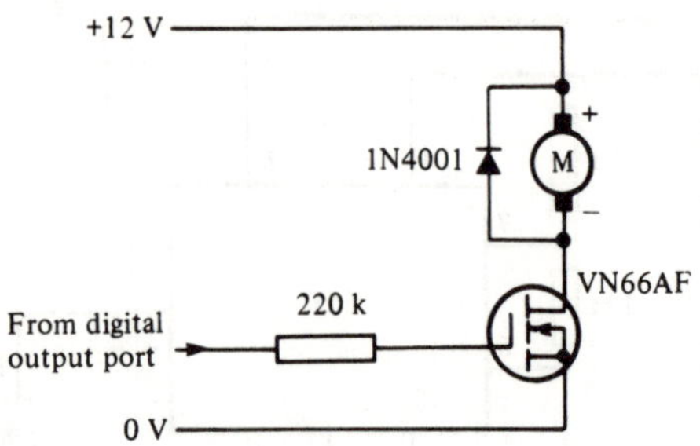

Figure 9.35 *VMOS FET d.c. motor driver*

voltage d.c. motor. This circuit is suitable for use with d.c. motors rated at up to 12 V with stalled currents of less than 1.5 A. A logic 1 from the output port operates the motor.

Output drivers

Where a number of output loads are to be driven from the same port, it is often expedient to make use of a dedicated octal driver chip rather than use eight individual driver circuits based on discrete components. Fortunately, a number of octal drivers are available and these invariably have TTL compatible inputs which makes them suitable for direct connection to an output port.

A simple, general-purpose byte-wide output driver can be based around a dedicated octal latch/driver of which the UCN5801A is a typical example. This device is directly bus compatible but may also be used in conjunction with a conventional parallel I/O port. The UCN5801A has separate CLEAR, STROBE and output ENABLE control lines coupled to eight bipolar Darlington driver transistors. This configuration provides an extremely low-power latch with a very high-output current capability.

The eight outputs of the UCN5801A are all open-collector, the positive supply voltage for which may be anything up to 50 V. Each Darlington output device is rated at 500 mA maximum however; if that should prove insufficient for a particular application then several output lines may be paralleled together subject, of course, to the limits imposed by the rated load current of the high-voltage supply.

Figure 9.36 shows a typical arrangement of the UCN5801 in which the load voltage supply is +12 V. The state of the bus is latched to the output of IC1 whenever the STROBE input is taken high; however, when used in conjunction with a conventional I/O device, this line can be tied high. A logic 0 present on a particular data line will turn the corresponding Darlington output device 'off' whilst a logic 1 will turn it

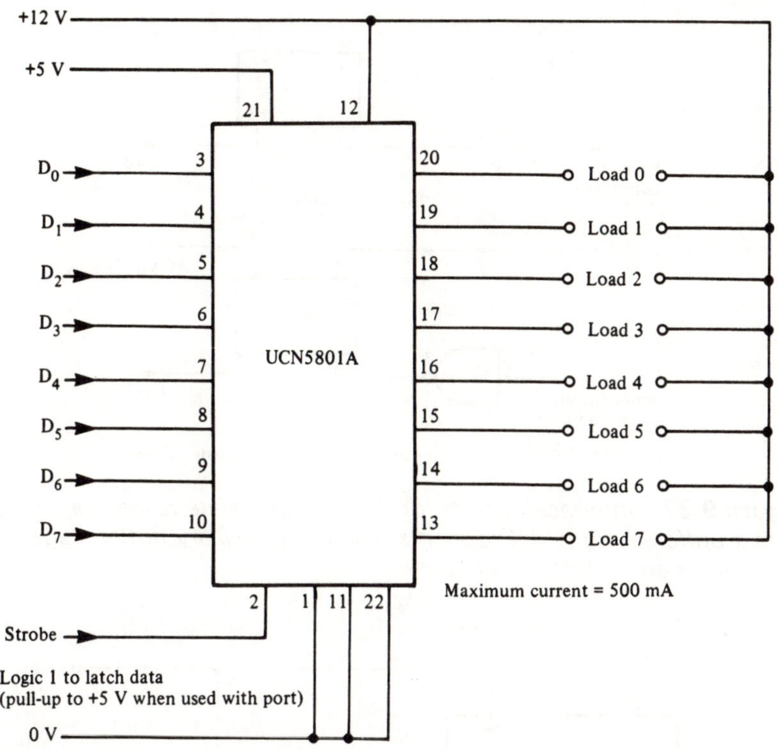

Figure 9.36 *Typical output driver arrangement based on the UCN5801*

'on'. It should also be noted that the output stages are protected against the effects of an inductive load by means of internal diodes. These are commoned at pin-12 and this point should thus be returned to the positive supply.

Driving mains connected loads

Control systems are often used in conjunction with mains connected loads. Modern solid-state relays (SSRs) offer superior performance and reliability when compared with conventional relays in such applications. SSRs are available in a variety of encapsulations (including DIL, SIL, flat-pack, and plug-in octal) and may be rated for RMS currents between 1 A and 40 A.

In order to provide a high degree of isolation between input and output, SSRs are optically coupled. Such devices require minimal input currents (typically 5 mA, or so, when driven from 5 V) and they can thus be readily interfaced with TTL devices. Figures 9.37(a)–(c) show how

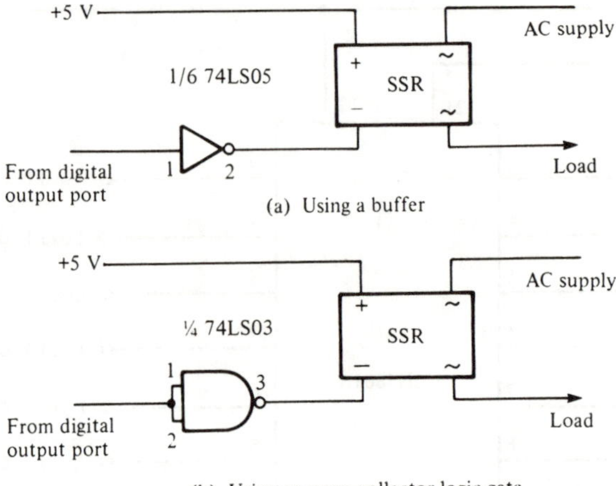

Figure 9.37 *Interface circuits for driving solid state relays: (a) driven from a buffered digital I/O port; (b) using an auxiliary buffer stage; (c) using an open-collector logic gate*

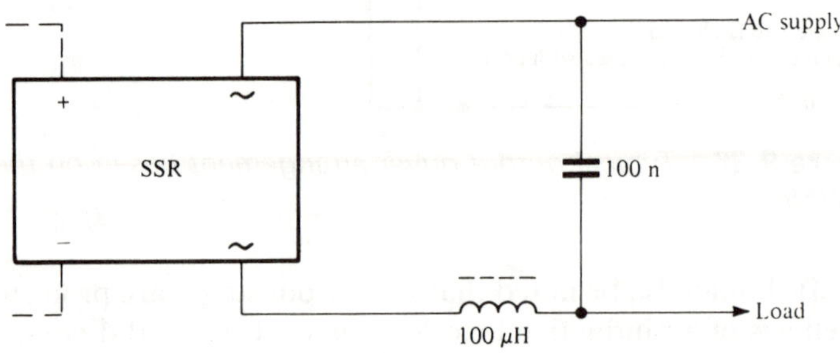

Figure 9.38 *Using a 'snubber' network with an inductive load*

an SSR can be driven from buffered and unbuffered I/O ports. It is important to note that, when an inductive load is to be controlled, a 'snubber' network should be fitted, as shown in Figure 9.38.

Driving solenoids and solenoid operated valves

Solenoids and solenoid operated valves are generally available with coils rated for 110 V/240 V a.c. or 12 V/24 V d.c. operation. The circuitry for interfacing solenoids will thus depend on whether the unit is rated for a.c. or d.c. operation. In the case of a.c.-operated units, a suitably rated SSR should be employed (See Figures 9.37(a) and 9.37(b)) while, in the case of d.c.-operated solenoids, interface circuitry should be identical to that employed with medium/high current relays (See Figure 9.31(c) and 9.31(d)).

Driving stepper motors

The complex task of interfacing a stepper motor to a PC-based system can be greatly simplified by using a dedicated stepper motor driver card. Alternatively, in many light-duty applications, a simple interface can be constructed based on a specialized stepper motor driver chip (such as the SAA1027). This device includes all necessary logic to drive a stepper motor as well as output drivers for each of the four phases. The chip operates from a nominal +12 V supply rail but its inputs are not directly TTL compatible and thus transistor or open-collector drivers will normally be required.

Figure 9.39 shows a typical stepper motor interface based on the SAA1027. The motor is a commonly available four-phase two-stator type having the following characteristics:

Supply voltage = 12 V
Resistance per phase = 47 Ω
Inductance per phase = 400 mH
Maximum working torque = 50 m Nm
Step rotation = 7.5°/step

The stepper motor interface requires only three port output lines which operate on the following basis:

- The STEP input is pulsed low to produce a step rotation.

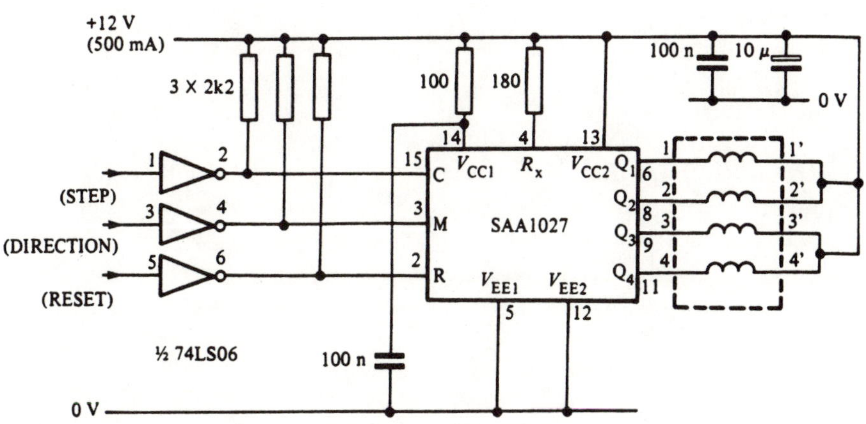

Figure 9.39 *Stepper motor interface based on the SAA1027*

- The DIRECTION input determines the sense of rotation. A low on the DIRECTION input selects clockwise rotation. Conversely, a high on the DIRECTION input selects anticlockwise rotation.
- The RESET input can be taken low to reset the driver.

The software routines for driving the stepper motor are quite straightforward and can be simply based on sequences of OUT (or equivalent) instructions which may be contained within loops where continuous rotation in a clockwise or anticlockwise direction is required.

10 Software packages

It should be self-evident that the effectiveness of any PC-based instrumentation or control system will be dependent not only upon the hardware employed but also upon the software which controls the system. Software has a vital role to play in that it provides an interface and acts as an intermediary between the user and the physical components of the system. Furthermore, the degree of control, flexibility, and ease of use will largely be dependent upon this software interface.

The newcomer to PC-based instrumentation and control systems can be forgiven for being baffled by the variety and complexity of software packages designed to assist him in his task. This chapter chategorizes software packages on a variety of grounds and provides details of several of the most popular software products. The aim has been that of providing a yardstick by which the control and instrumentation engineer can judge his or her current and future software requirements.

As an example, a stand-alone process controller may, for example, require relatively unsophisticated software in the form of a simple 'turnkey' program developed in a high-level programming language. A complex distributed factory instrumentation system, on the other hand, which requires frequent re-configuration and which may necessitate interfacing with several other applications programs, may require the services of an applications programming environment.

Selecting a software package

Several factors need to be considered when selecting any software package. These are:

- Ease of use – what level of expertise is required in order to make use of the package?

- Flexibility – can the package be easily adapted to differing requirements and can it be readily interfaced with other software?
- Performance – what performance criteria and specifications must be met?
- Functionality – does the package offer a suitable range of functions and will it interface correctly with the chosen hardware configuration?

To some extent, this last factor is paramount since, if the package cannot offer support for the particular hardware configuration in question, it may be of little use. Readers should be aware that a hardware specification will often be fixed before a full software specification has been developed.

Ease of use

The question of ease of use will largely depend upon the person (or persons) who will be using the system. In any event, a software package should be reasonably intuitive and the user should not be presented with outcomes which he or she did not expect. As an example, it should not be possible to quit from a package without being presented with a clear warning of the consequences (e.g. data loss). Similarly, on-screen controls and displays should, as far as possible, mimic those which the user will already be familiar with. All this may sound very obvious but programmers and software engineers often fail to identify with the level of expertise of the operator and he or she may be left to 'muddle through' by trial and error.

Fortunately, there is a trend towards making software 'user friendly' and this has been greatly aided by the availability of graphics-orientated operating system shells such as Microsoft's Windows and Digital Research's Graphics Environment Manager (GEM). These packages provide an excellent user interface with DOS which is based on windows, icons, and pull-down menus (WIMP).

The WIMP interfaces is also available to applications programs which are co-resident with the graphics shell. An application which wishes to make use of the WIMP environment can make calls to the graphics shell which, in turn, makes its own calls into DOS. The disadvantage of this system is that the graphics shell demands considerable RAM space and that operation (at least that which involves user interaction) is generally somewhat slower in comparison with an application which calls DOS directly. Anyone who may doubt this, is invited to compare two popular word processors: one running under a WIMP environment (e.g. First Word Plus running under GEM) and one running directly under DOS (e.g. Microsoft Word). The difference in the speed of certain operations can be quite stunning!

Flexibility

Some software packages (particularly those which may have been written for a particular application) tend to be somewhat rigid since they are generally based on a pre-defined model. In many cases, such programs will not provide the operator with an opportunity to configure the system or select from a range of choices (e.g. via a menu screen). If a change does become necessary, the software has to be modified at the source code level. Sometimes this task can be tackled by a keen control engineer but, more often than not, it will require the services of a programmer or software engineer.

More flexible software packages will allow the user to configure the software for a particular hardware system (e.g. by specifying the system components and expansion capability). It should be noted that the PC maintains an installed *equipment list* (in RAM) and it is possible for software to interrogate this and configure itself automatically. Furthermore, this process can be made transparent to the user.

As users become fully aware of the advantages of a PC-based instrumentation and control system, a further consideration which will often become increasingly important is that of interfacing with other software packages so that data can be exported and imported in a fashion which is largely transparent to the user. This is often advantageous where measurements must be made within one program before the captured data (stored in a disk file) is imported into another program for statistical and/or graphical analysis. This opens up a completely new scenario in which data acquired by one program can be made to yield all manner of new information when analysed by another package.

Performance

Performance of a software package is often somewhat difficult to gauge since determining factors may differ from one application to another. Processing speed, for example, will be all important in some applications but largely irrelevant in others.

To a large extent, processing speed will depend upon simultaneous demands placed on the processor. Where a system has to carry out many functions at the same time (such as regularly updating a graphics display, servicing interrupt requests generated by several expansion cards and responding to user input from a keyboard) it is hardly likely to run at what may be considered an acceptably high speed.

If speed is a paramount consideration and a great deal of 'number crunching' is expected, then a system with a high clock rate and numeric co-processor will normally be essential. If such a system still fails to deliver the speed which is a pre-requisite of a particular real-time application then it is probably questionable as to whether a PC-based system should have been adopted in the first place!

Provided that processing speed is identified as a premium requirement, custom software can usually be made to offer significant speed advantages over 'off-the-shelf' packages. The programmer or software engineer can elect to optimize his or her code for speed rather than data integrity. As an example, input range checking could be abandoned in favour of faster throughput of data or opt for post-acquisition rather than real-time display of data.

Functionality

Functionality is becoming increasingly important in software selection and it relates to the fitness and suitability of a program for a particular application and hardware configuration. Most of today's applications packages offer additional functionality which may not appear as part of an initial software specification. The ability to convert data files (e.g. to Lotus 1–2–3 format) is a prime example of this as is the availability of a built-in IEEE-488 command language.

Software classification

In order to provide a frame of reference, Table 10.1 shows the continuum which exists between custom-written 'turnkey' software at one extreme and graphics-orientated operating system shells at the other.

Custom-written software

Custom-written software has already been described at some length in earlier chapters and readers should refer to these for information concerning products such as Microsoft's MASM, QuickBASIC, and QuickC. You should not underestimate the task of developing your own 'turnkey' applications (particularly if starting from scratch); however, in some cases, this may be the only effective solution to a particular problem.

High-level language extensions

Most modern high-level language interpreters and compilers tend to provide a wide range of facilities suitable for those wishing to develop commercial and scientific applications. Unfortunately, the facilities offered are generally somewhat lacking when the software is to be used in conjunction with data acquisition or control hardware. Several

Table 10.1 Software classification

Ease of use	*Flexibility*	*Performance*	*Functionality*	*Examples*
Very complex (requires a high level of programming expertise)	Highly flexible and adaptable	Very fast and compact (where this is important)	As required to meet the needs	Custom written software (code developed in assembly language, C, BASIC, etc.)
Moderately complex (requires programming expertise)	Flexible and adaptable	Fast and efficient	Offers a high level of functionality	High level language extensions and drivers (e.g. LAB I/O and Soft500)
Average level of complexity (may require programming expertise)	Moderately flexible and adaptable to most situations	Reasonably fast and moderately efficient	Offers a fair degree of functionality but may not cover all eventualities	Programmable applications (e.g. ASYST)
Average level of complexity (does not require programming expertise but may require specialized knowledge)	Restricted to particular applications	Reasonably fast and moderately efficient	Offers functionality in particular areas	Tool kits (e.g. PC Tools and Norton Utilities)
Simple to use (requires no programming skills and little expertise on the part of the user)	Somewhat limited in flexibility and some packages may be rather specialized	May be slow in some circumstances	Levels of functionality vary from package to package	Applications packages (e.g. DADiSP, ASYSTANT, and LABTECH NOTEBOOK)
Very simple to use and easy to learn (highly intuitive)	Only provides standard DOS facilities	Not as efficient (in terms of speed and memory use) as DOS	Only provides functionality associated with the operating system	Graphics operating system shells (e.g. Windows and GEM)

software suppliers have realized this particular need and now supply extensions and drivers for popular languages. These provide the programmer with additional commands relating to external hardware devices and which may be used in conjunction with more complex data acquisition functions.

Programmable applications

Programming applications are software packages which allow the user to customize a program so that it may be employed in a dedicated (turnkey) application and, whilst such tools have something in common with a conventional language-orientated development systems, they tend to be based upon a limited number of simple commands which are incorporated into statements and executed automatically from within the application environment.

ASYST

ASYST is an integrated software environment which provides a full range of interactive and programmable functions. ASYST comprises four modules: System/Graphics/Statistics, Analysis, Data Acquisition, and GPIB/IEEE-488.

The first of ASYST's modules constitutes the base system (it is necessary for booting and running the ASYST system) which incorporates the data type specifications, data file format, and file I/O as well as basic arithmetic and statistical functions, graphics capabilities and ASYST's own programming language.

ASYST provides versatile array editing of one or two-dimensional arrays (any individual element within an array can be accessed by means of a cursor or by means of a typed command). Arrays can be reshaped, concatenated, laminated and sectioned as required. ASYST can read and write ASCII, DIF and binary formatted files. A special interface handles input and output of Lotus 1–2–3 files.

ASYST's second (analysis) module provides a manipulation and reduction capability for analysing data obtained through ASYST's own data acquisition system or otherwise read (or entered) into the package. The module also includes a number of analytical routines (e.g. Fast Fourier Transformation).

ASYST's applications include signal processing and waveform analysis as well as a full range of matrix and vector operations, polynomial mathematics and curve fitting. ASYST commands can be linked together in new word definitions which allow complex procedures to be executed from a single word in much the same manner as any extensible programming language.

ASYST's third module deals with data acquisition and provides control for a variety of A/D and D/A hardware. ASYST contains the necessary tools to create a wide range of highly specific data acquisition applications and supports foreground and background operation as well as multi-channel capability so that tasks operate concurrently with other ASYST functions. Acquisition can be triggered in various ways including signal threshold levels and time delays.

The last of ASYST's four modules provides an interface with the IEEE-488 bus. A full range of device dependent commands are provided for bus and instrument control.

ASYST requires a PC, PC-XT, PC-AT or any true compatible microcomputer fitted with minimum of 512 kilobytes of RAM (640 kilobytes recommended). The package requires DOS 2.0 (or later), an appropriate graphics adapter (e.g. IBM Enhanced Colour Graphics Card or Hercules Graphics Card) and preferably a hard disk. In addition, data acquisition hardware will be required for use with modules 3 and 4. A numeric co-processor (8087 or 80287, as appropriate) is required.

Applications packages

An applications package is simply a collection of applications programs (often linked together within an overall shell) designed to solve a particular problem or range of problems. Several powerful applications packages designed to simplify the acquisition, processing, and analysis of instrumentation and process control data have become available in recent years. Such packages tend to be very easy to use and generally provide the user with a graphical interface as well as the ability to import and export data from other packages.

Applications packages tend to be applicable to a wide variety of problems and may also incorporate direct data acquisition and control facilities (e.g. via the IEEE-488 bus).

ASYSTANT

ASYSTANT is a stand-alone, full-functioned software package for scientific data analysis, graphics, statistics and (optional) data acquisition. The package is completely menu-driven and requires no programming skills on the part of the user. ASYSTANT's analysis structure is based around a 'desk calculator' which contains a set of sophisticated data processing and analysis functions. These are presented to the user in the familiar form of a scientific calculator.

Two versions of the package are available: ASYSTANT, the basic package, and ASYSTANT+ which adds data acquisition facilities to the basic package.

ASYSTANT can display data in a variety of formats. Graphics menu options are listed across the bottom of the main graphics menu screen and functions such as on-screen labelling, a variety of window options, colour choices and manual or autoscaled axes allow the user to customize the presentation of graphics before sending plots to a digital plotter or graphics printer.

ASYSTANT also incorporates a waveform processor which allows the user to manipulate waveforms or waveform segments. The wave-processing screen display contains menu and status displays at the bottom of the screen, a main plotting area in the centre of the screen, and five windows at the top of the screen which constitute 'graphics memories'.

The desk calculator provided within ASYSTANT is an array-based device from which options are selected via four menu pages. Basic arithmetic operations are typed directly on a command line. Other operations are selected from the menu pages but may also be directly entered on the command line.

Fast Fourier Transformation (FFT), power spectra, matrix inversion and smoothing functions are provided as are curve fitting and statistical functions, differential equations, file and graphics based processing.

Each of the seven data-acquisition modes of ASYSTANT + emulates the function of a common laboratory data-collection instrument and operation is thus rendered largely intuitive. The mode of operation chosen for a particular application will depend upon a number of factors including sampling rate, need for a simultaneous graphics display, number of analogue-to-digital (A/D) channels required, etc. The modes may be summarized as follows:

- High-speed recording: This mode allows the user to acquire data at rates which approach the maximum supported by the data acquisition hardware.
- Signal averaging: The mode is similar to high-speed recording mode except that a running average is automatically calculated.
- Waveform scrolling: This mode is equivalent to that of a conventional strip-chart recorder (the RTDMS Graphics Accelerator Card is supported).
- Transient recording: This mode captures two windows of data surrounding a user-defined event trigger on up to eight channels with post-capture graphics display.
- X–Y recording: This mode emulates an X–Y recorder and provides a continuous display in a two-axis plot of two selected A/D channels with continuous throughput to disk.
- Data logging: This mode provides continuous low-speed data acquisition with throughput to disk and real-time print out to screen and/or printer. Up to four stages and six alarm conditions and messages can be defined. This mode has obvious applications in the area of process control and/or monitoring.

• Function generation:	This mode operates concurrently with several of the other data acquisition modes and may be used to provide an analogue output (D/A) from a waveform array.

ASYSTANT requires a PC, PC-XT, PC-AT or any true compatible microcomputer (e.g. HP Vectra) fitted with 640 kilobytes of RAM. The package requires DOS 2.0 (or later), an appropriate graphics adapter (e.g. IBM Enhanced Colour Graphics Card with 64–256 kilobytes of RAM, HP Multimode Video Adapter, AT&T High resolution Monochrom Graphics Card, or Hercules Graphics Card) and two (or more) disk drives including one double-sided floppy drive. A hard disk is recommended and an 8087 (or 80287) numeric co-processor is required.

DADiSP II

DADiSP II is a sophisticated package which can be used to analyse and display up to 64 windows from the same (or different) data files. The package offers a powerful graphing capability and contains a variety of standard and advanced mathematical and statistical functions ranging from standard arithmetic (addition, subtraction, multiplication and division) to hyperbolic functions, mean and standard deviations, integration and differentiation.

DADiSP is ideally suited to any application in which captured data is to be analysed and displayed in a graphical format. Typical examples might be X, Y and Z-axis strain developed in a structural member when subjected to repeated cycles of stress or variation in oscillator frequency, amplitude and noise when subjected to variations in temperature, pressure and humidity. In either case, the multiple windowing capability allows the user to form a complete picture of the performance of a system without recourse to a number of discrete graphs and displays.

DADiSP II supports a number of advanced mathematical functions including trigonometric and logarithmic functions. These can be applied to any valid signal, scalar or signal-scalar expression. When applied to a signal, they are applied successively to each point of the signal and the resulting signal is displayed in the current window. When applied to scalars (integers, real numbers or complex numbers) the resulting value is displayed at the bottom of the screen.

The calculus functions are provided for determining derivative and integral functions. Since the signals are discrete, DADiSP provides a means of performing left derivative and right derivative calculations. Four of these functions (DERIV, LDERIV, RDERIV and INTEG) take one signal as an argument and return a new one. AREA takes two additional arguments, the starting and ending points within a signal.

Note also that AREA returns a scalar whereas the other four calculus functions return signals. The algorithm for calculating the integral is a modification of Simpson's Rule and is more accurate than a simple trapezoidal approximation.

DADiSP's statistical functions provide information about a signal (or two signals in the case of LINREG2). MEAN and STDEV return appropriate values which can be nested in more complicated expressions. STATS does not return a signal but displays both the mean and standard deviation at the bottom of the screen. LINREG and LINREG2 display the regression coefficients and then create a new signal (i.e. the line generated by the linear regression) which is useful for over-plotting. AMPDIST generates a new signal which constitutes a bar graph distribution for a signal. The function accepts a real-number argument which is the incremental value (DELTA X).

DADiSP contains facilities for frequency domain analysis. Fourier analysis is provided with both the Discrete Transform and the much faster Fast Fourier Transform (FFT). PARTSUM creates a new signal which is equivalent to the partial sums of two input signals whilst SUMS adds any number of signals together. MOVAVG provides a smoothing function while AVG takes the point-by-point average of a group of signals.

A powerful range of signal editing functions are also provided. EXTRACT creates a new signal by extracting part of an existing signal while REVERSE simply changes the polarity of a given signal and CONCAT concatenates any number of signals. Signal generating functions are preceded by the letter G. An endless variety of waveforms can be synthesized through combination of functions (e.g. GASINH will generate the waveform of a hyperbolic arcsine function).

DADiSP runs on a PC, PC-XT, PC-AT or any true compatible microcomputer fitted with a minimum of 512 kilobytes of RAM. The package requires DOS 2.0 (or later), colour graphics, and at least one 360-kilobyte floppy drive together with a second drive (either floppy or hard disk). A numeric co-processor is optional.

DADiSP provides import and export utilities which may be used to achieve data file compatibility with Louts 1–2–3, ASYST, ASYSTANT, Soft500, and most ASCII formatted files.

Tool kits

Tool kits are aptly named utilities which provide the user with a range of facilities designed to maximize the efficiency of a system. Facilities provided by a tool kit generally include:

- File display and editing (in various formats).

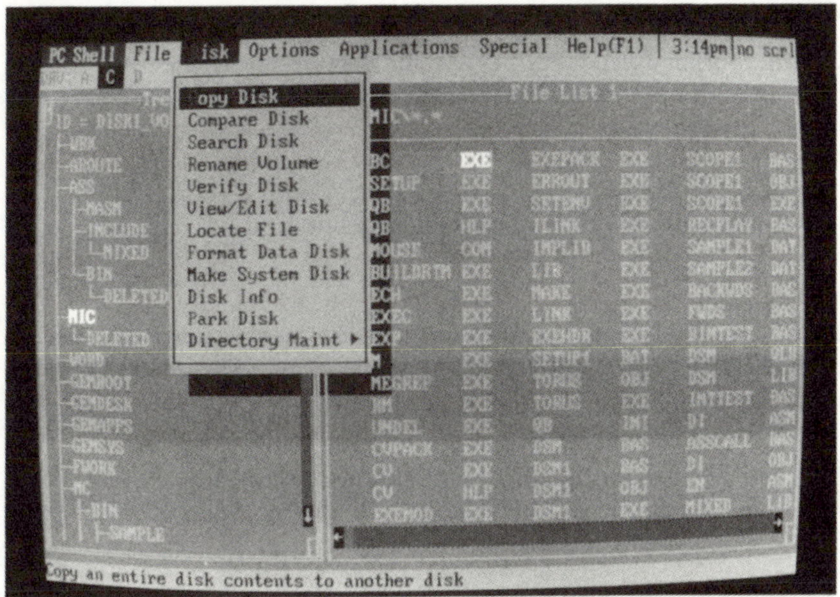

Plate 10.1 *PC tools shell program*

- Hard disk optimization.
- Hardware diagnostic facilities.
- DOS related facilities (e.g. directory listings).
- Data recovery.

Two of the most popular tool kits are Norton Utilities and PC Tools. Both sets of utilities can be highly recommended and indeed are considered invaluable by most users. Such tools would not, however, usually form part of an initial software specification since their use is generally restricted to troubleshooting and data recovery applications.

Norton utilities

Peter Norton's widely acclaimed package has been justifiably described as 'indispensable' by a number of reviewers. The package comprises a suite of programs which aid disk management and provide a means of recovering lost data.

PC tools

Central Point Software's popular PC Tools provides a range of useful utility programs together with an excellent DOS shell. The software includes a variety of disk management programs (including a disk optimizer and file recovery utilities).

Graphics-operating system shells

Graphics-orientated operating system shells provide the user with a friendly and largely intuitive interface with an operating system. The level of functionality associated with a graphics shell is generally no more than that which can be obtained using a conventional DOS command line interpreter (e.g. copying files, executing batch files, launching and running applications programs, etc.).

A graphics-orientated shell does, however, offer two significant advantages:

1 A more friendly and understandable interface with the operator or end user.
2 In-built routines which may be essential to the functioning of a particular graphics-based application.

If either (or both) of the foregoing can be identified as an important requirement, then it will be necessary to incorporate a graphics shell (e.g. Windows) into the software specification of a system (see page 144).

11 Applications

The PC is a potential prime mover in a huge variety of process control and instrumentation applications ranging from simple stand-alone machine controllers to fully integrated production control systems. This chapter aims to provide readers with an introduction to the procedure for selecting and specifying hardware and software for a PC-based instrumentation or process control system. In addition, several representative applications of PC-based systems are discussed.

Expansion cards

The range of PC expansion boards currently available from a large number of manufacturers includes:

- Analogue I/O cards with up to 16 analogue inputs (and up to four buffered analogue outputs.
- Digital I/O cards with direct TTL-compatible inputs and outputs.
- Digital I/O cards with opto-isolated inputs and outputs.
- Digital I/O cards with buffered I/O lines.
- Digital output cards fitted with reed relays.
- Digital output cards fitted with relays for a.c./d.c. power control or triads for a.c. power control.
- Expansion memory cards.
- Hard disk cards.
- EPROM programmers.
- IEEE-488/GPIB interface cards.
- Network adapter cards.
- Modem cards (to permit transfer of data via conventional telephone lines).
- Prototyping cards (these usually include the necessary PC expansion

bus interface logic and provide the user with an area for soldering components fitted into a 0.1″ matrix of plated through holes).
- Serial communications cards with up to four RS-232, RS-422, RS-423 or RS-485 serial ports.
- Stepper motor controllers.
- Multi-function I/O cards (offering mixed analogue and digital I/O facilities).
- Thermocouple interface cards.
- High-speed data acquisition cards.
- Bus expansion cards (which interface with external card frames or mother boards).
- Solid-state RAM/EPROM based disk drives.
- Specialized instrument cards (e.g. digital multimeters, counters/digital frequency meters, etc.).

In addition, the system builder is able to select from a large range of signal conditioning cards which provide the necessary interfacing circuitry for a wide range of popular sensors and output devices. It is thus eminently possible to construct a PC-based process control system simply by selecting 'off-the-shelf' modules. Only when dealing with very specialized applications is it necessary to manufacture one's own dedicated I/O cards and/or external signal conditioning boards. Appendix F list a number of major suppliers of PC expansion cards and signal conditioning circuitry.

Approaches

The system designer can select from a range of options depending upon the complexity and individual requirements of a particular application. The following general approaches are available:

- Stand-alone PC systems based on internally fitted expansion cards, rack modules, or separately enclosed units).
- PC systems based on standard PC expansion cards (and I/O processing cards, where appropriate) fitted into external card frame modules.
- Industrial PC systems (using a ruggedized PC functioning as a dedicated process controller or data-gathering device) fitted with internal or external expansion cards and housed in a rack or free-standing enclosure.
- RS-232 based systems with the PC as controller (peripheral hardware connected via an asynchronous serial link).
- IEEE-488 based systems with the PC as controller.

- Backplane bus-based (e.g. IEEE-1000) systems with a PC bus master/controller and standard bus I/O cards.
- Networked/distributed PC systems (e.g. based on Ethernet or Bitbus) with enclosures and expansion cards to meet local requirements.

PC instruments

In addition to the vast range of expansion cards currently available, have developed a range of 'PC instruments' which emulate conventional items of test equipment (such as digital voltmeters, digital frequency meters etc.).

A PC instrument offers many advantages over its conventional counterpart. It is flexible and adaptable and, in many cases, measurements may be automated under programmed control. Furthermore, considerable savings can be achieved from the elimination of redundant hardware (such as displays, operator controls etc.).

PC-based instruments can also offer very significant cost savings when compared with simple IEEE-488 bus-based instrumentation systems. A typical PC-based system for the acquisition of analogue voltages can, for example, be realized for less than 50% of the cost of a similarly specified system based on IEEE-488 hardware and software.

PC-based instruments are available in three general formats (Figure 11.1):

1 Using internally-fitted expansion cards (plugged into a free slot in the PC).
2 Using an external rack with plug-in PC expansion cards.
3 Using separately enclosed modules (which may, if desired, be stacked) based on RS-232 or IEEE-488 bus systems.

All three of these approaches have their own particular virtues and the system builder should include all three in his or her portfolio of engineering solutions.

Internally-fitted cards generally offer the lowest cost approach to building a PC instrument. The disadvantage of this technique is that it necessitates internal fitting and, since there may be a limited number of slots available, the expansion capability may be somewhat limited.

An external rack system allows the PC bus to be extended so that standard PC expansion cards may be fitted into an external card frame. This system is, however, relatively expensive and generally only appropriate where large-scale expansion is required. An alternative to that of extending the PC bus beyond the confines of the system enclosure

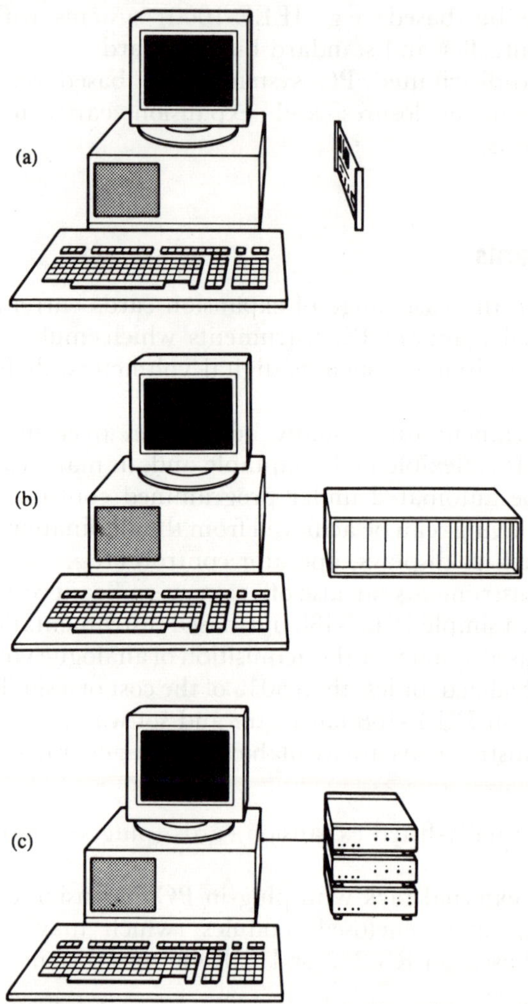

Figure 11.1 *Three basic approaches to PC-based instruments: (a) internally fitted PC expansion cards; (b) PC expansion cards in an external card frame; (c) separately enclosed PC instruments linked via the RS-232 or IEEE-488 bus*

is that of making use of a proprietary I/O bus such as Metrabyte's MetraBus. Such systems generally provide for between 1 and 32 I/O boards mounted in standard rack enclosures.

Separately enclosed modules (which may be interfaced to a PC by means or the RS-232 or IEEE-488 bus) provide the third of this trio of potential solutions. With the exception of a front panel display and controls, separate PC instrument modules usually resemble the conventional stand-alone instruments which they replace.

As an example, Siemens provide a range of individually enclosed PC instruments which include a multimeter, universal counter, transient recorders, digital input/output unit, function/pulse generator, voltage/current generator, and a scanner. Individual instruments can be combined to provide more complex instrumentation facilities. A data logger, for example, can be assembled from a scanner and multimeter and controlled flexibly from the PC.

PC instruments are ideal for making repetitive measurements during which data must be accumulated over a period time. The PC allows such measurements to be automated with the data acquired being sent to a file for future analysis.

As an example of the use of a PC instrument, consider an application in which the output frequency of an oscillator has to be monitored accurately over a long period of time. This task can be accomplished by means of a dedicated digital frequency meter with readings taken at appropriate intervals, logged on paper, and a graph showing the long-term variation of frequency can then be drawn. The alternative approach using a 'PC instrument' simply involves fitting a digital frequency meter expansion card (such as the Guide Technology GT200) to a standard PC-compatible and using simple software (in conjunction with the driver(s) supplied with the card) to automate the measurement and store the results in a data file for import into an analysis package (such as DADiSP). A typical application is discussed later in this chapter.

Industrial PC systems

Ruggedized PCs are the obvious choice for use in the harsh environment found in most industrial plants. Typical of such equipment is the immensely popular Hewlett Packard Vectra series of computers.

Industrial PCs usually offer the same range of facilities associated with conventional PCs and compatibles and invariably support the industry standard bus architecture. Hence an industrial PC will generally accept the same range of expansion cards as mentioned under the previous heading. Alternatively, where additional expansion beyond the limit imposed by the available free slots, industrial PCs may be fitted with bus extenders which are normally based on an external rack assembly.

In extremely harsh environments, it may become necessary to replace the floppy/hard disk drive system with a 'silicon disk'. Several manufacturers provide RAM/EPROM-based disk emulator cards which are ideal for use in dedicated diskless systems.

Backplane bus-based systems

A backplane bus system offers a reasonable compromise between a standard PC-based system at one extreme and a specialized industrial PC system at the other. Backplane bus systems are inherently flexible and reliable and can simply be fitted with a PC processor card in order to make use of standard PC software packages.

Networked/distributed PC systems

Networked or distributed PC systems are appropriate in large-scale applications where several processes are carried out concurrently. Each individual PC will be responsible for part of the process and data will be shared between the PCs by means of the network. As an example, consider the case of a packaging plant which manufactures and fills cardboard boxes on a continuous basis. One PC may be dedicated to the cutting and folding operation whilst another may be responsible for controlling glueing and stapling. A third PC would be responsible for filling and sealing the boxes. Data from all three PCs would then be collected by a fourth PC which oversees the entire process. Such a system provides an alternative to conventional solutions based on distributed programmable logic controllers (PLC).

Intel's BITBUS provides a simple and elegant solution to applications which require the services of a multi-drop network. BITBUS is a serial data bus based on the RS-485 physical and electrical interface standard (RS-485 is a multi-drop version of RS-422) and the datalink protocol employed is a subset of SDLC/HDLC.

BITBUS complements Manufacturing Automation Protocol (MAP) which has gained widespread recognition as the industrial standard for the upper level of factory data communications. At the machine and process level, however, where time critical data from sensors, actuators and alarms has to be transmitted, the response time of MAP, though guaranteed, is inadequate. BITBUS, on the other hand, is well suited to the transfer of short 'Field Data' messages.

BITBUS is configured as a single-master, multi-slave network and operates in one of two modes, synchronous and self-clocked. Synchronous operation permits speeds of up to 2.4 megabits/s but requires twin twisted-pair cables and is restricted to transmission over distances less than 300 m. Furthermore, since repeaters cannot be used in this mode, a maximum of 31 nodes is possible. Self-clocked mode, on the other hand, requires only single pair cable, can operate at either 62.5 or 375 kilobit/s and, with repeaters, can cater for up to 250 nodes at distances not exceeding 13 km.

Interfacing with BITBUS is usually made possible with the use of an

Intel 8044 micro-controller which implements the BITBUS protocol using an on-chip SDLC controller and ROM-based firmware. An interface of this type may be incorporated within a processor card or may be provided as part of an auxiliary communications interface.

Specifying hardware and software

When specifying hardware and software to be used in a given PC bus application, it is essential to adopt a 'top–down' approach. An important first stage in this process is that of defining the overall aims of the system before attempting to formalize a detailed specification. The aims should be agreed with the end-user and should be reviewed within the constraints of available budget and time. Specifications should then be formalized in sufficient detail for the performance of the system to be measured against them and should include such items as input and output parameters, response time, accuracy and resolution.

Having set out a detailed specification, it will be possible to identify the main hardware elements of the system as well as the types of sensor and output device required (see Chapter 9). The following checklist, arranged under six major headings, should assist in this process:

1 *Performance specification*

- What are the parameters of the system?
- What accuracy and resolution is required?
- What aspects of the process are time critical?
- What environment will the equipment be used in?
- What special contingencies should be planned for?
- What degree of fault-tolerance is required?

2 *I/O devices*

- What sensors will be required?
- What output devices will be required?
- What I/O and signal conditioning boards will be required?
- Will it be necessary to provide high-current or high-voltage drivers?
- Should any of the inputs or outputs be optically isolated?

3 *Displays and operator inputs*

- What expertise can be assumed on the part of the operator?
- What alarms and status displays should be provided?
- What inputs are required from the operator?
- What provision for resetting the system should be incorporated?

4 *Program/data storage*

- What storage medium and format is to be employed?
- How much storage space will be required for the operating system and/or control program?
- How much storage space will be required for data?
- How often will the control program need updating?
- Will stored data be regularly updated during program execution?
- What degree of data security and integrity must be achieved?

5 *Communications*

- What existing communications standards are employed by the end-user?
- Will a standard serial data link based on RS-232 be sufficient or will a faster, low-impedance serial data communications standard be needed?
- What data rates will be required?
- What distances are involved?
- Will it be necessary to interface with automatic test equipment?
- Will a networking capability be required?

6 *Expansion*

- What additional facilities are envisaged by the end-user?
- What additional facilities could be easily incorporated?
- Will expansion necessitate additional hardware, additional software, or both?
- What provision should be made for accommodating additional hardware?

Hardware design

Start by identifying the principal elements of the system including PC, card frame, power supply, etc. as appropriate. Then itemize the input devices (such as keypads, switches and sensors), and output devices (such as motors, actuators and displays). This process may be aided by developing a diagram of the system showing the complete hardware configuration and the links which exist between the elements. This diagram will subsequently be refined and modified but initially will serve as a definition of the hardware components of the system.

Having identified the inputs required, a suitable sensor or input device should be selected for each input (see Chapter 9). It should then be possible to specify any specialized input signal conditioning required with reference to the manufacturer's specification for the sensor concerned. Input signal conditioning should then be added to the system diagram mentioned earlier.

Next, a suitable driver or output interface should be selected for each output device present (see Chapter 9). Any additional output signal conditioning required should also be specified and incorporated in the system diagram.

Software design

Software design should mirror the 'top–down' approach adopted in relation to the system as a whole. At an early stage, it will be necessary to give some consideration to the overall structure of the program and identify each of the major functional elements of the software and their relationship within the system as a whole. It is important to consider the constraints of the system imposed by time critical processes and hardware limitations (such as the size of available memory). Furthermore, routines to cope with input and output may require special techniques (e.g. specialized assembly language routines).

The software should be designed so that it is easy to maintain, modify and extend. Furthermore, the programmer should use or adapt modules ported from other programs. These modules will already have been proven and their use should be instrumental in minimizing development time.

When developing software, it is advisable to employ only 'simple logic' (i.e. that which has been tried and understood). The temptation to produce untried and over-complicated code should be avoided. Simple methods will usually produce code which is easy to maintain and debug, even if the code produced requires more memory space or executes more slowly. If the process is time critical or memory space is at a premium then code can later be refined and optimized. It is also important to consider all eventualities which may arise, not just those typical of normal operation. The following are particularly important:

- Will the system initialize itself in a safe state – will there be momentary unwanted outputs during start-up?
- What will happen if the user defaults an input or if an input sensor becomes disconnected?
- What will happen if the power fails – will the system shut-down safely?
- What input validation checks are required – what steps should be taken if an 'out of range' input is detected?

Applications

The remainder of this chapter provides details of some representative PC-based applications. These applications are not particularly novel but they do address problems which are typical of those which face the instrumentation and control engineer. The applications have been chosen to illustrate contrasting aspects of design and, while it would be impossible to describe any of these applications in their entirety, they should provide a feel for various aspects within the process of designing and implementing a PC-based system.

Monitoring oscillator stability

The client is a manufacturer of synthesized HF radio transceivers and wishes to develop a prototype voltage-controlled oscillator (VCO) which operates in the range 40–60 MHz for use within the frequency-generating circuitry. Several circuits have been constructed and the client wishes to ascertain the short-term and long-term frequency stability of each unit.

Specification

The manufacturer requires that the output frequency is measured at appropriate intervals (e.g. every 100 ms for the short-term stability measurement and every 10 s for the long-term stability measurement). The results of each set of measurements are to be stored in an ASCII file for later graphical analysis. The software is, however, required to determine a number of simple performance indicators for each prototype unit including:

- Maximum frequency during the measurement period.
- Minimum frequency during the measurement period.
- Mean frequency over the measurement period.
- Total frequency drift during the measurement period.

The manufacturer also requires that the entire set of measurements and statistical calculations should be repeated at ambient temperatures of 0°C, 10°C, 20°C, 30°C and 40°C.

This task would require considerable manual effort if it were to be carried out using a conventional digital frequency meter. It is, however, an ideal candidate for automated measurement using a PC and appropriate expansion card.

Hardware

The Guide Technology GT200 Universal Counter was chosen to provide the frequency measuring facility in conjunction with a Samsung

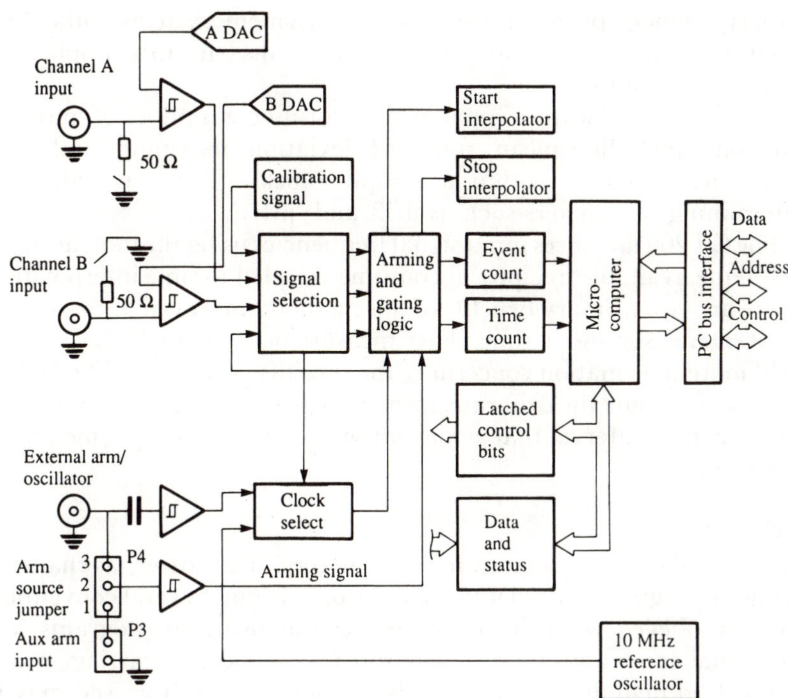

Figure 11.2 *Simplified block schematic of the GT200 digital frequency meter*

AT-compatible microcomputer which already resides in the client's RF laboratory. The GT200 takes the form of a full-size PC-compatible expansion card which is supplied together with a device driver (GT200.SYS) and virtual front panel software (VIRT.EXE) on a floppy disk. The simplified block schematic of the GT200 is shown in Figure 11.2.

The GT200 is supplied together with a device driver (GT200.STS) and virtual front panel software (VIRT.EXE) on floppy disk. The disk also contains software which assists with setting the base address switch and includes a program which allows users to test the GT200's programming commands.

The GT200 offers a variety of measuring facilities including frequency measurement (from d.c. to 100 MHz, with automatic pre-scaling above about 1 MHz), fast frequency measurement (a special mode for high-speed data acquisition which allows up to 2300 measurements per second), period (both single and multiple), and time interval (i.e. the elapsed time between 'start' and 'stop' events). In addition, a direct data acquisition mode places measurements into a memory array without the usual overheads required to communicate results back to an application program via DOS).

The GT200 software is capable of performing a number of statistical functions (including mean, standard deviation, maximum and minimum measurements within a sample block). These are ideal for determining parameters such as drift and 'jitter'.

The GT200 measures input signal frequencies using the most accurate technique available, reciprocal counting coupled to time interpolation. There are two primary benefits of this method: improved accuracy and reduced measurement time. Fast measurements with high accuracy yield more information concerning the stability of a signal. The GT200 is able to compute the drift rate, mean and peak–peak jitter of a signal in the same time interval that a conventional counter is simply measuring frequency.

Software

The control program sends commands to the GT200 driver as character strings through standard DOS file write operations. Several conventions must be obeyed when incorporating commands into programs (e.g. individual commands must be separated by semicolon, carriage return or line feed delimiters). Commands are not case sensitive and may be abbreviated for convenience. The minimum acceptable abbreviations for each command are listed in the manual. As an example, FREQ may be used instead of FREQUENCY, FU instead of FUNCTION, and so on.

GT200 commands are incorporated in normal program statements such as:

```
PRINT #1, "fu freqa; gate 0.01"                    (BASIC)
fprintf(COUNTER, "fu freqa; gate 0.01");            (C)
```

Note that the foregoing have an identical effect on the GT200 card.

Figure 11.3 shows how a simple program can be easily developed to meet the client's requirements for the long term stability measurement (involving 100 readings taken at 10-s intervals).

Three sub-programs, max(), min(), and mean() are declared at the beginning of the program. The array, freq(), (which will contain the returned data from the GT200 card) is then dimensioned for a total of 100 values.

The user is then prompted to enter the oscillator reference (which is truncated to include only the first six characters) and the ambient temperature used for the measurement.

The GT200 digital frequency meter is then associated with channel 1 for output and 2 for input by means of the OPEN statements. The

```
REM Oscillator test program
REM Declare sub-programs
DECLARE SUB max ()
DECLARE SUB min ()
DECLARE SUB mean ()

REM Dimension array
DIM freq(100)

REM Get oscillator reference
CLS
INPUT "Enter oscillator reference: "; ref$
LET ref$ = LEFT$(ref$, 6)
INPUT "Enter ambient temperature:  "; temp$
osc$ = ref$ + temp$
REM Initialise digital frequency meter
OPEN "GT200$" FOR OUTPUT AS #1
OPEN "GT200$" FOR INPUT AS #2
PRINT #1, "init; timeout 4; function frequency A; gate 0.2"

REM Start collecting readings
PRINT "Hit <RETURN> to start measurement..."
WHILE r$ = ""
  r$ = INKEY$
WEND
FOR time% = 0 TO 99
  PRINT #1, "reset"
  INPUT #2, freq(time%)
  PRINT "Time = "; 10 * time%; " sec.  Frequency = "; freq(time%); " Hz"
  PRINT #1, "wait 10"
NEXT time%
CLOSE #1
CLOSE #2

REM Calculate and print statistics
LPRINT
LPRINT "Performance data for oscillator ref: "; ref$
LPRINT "Performance measured at:             "; temp$; " deg.C"
max
LPRINT "Maximum frequency: "; maxfreq; " Hz"
min
LPRINT "Minimum frequency: "; minfreq; " Hz"
mean
LPRINT "Mean frequency:    "; ave; " Hz"
LPRINT "Frequency drift:   "; maxfreq - minfreq; " Hz"
LPRINT

REM Save data in ASCII file
LET file$ = osc$ + ".DAT"
OPEN file$ FOR OUTPUT AS #3
FOR time% = 0 TO 99
  PRINT #3, freq(time%)
NEXT time%
CLOSE #3
END

SUB max
SHARED freq()
SHARED maxfreq
maxfreq = 0
FOR i% = 0 TO 99
```

Figure 11.3 *Oscillator test program for long term stability*

```
 IF freq(i%) > maxfreq THEN maxfreq = freq(i%)
NEXT i%
END SUB

SUB mean
SHARED freq()
SHARED ave
total = 0
FOR i% = 0 TO 99
  total = total + freq(i%)
NEXT i%
ave = total / 100
END SUB

SUB min
SHARED freq()
SHARED minfreq
minfreq = 1E+09
FOR i% = 0 TO 99
IF freq(i%) < minfreq THEN minfreq = freq(i%)
NEXT i%
END SUB
```

Figure 11.3 *(cont.)*

instrument is initialized to measure frequency using input A with a timeout and gate times of 4 s and 0.2 s, respectively.

The program then waits for the user to indicate that he or she is ready to begin a measurement by hitting the RETURN key. Once the key has been hit, the program takes 100 readings of frequency, placing each returned reading into the freq() array. The time between readings is set at 10 s by means of the wait command. Times and corresponding frequency readings are displayed on the screen on each pass through the main FOR . . . NEXT loop so that the user is kept informed of the current state of measurement.

When the main loop has been completed, the two communications channels are closed. Thereafter, the performance data for the oscillator in question is printed with calls to the three sub-programs which determine the maximum, minimum and mean frequency values. The total frequency drift is calculated simply by subtracting the minimum frequency from the maximum frequency.

The three sub-programs, max(), min(), and mean(), are quite straightforward and need no comment. A typical résumé of oscillator performance (printed by the program) is shown in Figure 11.4.

Finally, the data is stored in an ASCII file. Note that the filename is constructed from the concatenation of the first six (or less) characters of the oscillator reference and the ambient temperature which was entered

```
Performance data for oscillator ref: HXO
Performance measured at:              20 deg.C
Maximum frequency:   5.016576E+07  Hz
Minimum frequency:   5.015442E+07  Hz
Mean frequency:      5.015941E+07  Hz
Frequency drift:     11336  Hz
```

Figure 11.4 *Sample printed oscillator performance data generated by the program shown in Figure 11.3*

by the user, together with the file extension, .DAT. The file is opened for output (via channel 3) and all 100 values stored in the array are written to it. The channel is then closed.

Testing crystal filters

The clients is a manufacturer of RF passive components. Part of the company's product range includes 10.7-MHz crystal filters of various types which are manufactured to close tolerance in a batch process. Each filter is checked (on a test jig) to determine whether it meets the design specification which includes bandwidth (measured at −6 dB and −40 dB) and pass-band ripple. It is also considered desirable to display the response of the filter graphically in order that the ultimate stop-band attenuation can be gauged. Figure 11.5 shows a typical filter response characteristic.

The company wishes to automate the process of filter measurement and, at the same time, generate statistical information which can be used to check the manufacturing process.

Hardware

This application is ideally suited to an IEEE-488 based system (based on test instruments fitted with the requisite IEEE-488 interface which are already available in the company's test department). Apart from the PC controller (which will require an IEEE-488/GPIB interface card) the two instruments required are:

- An RF voltmeter (Marconi 2610 with GPIB interface).
- An RF signal generator (Marconi 2018A with GPIB module).

The RF signal generator will be configured as a 'listener' whilst the RF voltmeter will be a 'talker'. A test jig will have to be constructed to accommodate the filter under test. Furthermore, since the filter source

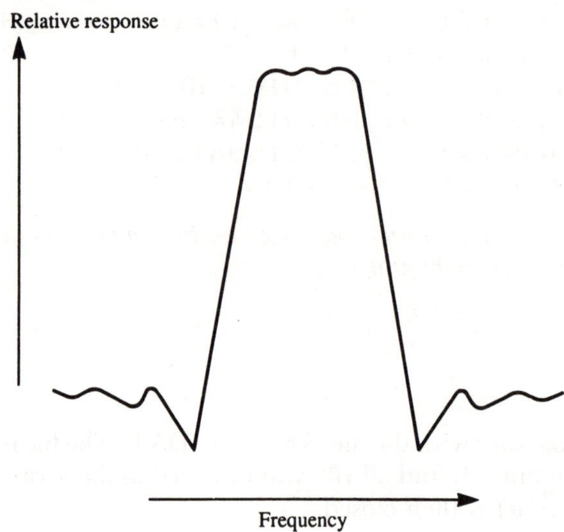

Figure 11.5 *Typical crystal filter response characteristic*

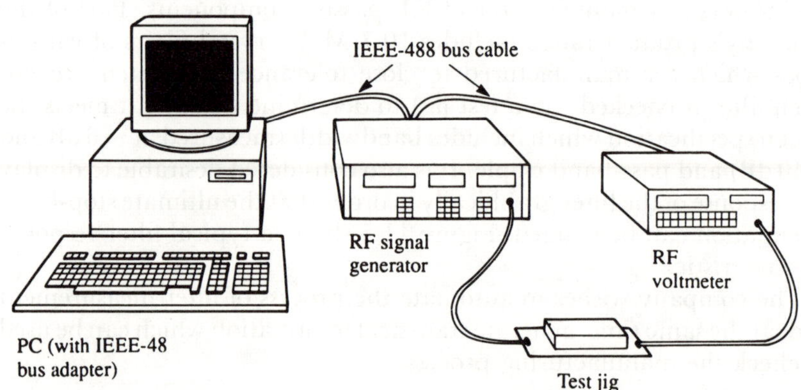

Figure 11.6 *Crystal filter test hardware configuration*

and load impedances are critical, appropriate matching components must be incorporated into the test jig. The simplified block schematic of the hardware is shown in Figure 11.6.

Software

The control software is again easily written in QuickBASIC (or equivalent) and the required program can be based on the following algorithm (expressed in a form of structured English):

```
INITIALISE-SYSTEM
DISPLAY-WELCOME-SCREEN
DO
  GET-SYSTEM-PARAMETERS
  CONFIGURE IEEE-SYSTEM
  ENTER-FILTER-REFERENCE
  DO
    READ-VOLTAGE-LEVEL
    INCREMENT GENERATOR-FREQUENCY
  LOOP UNTIL FINAL-FREQUENCY
  CALCULATE-FILTER-SPEC
  DISPLAY-FILTER-SPEC
  STORE-FILTER-SPEC
  PRINT-FILTER-SPEC
  PRINT-FILTER-LABEL
LOOP UNTIL LAST-FILTER
END
```

Most of the statements within the algorithm are coded as procedures. As an example, the procedure which prompts the user for values which will be used to set the system parameters (GET_SYSTEM_PARAMETERS) is itself described by the algorithm:

```
PROCEDURE GET-SYSTEM-PARAMETERS
  GET-INITIAL-FREQUENCY
  GET-FINAL-FREQUENCY
  GET-FREQUENCY-INCREMENT
  GET-RF-LEVEL
END PROCEDURE
```

Having decomposed each procedure, it is possible to translate each structured English statement into equivalent BASIC program statements. As an example, GET_INITIAL_FREQUENCY could be coded (in minimal form) as follows:

```
INPUT "Start frequency (kHz): "; start
```

In practice, a range check is desirable on this input since the normal range of start frequencies will lie within the range 400–450 kHz. The final code for GET_INITIAL_FREQUENCY was therefore:

```
DO
  INPUT "Start frequency (kHz): ";start
LOOP WHILE start<400 OR start>450
```

A speech annunciator

The client is a manufacturer of 'user friendly' data entry devices and requires a low-cost system capable of recording and playing back analogue speech signals. This system will then be incorporated into an existing terminal based on a PC-XT compatible motherboard and fitted with a 20-megabyte hard disk. The prototype is shown in Figure 11.7.

Plate 11.1 *Prototype speech annunciator card*

Specification

The client requires that speech of up to 30 s duration and nominal bandwidth 6 kHz be available within the system. The speech signal (input from a microphone) is to be converted to digital information and stored in one or more data files within a reserved partition on the hard disk. The speech data is then to be made available for replay (as required) by the terminal control program.

Hardware

This system requires a fast A/D and D/A interface together with additional analogue signal filtering in order to reduce the effects of aliasing. No card of this type is available 'off the shelf' and thus a board must be prototyped from scratch. A suitable board is the PC-35 (available in the UK from Amplicon Liveline, see Appendix F). This full-size PC expansion card is supplied with the necessary bus interface logic and provides access to the standard PC/PC-XT bus.

The need for A/D and D/A conversion can be realized by using a complete analogue I/O system in the form of the Analogue Devices AD7569. This unit offers 8-bit resolution (adequate for this simple speech application) coupled with a 2 μs ADC track/hold time, and on-chip bandgap 1.25 V voltage reference. The device is fabricated in

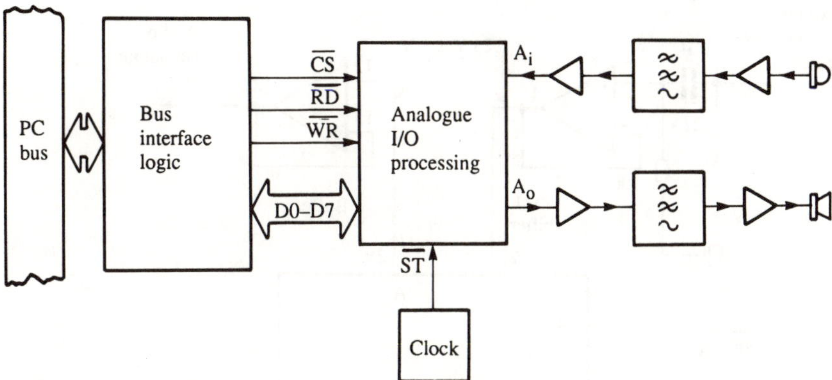

Figure 11.7 *Simplified block schematic of the prototype speech annunciator*

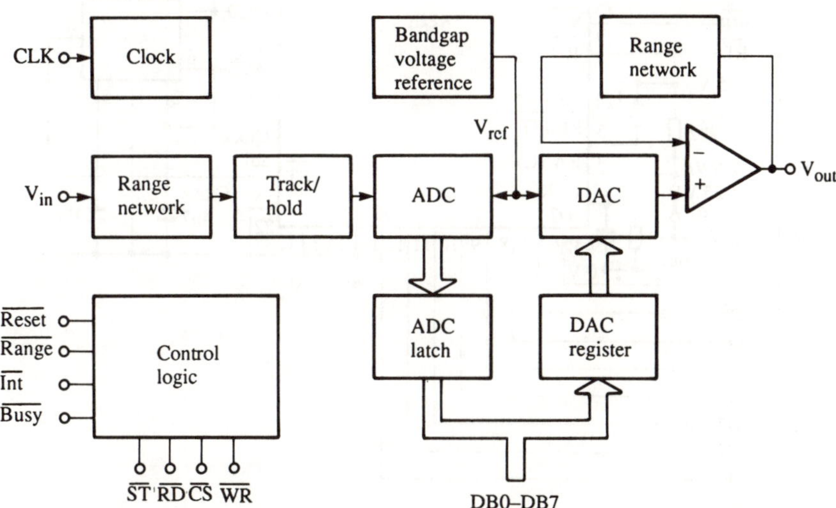

Figure 11.8 *Internal arrangement of the Analog Devices AD7569 8-bit analogue I/O system*

linear-compatible CMOS (LC²MOS) and is supplied in a 24-pin 'skinny' DIP package. The internal architecture of the AD7569 is shown in Figure 11.8 while the simplified circuit of the prototype interface card is shown in Figure 11.9.

Software

The software for the speech annunciator can usefully take advantage of the mixed language interface which is provided within the Microsoft

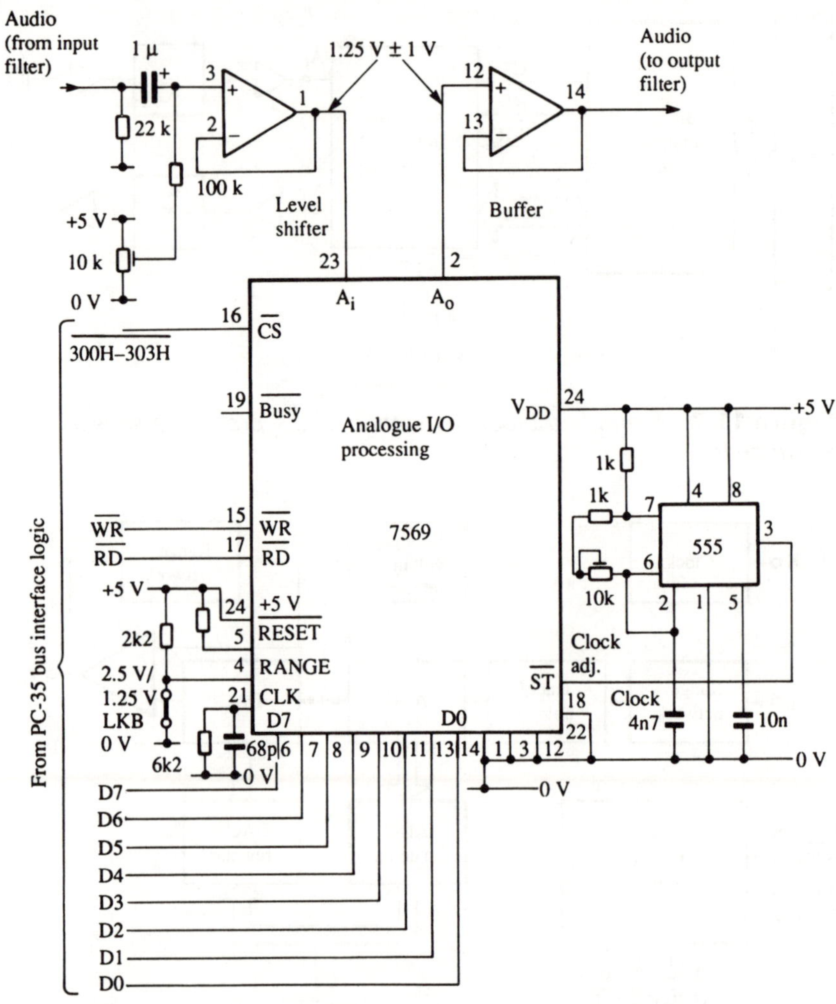

Figure 11.9 *Simplified circuit of the annunciator card*

suite of programming languages. Time critical routines (such as those which drive the ADC and DAC) can be written in assembly language while those which deal with disk filing and screen displays can be quickly and easily developed in QuickBASIC.

Figure 11.10 shows the assembly language modules which are responsible for the recoding and playback process. These routines are liberally commented and are thus reasonably self-explanatory.

```
.MODEL  MEDIUM
.STACK  100H
.CODE
        ; This routine records data from the ADC in a
        ; 128k byte buffer starting at 70000H
        ; Registers used: AX,BX,CX,DX,DI,DS
        ; Parameters passed in: 16-bit delay in stackframe
        ; Parameters returned : none

        PUBLIC Rec
Rec PROC
        PUSH     BP        ; save old base pointer
        MOV      BP,SP     ; set stack framepointer
        MOV      BX,[BP+6]       ; get argument passed
        MOV      AX,[BX] ; and preserve in BX
        MOV      BX,AX

        PUSH     SI
        PUSH     DI
        PUSH     SS
        PUSH     DS

        MOV      DX,0300H     ; analogue input port
        MOV      AX,7000H; block 0 is at 70000H
        MOV      DS,AX     ;
        MOV      DI,0      ; first location
        MOV      CX,0FFFFH   ; buffer size 64K
Rloop1: IN       AL,DX     ; get a byte
        MOV      [DI],AL ; and save it in the buffer
        INC      DI        ; point to next location
        CALL     Sdelay    ; sampling delay
        LOOP     Rloop1    ; go back for more
        MOV      AX,8000H; block 1 is at 80000H
        MOV      DS,AX     ;
        MOV      DI,0      ; first location
        MOV      CX,0FFFFH    ; buffer size 64K
Rloop2: IN       AL,DX     ; get a byte
        MOV      [DI],AL ; and save it in the buffer
        INC      DI        ; point to next location
        CALL     Sdelay    ; sampling delay
        LOOP     Rloop2    ; go back for more

        POP      DS
        POP      SS
        POP      DI
        POP      SI

        POP      BP
        RET      2         ; bye!
Rec ENDP
        ; This routine outputs data to the DAC from a
        ; 128k byte buffer starting at 70000H
        ; Registers used: AX,BX,CX,DX,DI,DS
        ; Parameters passed in: 16-bit delay in stackframe
        ; Parameters returned : none
```

Figure 11.10 *Assembly language recording and playback routines used in the annunciator software*

```
        PUBLIC Playb
Playb PROC
        PUSH    BP          ; save old base pointer
        MOV     BP,SP       ; set stack framepointer
        MOV     BX,[BP+6]          ; get argument passed
        MOV     AX,[BX] ; and preserve in BX
        MOV     BX,AX

        PUSH    SI
        PUSH    DI
        PUSH    SS
        PUSH    DS

        MOV     DX,0300H        ; analogue input port
        MOV     AX,7000H; block 0 is at 70000H
        MOV     DS,AX     ;
        MOV     DI,0      ; first location
        MOV     CX,0FFFFH    ; buffer size 64K
Ploop1: MOV     AL,[DI]      ; get a byte
        OUT     DX,AL     ; and output it
        INC     DI        ; point to next location
        CALL    Sdelay    ; sampling delay
        LOOP    Ploop1    ; go back for more
        MOV     AX,8000H; block 1 is at 80000H
        MOV     DS,AX     ;
        MOV     DI,0      ; first location
        MOV     CX,0FFFFH    ; buffer size 64K
Ploop2: MOV     AL,[DI]      ; get a byte
        OUT     DX,AL     ; and output it
        INC     DI        ; point to next location
        CALL    Sdelay    ; sampling delay
        LOOP    Ploop2    ; go back for more

        POP     DS
        POP     SS
        POP     DI
        POP     SI

        POP     BP          ; restore base pointer
        RET     2           ; bye!
Playb ENDP
; delay routine to determine sampling rate
; called by Rec and Playb
; Registers used: BX,CX
; Parameters passed:  none

Sdelay: PUSH    CX          ; save current byte count
        MOV     CX,BX       ; sets time delay
Sloop:  LOOP    Sloop       ;
        POP     CX          ; restore byte count
        RET                 ; back to the main loop
        END
```

Figure 11.10 *(cont.)*

Strain measurement and display

The client is a manufacturer of aircraft undercarriage components and wishes to carry out a series of strain measurements on structures when a stress is suddenly applied. In addition, the company wishes to display the response to an impulse force in real-time using a conventional oscilloscope-type display on the screen of a PC.

The measurement interval is to range from approximately 200 ms to 3 s and the strain gauges and associated signal conditioning circuitry are expected to produce signals in the range +250 mV. Eight sets of strain gauges are fitted to the structural member under test.

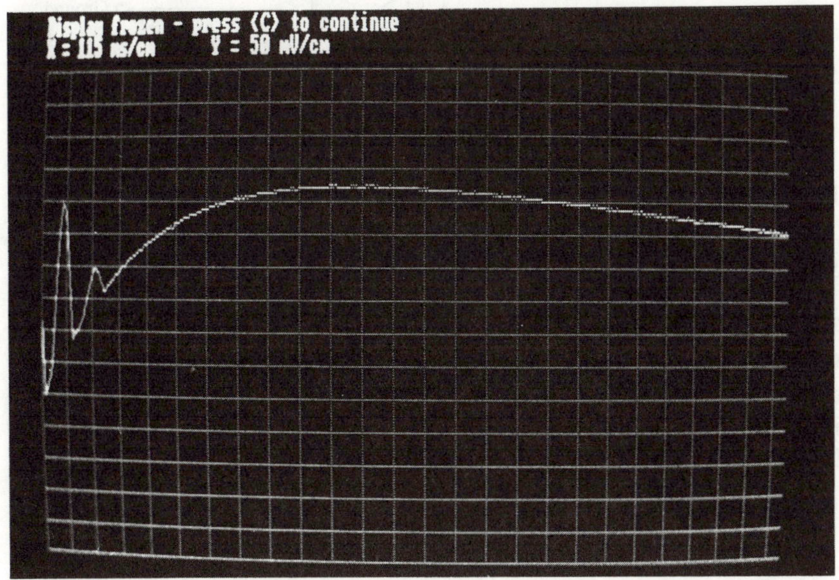

Plate 11.2 *Typical display produced by the oscilloscope program shown in Figure 11.12 (strain plotted against time)*

Hardware

The system can be based on a PC fitted with almost any 8-channel analogue input card (e.g. MetraByte DAS-8PGA, Deer Mountain PO11, etc.). The signal conditioning circuitry (replicated eight times) is based on a conventional temperature compensated half-bridge with operational amplifiers to provide voltage gain (variable from approximately 500 to 1500). To minimize noise, the input cable from each strain gauge bridge is balanced and shielded. Figure 11.11 shows the signal conditioning circuitry associated with each strain gauge bridge.

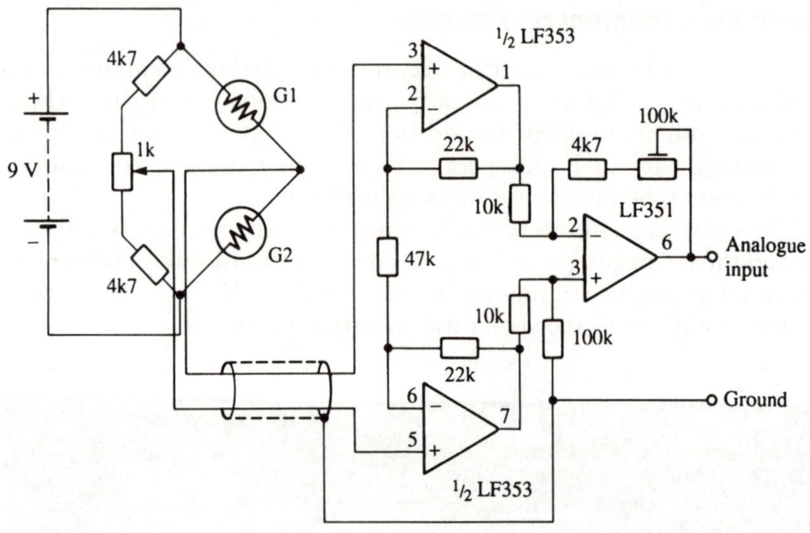

Figure 11.11 *Strain gauge input signal conditioning circuitry*

Software

The quasi-real-time oscilloscope display can easily be developed in C or QuickBASIC. An unrefined (but nevertheless functional) routine is shown in Figure 11.12. The routine displays the analogue signal returned from the strain gauge fitted to channel 0 (I/O address 300 hex.).

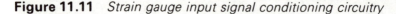

```
' Oscilloscope display
' Microsoft QuickBASIC

DECLARE SUB delay (count%)

' Set up the screen and graphics viewport
SCREEN 8
VIEW (0, 20)-(639, 199)

' Set initial timebase rate
dly% = 50

' Get initial voltage level
v% = INP(&H300)
IF v% > 127 THEN v% = v% - 256
q% = 350 + v% - 255
```

Figure 11.12 *Oscilloscope display routine*

```
DO
        CLS
        ' Plot the axes
        LINE (0, 0)-(0, 179), 5
        LINE (0, 179)-(640, 179), 5
        ' and the 1cm grid
        FOR i% = 0 TO 179 STEP 12
         LINE (0, i%)-(640, i%), 5
        NEXT i%
        FOR i% = 0 TO 640 STEP 25
         LINE (i%, 179)-(i%, 0), 5
        NEXT i%
        ' Update the status display
        LOCATE 2, 1
        PRINT "X = ";
        SELECT CASE dly%
                CASE 0
                        PRINT "30";
                CASE 25
                        PRINT "50";
                CASE 50
                        PRINT "80";
                CASE 75
                        PRINT "100";
                CASE 100
                        PRINT "115";
                CASE ELSE
                        BEEP
                END SELECT
                PRINT " ms/cm"
        LOCATE 2, 20
        PRINT "Y = 50 mV/cm"
        LOCATE 1, 1


        PRINT "Press <SPACE> to abort, <X> to freeze, ";
        PRINT "<+> or <-> to change timebase setting"
        ' Get initial voltage level and plot the starting point
        v% = INP(&H300)
        IF v% > 127 THEN v% = v% - 256
        q% = 85 + v%
        PSET (0, q%), 10
        ' Scan across the screen from left to right
         FOR x% = 0 TO 639
             v% = INP(&H300)
             IF v% > 127 THEN v% = v% - 256
             q% = 85 + v%
             LINE -(x%, q%), 10
             CALL delay(dly%)
        NEXT x%
        ' Check to see whether the user wishes to alter the scan rate
        r$ = INKEY$
        IF r$ = "+" THEN dly% = dly% - 25
        IF r$ = "-" THEN dly% = dly% + 25
        IF dly% < 0 THEN dly% = 0
        IF dly% > 100 THEN dly% = 100
        ' Check to see if the user wishes to freeze the screen
        WHILE r$ = "X" OR r$ = "x"
                ' Erase previous status line
                LOCATE 1, 1
                PRINT STRING$(80, 32)
```

Figure 11.12 (*cont.*)

```
                ' Tell the user how to resume
                LOCATE 1, 1
                PRINT "Display frozen - press <C> to continue"
                DO
                        r$ = INKEY$
                LOOP UNTIL r$ = "C" OR r$ = "c"
        WEND
LOOP UNTIL r$ = " " ' Does the user wish to quit?

SUB delay (count%)
FOR z% = 0 TO count%
NEXT z%
END SUB
```

Figure 11.12 *(cont.)*

12 Reliability and fault-finding

The principal goal of the designer of an instrumentation or process control system is that of optimizing system performance within the constraints imposed by time and a given budget. At the same time, he or she will not wish to compromise the overall quality or reliability of the system. This final chapter deals with quality and reliability in the context of PC-based instrumentation and process control systems and also sets out to examine some basic fault-finding and troubleshooting techniques which can be instrumental in reducing system down-time.

Quality procedures

In a general engineering context, quality is often defined as the degree to which a product or its components conform to the standards specified by the designer. Such standards generally relate to identifiable characteristics relating to materials, dimensions, tolerances, performance and reliability. In a production engineering environment, the degree of effectiveness in meeting these standards can be assessed by conventional acceptance tests, sampling, and statistical analysis. In the case of a one-off process control system, quality control procedures will generally involve the following tests:

- Functional tests under normal (or simulated normal) operating conditions.
- Functional tests under extreme (or simulated extreme) operating conditions.
- Overload tests to determine the behaviour of the system under abnormal or totally unexpected operating conditions.
- Environmental testing to determine the performance of the system under various extreme conditions of humidity, temperature, vibration, etc.

The instrumentation and process control specialist must inevitably undertake some or all of the functions performed by the quality engineer in a production environment. Not only will he be involved with specifying, designing, building and installing a system but he must also ensure that the overall quality of the system is assured and that the system meets the standard and criteria laid down in the initial specification. The quality assurance function requires an ongoing involvement with the project from design to subsequent installation and use.

Reliability and fault-tolerance

Reliability of a process control system is often expressed in terms of its percentage 'up-time'. Thus, a system which is operational for a total of 950 hours in a period of 1000 hours is said to have a 95% up-time. An alternative method of expressing reliability involves quoting a 'mean time before failure' (MTBF). The MTBF is equivalent to the estimated number of hours that a system is expected to operate before it encounters a failure requiring a period of 'down-time'.

Various techniques can be used to make PC-based instrumentation and process control systems inherently fault-tolerant. Such techniques can be classified under the general categories of 'hardware' and 'software'. We shall discuss these techniques separately.

Hardware techniques

Hardware methods generally involve the use of a 'watchdog methods'. These are based upon hardware devices for monitoring the performance of the system. Typical techniques include:

1 Configuring external hardware such that it generates a status byte which is periodically read (typically every 2–10 s) by the control program in order to ascertain the state of the system. If the status byte is not read within a pre-determined period, the PC controller assumes that a fault condition has been encountered and then takes appropriate action (such as generating an error message, sounding an alarm, or invoking redundant backup hardware). Watchdog techniques can be useful in overcoming a system 'hang' which may occur when the PC fails to access a malfunctioning item of peripheral hardware.

2 Monitoring a power rail and generating appropriate signals when the voltage present fails to meet the defined tolerance limits for the rail concerned. Typical actions involve closing down the system in an orderly fashion or invoking the changeover to a backup supply.

3 Fitting an uninterruptible power supply to the PC and important items of peripheral hardware.
4 Using a backup control system and, where necessary, duplicating critical I/O circuitry attached to independent signal conditioning boards.

Software techniques

Software techniques generally involve incorporating software routines, procedures, or functions which will:

1 Perform full system diagnostic tests during initialization.
2 Perform periodic diagnostic tests during program execution (e.g. periodically reading a status byte).
3 Ensure that out of range indications are recognized and erroneous data is ignored.
4 Generate error and warning messages to alert the user to the presence of a malfunction.
5 Log faults as they occur together, where possible, with sufficient information (including date and time) so that the user can determine the point at which the fault occurred and the circumstances prevailing at the time.

The resident system software invariably incorporates simple diagnostic routines of the type mentioned in (1). These routines check the major hardware components within the PC (including ROM and RAM) and are described in further detail later. Where necessary (particularly when a system is in constant operation) it may be desirable to make further checks of the system available as a menu option. The necessary routines are quite straightforward.

As an example, a ROM checksum can be produced simply by reading each byte in turn, adding the values returned (ignoring any overflow) and comparing the result with the known checksum for the ROM. Any difference will indicate a ROM error and appropriate action can be taken. In the case of the RAM, a somewhat different technique is employed. Here the process involves writing and reading each byte of RAM in turn, checking, in each case, that the desired change has been effected. Where a particular bit refuses to be changed, the diagnostic procedure is temporarily halted and an appropriate error message is generated (this may also provide sufficient information for it to be possible to locate the individual device which has failed).

It is, of course, desirable that RAM diagnostics can also be carried out on a non-destructive basis. In such cases, the byte read from RAM is replaced immediately after each byte has been tested. It is thus possible to perform a major RAM diagnostic routine without destroying data stored in read/write memory.

Table 12.1 Error codes used in the PC-AT POST

Code	*Functional element*
1xx	System board
20x	Memory board
30x	Keyboard
4xx	Display (monochrome)
5xx	Display (colour)
6xx	Disk drive
7xx	Maths co-processor
9xx	Serial/parallel adapter – parallel port
10xx	Alternative serial/parallel adapter – parallel port
11xx	Serial/parallel adapter – serial port
12xx	Alternative serial/parallel adapter – serial port
13xx	Games control adapter
14xx	Graphics printer
15xx	SDLC adapter
17xx	Fixed disk drive
20xx	BCS adapter
21xx	Alternative BSC adapter
29xx	Colour printer

Note: x = any digit.

The power on self test (POST)

The IBM Power On Self Test (POST) checks the hardware system during initialization and performs the following checks:

- System board
- Memory expansion adaptor
- Fixed disk and disk drive adapter
- Keyboard
- Disk drives
- Fixed disk drive

The error codes shown in Table 12.1 are generated by the PC-AT POST. For further information, consult the relevant IBM publication.

RAM fault-finding

Parity bits are written to memory during a memory write cycle and read from memory during a memory read cycle. A non-maskable interrupt (NMI) is generated if a parity error is detected and thus uses are notified if RAM faults develop. When a RAM fault is detected an error code is

Table 12.2 First two digits of error codes used in the PC-XT RAM diagnostic

First two digits of error code	*System board RAM location*
00	Bank 0
01	Bank 1
02	Bank 2
03	Bank 3

Table 12.3 Last two digits of error codes used in the PC-XT RAM diagnostic

Last two digits of error code	*RAM number*
00	P
01	0
02	1
04	2
08	3
10	4
20	5
40	6
80	7

Table 12.4 First two digits of PC-AT RAM diagnostic error code

First two digits of error code	*RAM location*
00–03	System board – bank 0
04–07	System board – bank 1
08–09	128K memory adaptor
10–17	First 512K memory adaptor
18–1F	Second 512K memory adaptor
20–27	Third 512K memory adaptor
28–2F	Fourth 512K memory adaptor
30–37	Fifth 512K memory adaptor

generated and displayed on the screen. The first two digits of this code indicate where in RAM the fault lies, as shown in Table 12.2 which relates to the PC-XT. The last two digits of the error code indicate the exact RAM device which has failed as shown in Table 12.3.

In the case of the PC-AT (which has more RAM capability), the first two digits of the error code indicate where in RAM the fault lies, as shown in Table 12.4. The last four digits of the error code shown in Table 12.5 indicates the exact RAM device which has failed.

Table 12.5 Last four digits of error code in PC-AT RAM diagnostic

Last four digits of error code	*RAM number*
0000	P
0001	0
0002	1
0004	2
0008	3
0010	4
0020	5
0040	6
0080	7
0100	8
0200	9
0400	10
0800	11
1000	12
2000	13
4000	14
8000	15

Note: Replace both parity RAM devices in a failing bank if a defective parity RAM is indicated.

Advanced diagnostic tests

IBM and other manufacturers provide diagnostic disks which perform more exhaustive tests. IBM recommend that the Advanced Diagnostic disk is used whenever the equipment is serviced or whenever new options are added. The Advanced Diagnostic disk is normally used in conjunction with Problem Isolation Charts (PIC) in order to locate faults such as failed RAM devices.

Problem Isolation Charts (PIC) provide a logical sequence of fault-finding steps and screen menus that require the user to perform basic checks and provide suitable responses. The combined result of using Problem Isolation Charts in conjunction with the Advanced Diagnostic disk is that of identifying Field Replaceable Units (FRU). With such aids, very little skill and minimal test equipment is required to successfully localize the majority of faults. Fault location down to component level is, however, invariably a somewhat more demanding task which requires a great deal of expertise coupled with the resources offered by a well-equipped workshop!

Finally, it is important to mention that, wherever possible, to refer to the service information provided by individual manufacturers before embarking on systematic fault finding. A selection of service manuals can be a very worthwhile investment which will be repaid many times over.

Fault-finding and troubleshooting techniques

A popular misconception concerning electronic fault finding is that good troubleshooters are borne and not made. The implication of this is that the skills of a service or test engineer cannot be acquired unless the person concerned happens to possess the equivalent (in electronic terms) of 'green fingers'. Nothing could be further from the truth – indeed it is quite possible for anyone of moderate intelligence and manual dexterity to successfully locate faults on even the most complex systems. The secret lies with adopting the correct *approach* to troubleshooting. This is the real key to successful fault finding.

With experience, the right technique will come as second nature. Indeed, a practised service engineer may not even be conscious of the technique which he or she is applying when tackling a fault. They may appear to get right to the cause of the problem without even thinking – by applying a little logic and reasoning, you can do the same.

Fault finding is a disciplined and logical process in which 'trial fixing' should never be contemplated. The generalized process of fault finding is illustrated in the flow chart of Figure 12.1. The first stage is that of identifying the defective equipment and ensuring that the equipment really is defective! This may sound rather obvious but in some cases a fault may simply be attributable to maladjustment or misconnection. Furthermore, where several items of equipment are connected together, it may not be easy to pinpoint the single item of fault equipment. For example, take the case of a process control system in which the user simply states that there is 'no output'. The fault could be almost anywhere in the system – computer, display, printer or any one of several connecting cables.

The second stage is that of gathering all relevant information. This process involves asking questions such as:

- In what circumstances did the equipment fail?
- Has the equipment operated correctly before?
- Exactly what has changed?
- Has there been a progressive deterioration in performance?

The questions used are crucial and they should explore all avenues and eventualities (particularly when the repairer has no previous experience of the equipment in question). The answers to the questions

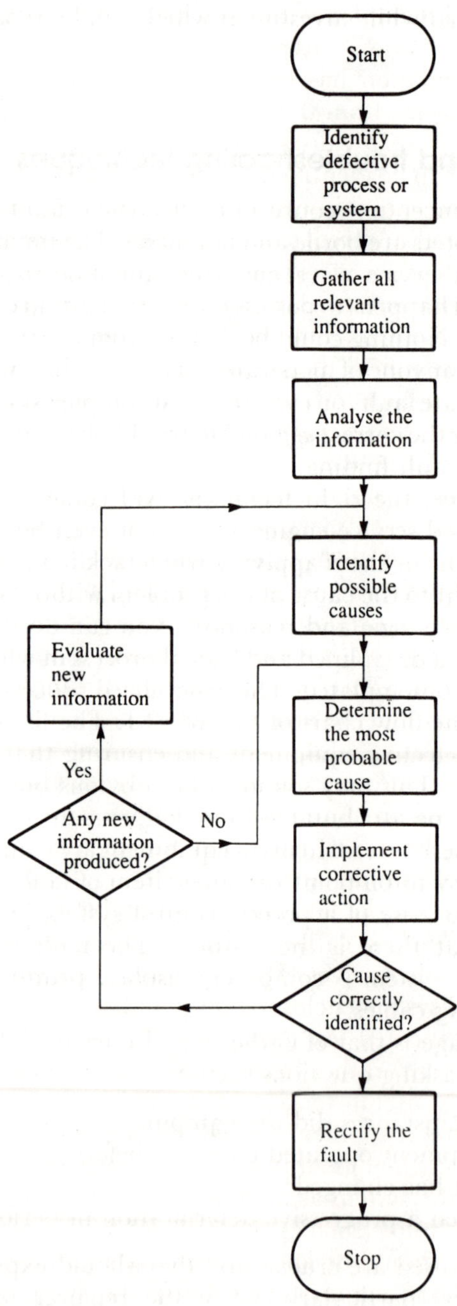

Figure 12.1 *Flow chart to illustrate the generalized approach to troubleshooting*

will help to build a conceptual model of the symptoms – before and after the fault occurred. Coupled with knowledge of the equipment (e.g. its performance specification) this model can often point to a unique cause.

Once the information has been analysed, the next stage involves separating the 'effects' from the 'cause'. Here the aim is simply that of listing each of the *possible* causes. Once this has been accomplished, the *most probable* case can be identified and focussed upon. Corrective action should be applied (to this cause alone). Such action may require component removal and replacement, adjustment or alignment, etc.

Next it is necessary to decide whether the fault has been correctly identified. A component may have failed (open circuit or short circuit) or a fuse may have blown. This will confirm that the cause has, in fact, been correctly identified. If so, the fault can be rectified and the equipment brought back into service. If not, any new information that has been generated can be evaluated before reverting to the selection of the next most probable cause. In practice, the loop may have to be executed several times until the fault is correctly identified and rectified.

Instrumentation and process control specialists will rarely wish to deal with fault-finding down to component level. In order to avoid a prohibitive investment in test equipment and technical expertise, it is generally considered more cost-effective to have such repairs carried out by specialists. Despite this, it is sometimes essential to minimize the time taken to correct the failure of a PC-based instrumentation or process control system. An ability to make on-site repairs, at least to board level, is thus highly desirable.

At first sight, the prospect of fault-finding a PC-based instrumentation or control system can be somewhat daunting. This is especially true when those having to carry out the repairs may be relatively unfamiliar with electronic circuitry. However, in the author's experience, the vast majority of faults are attributable to failure of external devices (such as sensors, cables and connectors) rather than with the board and cards themselves. Furthermore, even when dealing with boards within the system enclosure, most faults can be detected without recourse to sophisticated test gear.

When component rather than board level servicing has to be undertaken, it is useful to obtain a circuit diagram and service information on the equipment before starting work. This information will be invaluable when identifying components and establishing their function within the system as a whole.

Certain 'stock faults' (such as chip failure) may be prevalent on some boards and these should be known to manufacturers and their service agents. A telephone enquiry, describing the symptoms and clearly stating the type and version number of the card or board, will often save much time and effort. Furthermore, manufacturers are usually very receptive to information which leads to improvement of their products and may also be prepared to offer retrofit components and/or circuit modifications to overcome commonly identified problems.

Test equipment

A few items of basic test gear will be required by anyone attempting to perform fault location on bus systems. None of the basic items is particularly costly and most will already be available in an electronics laboratory or workshop. For the benefit of the newcomer to electronics we will briefly describe each item and explain how it is used in the context of PC-based system fault-finding.

Multi-range meters

Multi-range meters provide either analogue or digital indications of voltage, current and resistance. Such instruments are usually battery-powered and are thus eminently portable. Connection to the circuit under test is made via a pair of test leads fitted with probes or clips. The following specification is typical of a modern digital multi-range meter:

d.c. voltage	200 mV, 2 V, 20 V, 200 V, and 1.5 kV full-scale Accuracy ±0.5% Input resistance 10 MΩ
a.c. voltage	2 V, 20 V, 200 V, and 1 kV full-scale Accuracy ±2% Input resistance 10 MΩ
d.c. current	200 μA, 2 mA, 20 mA, 200 mA, and 2 A full-scale Accuracy ±1%
a.c. current	200 μA, 2 mA, 20 mA, 200 mA, and 2 A full-scale Accuracy ±2%
resistance	200 Ω, 2 kΩ, 20 kΩ, 200 kΩ, 2 MΩ full-scale Accuracy ±2%

A typical application for a multi-range meter is that of checking the various supply voltages present within the PC. For an operational system the supply voltages should be within the range given below:

Nominal value	*Acceptable value*		*Connector pin number*
	minimum	*maximum*	
+5 V	+4.75 V	+5.25 V	B3 and B29
−5 V	−4.75 V	−5.25 V	B5
+12 V	+11.4 V	+12.6 V	B9
−12 V	−11.4 V	−12.6 V	B7

Multi-range meters may also be used for checking the voltages present on the supply rails within individual expansion cards. Particular points of interest will be those associated with the supplies to individual chips. In such cases, PC bus extension card frames can be employed in order to gain access to a 'live' expansion card. Alternatively, the expansion card in question can be fitted to the left-most slot within a PC in order to provide access to the printed wiring of the card.

Multi-range meters may even be used to display logic states on signal lines which remain static for long periods. This is often the case when dealing with I/O lines however, in situations where logic levels are continuously changing, a multi-meter cannot provide a reliable indication of the state of a line.

Where logic levels do remain static for several seconds, the multi-range meter may be used on the d.c. voltage ranges to sense the presence of logic 0 or logic 1 states according to the following table which gives the conventional voltage levels associated with TTL logic:

Logic level	*Voltage present*
1	>2.0 V
0	<0.8 V
indeterminate	0.8–2.0 V

It should be noted that an 'indeterminate' logic level may result from a tri-state condition in which bus drivers are simultaneously in a high impedance state. Modern high-impedance instruments will usually produce a misleading fluctuating indication in such circumstances and this can sometimes be confused with an actively pulsing bus line.

Logic probes

The simplest and most convenient method of tracing logic states involves the use of a logic probe rather than a multi-range meter. This invaluable tool comprises a hand-held probe fitted with LEDs to indicate the logical state of its probe tip.

Unlike multi-range meters, logic probes can generally distinguish between lines which are actively pulsing and those which are in a permanently tri-state condition. In the case of a line which is being pulsed, the logic 0 and logic 1 indicators will both be illuminated (though not necessarily with the same brightness) whereas, in the case of a tri-state line neither indicator should be illuminated.

Logic probes generally also provide a means of displaying pulses having a very short duration which may otherwise go undetected. A pulse stretching circuit is usually incorporated within the probe circuitry so that an input pulse of very short duration is elongated sufficiently to produce a visible indication on a separate pulse LED.

Logic probes invariably derive their power supply from the circuit under test and are connected by means of a short length of twin flex fitted with insulated crocodile clips. While almost any convenient connecting point may be used, the leads of an electrolytic +5 V rail decoupling capacitor fitted to an expansion card make ideal connecting points which can be easily identified.

Logic pulsers

It is sometimes necessary to simulate the logic levels generated by a peripheral device or sensor. A permanent logic level can easily be generated by pulling a line up to +5 V via a 1 kΩ resistor or by temporarily tying a line to 0 V. However, on other occasions, it may be necessary to simulate a pulse rather than a permanent logic state and this can be achieved by means of a logic pulser.

A logic pulser provides a means of momentarily forcing a logic level transition into a circuit regardless of its current state and thus overcomes the need to disconnect or desoldering any of the devices. The polarity of the pulse (produced at the touch of a button) is adjusted so that the node under investigation is momentarily forced into the opposite logical state. During the period before the button is depressed and for the period after the pulse has been completed, the probe tip adopts a tri-state (high impedance) condition. Hence the probe does not permanently affect the logical state of the point in question.

Pulsers derive their power supply from the circuit under test in the same manner as logic probes. Here again, the leads of an electrolytic decoupling capacitor or the +5 V and GND terminals fitted to an expansion card make suitable connecting points.

Oscilloscopes

The use of an oscilloscope in the examination of time-related signals (waveforms) is well known. Such instruments provide an alternative means of tracing logic states present in a PC-based system and may also be used for detecting noise and unwanted a.c. signals which may be present on power-supply rails. It must, however, be stressed that, since low-cost oscilloscopes generally do not possess any means of storing incoming signals, severe triggering problems arise when signals are non-repetitive. This is an important point since many of the digital signals present on a bus are both asynchronous and non-repetitive.

Apart from displaying the shape of waveforms present in a bus system, oscilloscopes can also be used to make reasonably accurate measurements of voltage and time. In such cases, measurements are made with reference to a graticule fitted to the CRT and scale factors are applied using the time and voltage range switches. However, before attempting to take measurement from the graticule it is essential to check that any variable front panel controls are set to the calibrate (CAL) position. Failure to observe this simple precaution may result in readings which are at best misleading or at worst grossly inaccurate.

Since modern oscilloscopes employ d.c. coupling throughout the vertical amplifier stages, a shift along the vertical axis will occur whenever a direct voltage is present at the input. When investigating waveforms in a circuit one often encounters a.c. signals superimposed on d.c. levels; the latter may be removed by inserting a capacitor in series with the input using the 'AC–GND–DC' switch. In the 'AC' position the capacitor is inserted at the input, whereas in the 'DC' position the capacitor is shorted. If 'GND' is selected the vertical input is taken to common (0 V) and the input terminal is left floating. In order to measure the d.c. level of an input signal, the 'AC–GND–DC' switch must first be placed in the 'GND' position. The 'vertical position' is then adjusted so that the trace is coincident with the central horizontal axis. The 'AC–GND–DC' switch is then placed in the 'DC' position and the shift along the vertical axis measured in order to ascertain the d.c. level.

Most dual-beam oscilloscopes incorporate a 'chopped-alternate' switch to select the mode of beam splitting. In the 'chopped' position, the trace displays a small portion of one vertical channel waveform followed by an equally small portion of the other. The traces are thus sampled at a fast rate so that the resulting display appears to consist of two apparently continuous traces. In the 'alternate' position, a complete horizontal sweep is devoted to each channel on an alternate basis.

Chopped mode operation is appropriate to signals of relatively low frequency (i.e. those well below the chopping rate) where it is important that the display accurately shows the true phase relationship between the two displayed signals. Alternate mode operation, on the other hand, is suitable for high-frequency signals where the chopping signal would otherwise corrupt the display. In such cases it is important to note that the relative phase of the two signals will not be accurately displayed.

Most modern oscilloscopes allow the user to select one of several signals for use as the timebase trigger. These 'trigger source' options generally include an internal signal derived from the vertical deflection system, a 50 Hz signal derived from the a.c. mains supply, and a signal which may be applied to an 'external trigger input'. As an example, the 50 Hz trigger source should be selected when checking for mains-borne noise and interference whereas the external trigger input may usefully be derived from a processor clock signal when investigating the synchronous signals present within the PC expansion bus.

Fault location procedure

To simplify the process of fault location on a PC and associated expansion bus, it is useful to consider the system as a number of interlinked sub-systems. Each sub-system can be further divided into its constituent elements. Fortunately, the use of a standard expansion bus makes fault finding very straightforward since it is eminently possible to isolate a fault to a particular part of the system just by removing a suspect board and substituting one which is known to be functional.

The following eight-point checklist may prove useful; the questions should be answered *before* attempting to make any measurements or remove any suspect boards.

1 Has the system operated in similar circumstances without failure? Is the fault inherent in the system?
2 If an inherent fault is suspected, why was it not detected by normal quality procedures?
3 If the fault is not considered inherent and is attributed to component failure, in what circumstances did the equipment fail?
4 Is the fault intermittent or is it present at all times?
5 If the fault is intermittent, in what circumstances does it arise? Is it possible to predict when the fault will occur?
6 To facilitate testing and diagnosis, can conditions be reproduced so that the fault manifests itself permanently?
7 What parts of the equipment are known to be functioning correctly? Is it possible to isolate the fault be isolated to a particular part?
8 Is the fault a known 'stock fault'? Has the fault been documented elsewhere?

Having answered the foregoing questions, and assuming that one is confronted with a system which is totally unresponsive, the first step is that of checking the power-supply rails using a multi-range meter. Where any one of the supply rails is low (or missing altogether) the power supply should be disconnected from the backplane and the measurement should be repeated in order to establish whether the absence of power is due to failure of the power supply or whether the fault can be attributed to excessive loading. This, in turn, can either be due to a short-circuit component failure within an expansion card or a similar fault within the system mother board.

The system power supply employs switched mode techniques and it should be borne in mind that such units generally require that a nominal load be present on at least one of their output rails before satisfactory regulation can be achieved. Failure to observe this precaution can lead the unsuspecting test engineer to conclude that a unit is not regulating correctly when it has been disconnected from a system. In any event, it is advisable to consult the manufacturer's data before making measurements on individual supplies.

Having ascertained that the system is receiving its correct power supply voltages, the next stage is that of activating the system reset switch and noting whether any changes are produced. After each of the initial diagnostic procedures are completed an appropriate message is printed on the screen. Furthermore, once the initial procedures have been complete, any disk drive fitted to the system will normally become active as the system is 'booted'. If neither of these indications is produced, the system motherboard must be suspect as the fault will probably be attributable to failure of the CPU or one of the major VLSI support devices present.

At this stage it may be worth replacing the system motherboard with a known functional unit. If this is not possible, checks should be performed on each of the VLSI devices starting with the CPU. Table 12.6 shows typical CPU logic probe indications for a PC or PC-XT microcomputer.

The state of the expansion bus lines can often be instrumental in pointing to a faulty board. However, where a fault is intermittent (e.g. the system runs for a time before stopping) it is worth checking connectors and also investigating the cleanliness of the supply. Connectors are often prone to failure and, if the principal chips are socketed these, too, can cause problems. Intermittent faults can sometimes be corrected simply by pressing each of the larger chips into its socket. In some cases it may be necessary to carefully remove the chips before replacing them; the action of removal and replacement can sometimes be instrumental in wiping the contacts clean.

Where a fault is permanently present and one or more of the supply rail voltages is lower than normal, chip failure may be suspected. In such an event, the system should be left running for some time and the centre of each chip should be touched in turn in order to ascertain its working temperature. If a chip is running distinctly hot (i.e. very warm or too hot to comfortably touch) it should be considered a prime suspect. Where possible the temperature should be compared with that generated by a similar chip fitted in the same board or that present in another functional module. Where the larger chips have been fitted in sockets, each should be carefully removed and replaced in turn (disconnecting the power, of course, during the process) before replacing it with a known functional device.

Expansion cards are invariably fitted with links which provide selection of base addresses. These links *must* be configured so that no conflicts occur. This precaution is particularly important when new or replacement cards are fitted to a system. This is, perhaps, a rather obvious precaution but it is nevertheless one which is easily forgotten!

Finally, in a perfect world there would be no uncertainty nor any

Table 12.6 IBM PC/PC-XT CPU logic probe indications

Pin number	*Signal*	*State*
1	GND	0V
2	A14	Pulsing
3	A13	Pulsing
4	A12	Pulsing
5	A11	Pulsing
6	A10	Pulsing
7	A9	Pulsing
8	A8	Pulsing
9	A7/D7	Pulsing
10	A6/D6	Pulsing
11	A5/D5	Pulsing
12	A4/D4	Pulsing
13	A3/D3	Pulsing
14	A2/D2	Pulsing
15	A1/D1	Pulsing
16	A0/D0	Pulsing
17	NMI	Low
18	INTR	Low
19	Clock	Pulsing
20	GND	0V
21	RESET	Low
22	READY	Pulsing
23	$\overline{\text{TEST}}$	High
24	QS1	Pulsing
25	QS0	Pulsing
26	$\overline{\text{S0}}$	Pulsing
27	$\overline{\text{S1}}$	Pulsing
28	$\overline{\text{S2}}$	Pulsing
29	$\overline{\text{LOCK}}$	Pulsing
30	$\overline{\text{RQ}}/\overline{\text{GT1}}$	High
31	$\overline{\text{RQ}}/\overline{\text{GT0}}$	High
32	$\overline{\text{RD}}$	Pulsing
33	MN/$\overline{\text{MX}}$	0V
34	SSO	High
35	A19/S6	Pulsing
36	A18/S5	Pulsing
37	A17/S4	Pulsing
38	A16/S3	Pulsing
39	A15	Pulsing
40	Vcc	+5V

ambiguity about the logic levels present in a digital system. Unfortunately, this is seldom the case since spurious signals (or 'noise') are invariably present to some degree. The ability to reject noise is thus an important requirement of PC-based control systems. This is particularly true where a system is to be used in a particularly noisy environment (such as a shipyard or steelworks). In such a situation, special precautions may be necessary in order to avoid corruption of signals and data and one or more of the following techniques may be applied:

- Using a 'clean' a.c. supply for the PC controller and peripheral devices (where appropriate). If such a supply is not available, a supply filter or a.c. power conditioner should be fitted.
- Screening all signal cables (particularly those used to connected remote transducers) and returning the outer braid screen to earth (note that noise rejection is sometimes enhanced if the screen is only earthed at one point).
- Ensuring that the PC system enclosure is adequately earthed and that none of the outer panels or metal chassis parts of external card frames or enclosures are allowed to 'float'.
- Decoupling supply rails at the point at which they enter each external signal conditioning board (where appropriate).
- In extreme cases, making use of optical fibres (and appropriate interface hardware) rather than twisted pairs or co-axial cables for the asynchronous transmission of digital signals.

13 System configuration

The configuration of a system can be crucial in determining its overall level of performance in any particular application. It is therefore essential to ensure that a PC is correctly set up at the outset. DOS provides various means of configuring a system including:

- Commands entered from within the DOS command processor.
- Batch files containing DOS commands.
- Hardware device drivers.
- Configuration files; CONFIG.SYS and AUTOEXEC.BAT.

This chapter will provide you with an understanding of various techniques that can be used to optimize a system. It shows you how to install and configure device drivers, how to set up a disk cache or a RAM drive, and how to make use of extended and/or expanded memory. The chapter also deals with the configuration of displays and printers.

Device drivers

DOS provides a number of device drivers and utility programs which can be installed from CONFIG.SYS. The drivers in the table opposite will allow you to make the most of your own particular hardware configuration and to modify it to cater for hardware upgrades.

Disk caching

A disk cache provides improved file access times and helps to reduce the number of physical disk accesses made by a program which makes regular use of disk files. Data is initially read from the disk into the cache.

Function	*Device driver names*
Disk caches	IBMCACHE.SYS, SMARTDRV.SYS
RAM drives	RAMDRIVE.SYS, VDISK.SYS
Non-standard or additional disk drives	DRIVER.SYS
Memory management	XMAEM.SYS, EMM386.SYS, EMM386.EXE
Display adapter configuration	DISPLAY.SYS
Printer configuration	PRINTER.SYS
Multimedia extension	MSCDEX.SYS

Subsequent file accesses make use of the cache rather than the disk itself. At some later time, data is written back to the disk. Generally, a cache will hold more information than is likely to be requested at any one time by the program. Redundant disk accesses are eliminated as data in the cache can be manipulated directly without having to access data on the disk.

A disk cache remembers what sections of the disk have been used most frequently. When the cache must be recycled, the program keeps the more frequently used areas and discards those that are used less frequently. Some experimentation will normally be required in order to obtain the optimum size of a disk cache. This also varies according to the requirements of a particular application program.

Because of its 'intelligent' features, a disk cache will normally outperform a disk buffer (created using the BUFFERS directive). Furthermore, a disk cache offers some advantages over a RAM drive, since its operation is 'transparent' and you are less likely to lose all your data due to power failure. If your applications program makes frequent use of disk files, it is well worth setting up a cache.

IBMCACHE.SYS is a file that accompanies any IBM PS/2 machine which uses extended memory (this excludes Models 25 and 30). The program is installed by inserting from the Reference Disk. The cache can be changed after installation either by re-running IBMCACHE or by editing the reference to IBMCACHE in CONFIG.SYS.

Example:

```
DEVICE = C:\IBMDOS\IBMCACHE.SYS 32 /E /P4
```

configures the cache size to 32 kilobyte using extended memory, and allows four sectors to be read at a time. IBMCACHE.SYS is resident in the IBMDOS directory.

MS-DOS and DR-DOS offer an alternative disk-caching utility

called SMARTDRV.SYS installed by means of an appropriate directive within CONFIG.SYS.

Example:

```
DEVICE = C:\WINDOWS\SMARTDRV.SYS 512 /A
```

configures the cache size to 512 kilobyte using expanded memory with SMARTDRV.SYS resident in the WINDOWS directory. (Note that 256 kilobyte is the default cache size for SMARTDRV.)

RAM drives

A RAM drive is a virtual disk drive that is created in RAM. The size of the virtual disk drive is limited by the amount of available memory. For a system with a basic system memory (640 kilobyte) there will be obvious limitations on the size of RAM that can be created. Despite this, a RAM drive can be useful in a number of occasions and particularly where you may wish to avoid a large number of repeated disk accesses involving data files of relatively small size.

A RAM drive can be established in conventional memory, extended memory, or expanded memory, as appropriate. The simulated drive behaves exactly like a conventional drive but with vastly improved access time.

RAMDRIVE.SYS (or VDISK.SYS) is used to simulate a disk drive in RAM. RAMDRIVE.SYS is installed using the DEVICE command from within CONFIG.SYS. The size of the RAM drive can be specified within the command as can the number of bytes per sector. Switches can be used to specify the use of extended or expanded memory (if no switch is specified the drive is established within base memory).

Examples:

- `DEVICE = C:\DOS\VDISK.SYS`

  establishes a RAM drive of 64 kilobyte (the default size) with 128 byte per sector (the default sector size) and 64 directory entries (the default number of directory entries) within *conventional base memory*.

- `DEVICE = C:\DOS\RAMDRIVE.SYS 128 512 96 /X`

  establishes a RAM drive of 128 kilobyte with 512 byte per sector and 96 directory entries within *expanded memory*.

- `DEVICE = C:\DOS\VDISK.SYS 256 256 128 /E`

  establishes a RAM drive of 256 kilobyte with 256 byte per sector and 128 directory entries within *extended memory*.

- `DEVICE = C:\DRIVERS\RAMDRIVE.SYS 512 256 96 /A`

  creates a 512 kilobyte RAM disk with 256-byte sectors and 96 directory entries in *expanded memory*.

- `DEVICE = C:\DRIVERS\RAMDRIVE.SYS 1024 512 128 /E`

  creates a 1 megabyte RAM disk with 512-byte sectors and 128 directory entries in *extended memory*.

Where RAM space is at a premium it can be beneficial to make use of a small sector size (e.g. 128 byte). This works because DOS does not allow files to share the same sector – a 513-byte file would occupy *two* sectors with 511 byte wasted. If, however, memory is not limited, it is better to use a relatively large sector size in order to reduce access time.

Before we move on, it is vitally important to remember that the data stored in a RAM drive will be lost when the power is removed from the system. It is thus absolutely *vital* to back-up your data before switching off!

Using a RAM drive to copy disks

A RAM drive can provide a quick method of copying on a system which has only one disk drive. You simply use VDISK or RAMDRIVE to establish a RAM drive with identical features to your existing (single) drive and then perform a DISKCOPY. The parameters for standard drives are:

Drive type	*RAMDRIVE or VDISK parameters*
5.25 in., 360 kilobyte	360 512 112 /E or /A, as required
5.25 in., 1.2 megabyte	1200 512 224 /E or /A, as required
3.5 in., 720 kilobyte	720 512 112 /E or /A, as required
3.5 in., 1.44 megabyte	1440 512 224 /E or /A, as required
3.5 in., 2.88 megabyte	2880 512 240 /E or /A, as required

Adding or using non-standard disk drives

Most PC disk drives conform to one of the five specifications listed above. On some occasions, however, it may be necessary to configure a system for a non-standard disk drive. DRIVER.SYS can be used to configure a system for use with a non-standard internal or external disk drive. The driver will allow you to specify the physical drive number, the number of sectors per track, and the number of tracks.

Examples:

- `DEVICE = C:\DOS\DRIVER.SYS /D:2 /F:0 /S:9 /T:40`

  designates a third floppy disk drive in a system which already has two disk drives (the parameter following the /D switch is the 'physical drive number') having a capacity of 360 kilobyte, 9 sectors per track, and 40 tracks.

- ```
  DEVICE = C:\DOS\DRIVER.SYS /D:2 /F:0 /S:9 /T:40
  DEVICE = C:\DOS\DRIVER.SYS /D:2 /F:0 /S:9 /T:40
  ```

  specifies two further disk drives (D: and E:), each of 360 kilobyte and 9 sectors per track, with 40 tracks. Each time the driver is loaded, the physical disk drive is assigned an additional valid drive letter automatically (D: the first time, E: the second time, and so on).

- `DEVICE = C:\DOS\DRIVER.SYS /F:1 /T:80 /S:15 /H:2 /C`

  specifies a conventional 1.2 megabyte, 5.25 inch floppy disk drive (incorporating a disk change line).

- `DEVICE = C:\DOS\DRIVER.SYS /F:2 /T:80 /S:9 /H:2 /C`

  specifies a conventional 720 kilobyte, 3.5 inch floppy disk drive (incorporating a disk change line).

- `DEVICE = C:\DOS\DRIVER.SYS /F:7 /T:80 /S:18 /H:2 /C`

  specifies a conventional 1.44 megabyte, 3.5 inch floppy disk drive (incorporating a disk change line).

Memory managers

The EMM386 memory manager requires a 386 (or later) CPU and provides a means of accessing the unused parts of the upper memory area. The memory manager also allows you to use your system's extended memory to simulate expanded memory. Note that EMM386 is provided as EMM386.SYS with MS-DOS 4.0 and DR-DOS 6.0 but as EMM386.EXE with MS-DOS 5.0 (and later).
  ```

Examples:

- `DEVICE = C:\DOS\EMM386.SYS`

  installs the memory manager and takes 256 kilobyte (the default) of extended memory to provide expanded memory (MS-DOS 4.0).

- `DEVICE = C:\DOS\EMM386.SYS 1024`

  installs the memory manager and takes 1 megabyte of extended memory to provide expanded memory (MS-DOS 4.0).

- `DEVICE = C:\DOS\EMM386.EXE 1024 RAM`

  provides access to the upper memory area and also uses 1024 kilobyte of the computer's extended memory as expanded memory (MS-DOS 5.0 and later).

- `DEVICE = C:\DOS\EMM386.EXE NOEMS`

  provides access to all available portions of the upper memory area but without functioning as an expanded memory emulator (MS-DOS 5.0 and later).

- `DEVICE = C:\DRDOS\EMM386.SYS /F=NONE /B=AUTO /E=E800-FFFF`

  autoscans upper memory from C000 to FFFF (the default) but specifically excludes the area from E800 to FFFF. No LIM page frame is set up and the DOS kernel is loaded into upper memory or, if there is not enough upper memory, it is loaded into high memory (DR-DOS 6.0).

- `DEVICE = C:\DRDOS\EMM386.SYS /FRAME=C400 /KB=2048 /BDOS=FFFF`

  autoscans upper memory from C000 to FFFF (the default) and sets up a LIM window with 2 megabyte available. BDOS is located to segment address FFFF in high memory (DR-DOS 6.0).

Configuring the printer and display

DISPLAY.SYS enables you to switch code pages without restarting DOS, and PRINTER.SYS enables you to download a font table to supported printers so that they can print non-English language and graphic characters.

The DISPLAY.SYS utility caters for different display types and for specified code pages. You do not have to include DISPLAY.SYS if you do not use code-page switching. Valid display types are:

Display type	*Meaning*
MONO	Monochrome adapter
CGA	Color-graphics adapter
EGA	Enhanced color-graphics adapter or PS/2 or VGA
LCD	Convertible LCD

Valid code pages include:

Code page	*Country*
437	USA
850	Multilingual
860	Portugal
863	Canadian–French
865	Norway

The number specified within the DISPLAY.SYS command is the maximum number of *additional code pages* that the adapter can use.

Examples:

- `DEVICE = C:\DOS\DISPLAY.SYS CON: = (EGA,850,2)`

  specifies an EGA (or VGA) display using the multilingual code page 850 with two code pages.

- `DEVICE = C:\SYS\DISPLAY.SYS CON = (EGA,437,1)`

  specifies an EGA (or VGA) display using 437 as the starting code page together with one additional code page.

- `DEVICE = C:\SYS\DISPLAY.SYS CON = (EGA,850,863,2)`

  selects an EGA (or VGA) display and a code page of 863 (French-Canadian) together with the multilingual code page, 850.

Finally, if your existing code page is not 437 (USA or UK), you should use 850 as the new code page and (2) as the number of additional code pages.

Making changes to CONFIG.SYS

It is very important to note that there is an optimum sequence in which device drivers should appear within your CONFIG.SYS. The recommended sequence is as follows:

1 HIMEM.SYS.
2 An expanded memory manager (if your system is fitted with expanded memory).
3 Any device drivers that use extended memory.
4 EMM386.EXE (do not use EMM386 if you are using an expanded memory manager).
5 Any device drivers that use expanded memory.
6 Any device drivers that use the upper memory area.

It is also worth noting that, if you use ANSI.SYS and DISPLAY.SYS together in your CONFIG.SYS file, the ANSI.SYS command must be used *before* the DISPLAY.SYS directive. ANSI.SYS will not take affect when the order is reversed.

Using PRINTER.SYS

You can use PRINTER.SYS to print the required code page characters on supported printers (or on printers that provide emulation for the supported types). Note that MS-DOS 3.3 supports code pages on only two printers:

- IBM ProPrinter Model 4201, and
- IBM Quietwriter III Model 5202.

MS-DOS 4.0 and MS-DOS 5.0 provide additional support for the following printers:

- IBM ProPrinter Model 4202,
- IBM ProPrinter Model 4207, and
- IBM ProPrinter Model 4208.

Valid code pages are:

Code page	*Country*
437	USA
850	Multilingual
860	Portugal
863	Canadian–French
865	Norway

Examples:

- `DEVICE = C:\DOS\PRINTER.SYS LPT1=(4201,850,2)`

  specifies the IBM ProPrinter Model 4201 connected to the parallel printer port and using the multilingual code page as the built-in character set. You can specify two code pages for the printer and then switch between them.

Using AUTOEXEC.BAT

The AUTOEXEC.BAT file allows you automatically to execute a series of programs and DOS utilities to add further functionality to your system when the system is initialized. AUTOEXEC.BAT normally contains a sequence of DOS commands, but in addition it can also contain the name of an application or shell that will be launched automatically when the system is booted. This is a useful facility if you alway use the same shell or application whenever you power-up your system or if you wish to protect the end-user from the need to remember rudimentary DOS commands (such as CD, etc.).

AUTOEXEC.BAT is typically used to:

1 Set up the system prompt (see page 120).
2 Define the path for directory searches (using SET PATH, see page 120).
3 Execute certain DOS utilities (e.g. SHARE).
4 Load a mouse driver (e.g. MOUSE.COM), but note that Microsoft's Windows and some DOS programs (e.g. AutoCAD) have their own built-in mouse drivers and can thus communicate directly with the mouse.
5 Change directories (e.g. from the root directory to a 'working' directory).
6 Launch an application or shell program (e.g. PCSHELL and Windows).

The Example CONFIG.SYS and AUTOEXEC.BAT files which follow should give you plenty of food for thought but note that you should always keep back-up copies of your original files (saved as CONFIG.BAK, CONFIG.OLD, etc.) before attempting to make use of your modified configuration files.

Examples of CONFIG.SYS and AUTOEXEC.BAT files

Two-megabyte notebook for Windows applications, MS-DOS 5.0

File	*Content*	*Notes*
CONFIG.SYS		
	DEVICE=C:\WINDOWS\HIMEM.SYS	Load the Windows high memory device driver.
	DOS=HIGH	Load DOS into high memory.
	FILES=40	Allow for up to 40 files.
	BUFFERS=10	Provide 10 file buffers.
	STACKS=9,256	Create 9 stacks each of 256 byte.
AUTOEXEC.BAT		
	PROMPT PG	Prompt with current directory path.
	SET PATH=C:\DOS;C:\WINDOWS	Search DOS and Windows directories for executable files.
	SET TEMP=C:\TEMP	Directory for Windows swap files.
	C:\DOS\SHARE.EXE	Permits file sharing and locking.
	C:\WINDOWS\SMARTDRV.EXE	Establish a disk cache.
	CD WINDOWS	Change to the Windows directory . . .
	WIN	and run Windows!

Comment: Windows performance will be disappointing with limited memory; however, the disk cache will improve disk access time.

Four-megabyte general purpose laboratory PC for Windows applications, MS-DOS 6.2

File	*Content*	*Notes*
CONFIG.SYS		
	DEVICE=C:\WINDOWS\HIMEM.SYS	Load the Windows high memory device driver.
	DOS=HIGH	Load DOS in high memory.
	FILES=64	Allow for up to 64 files.
	BUFFERS=16	Provide 16 file buffers.
	STACKS=9,256	Set up 9 stacks each of 256 bytes.
	SHELL=C:\DOS\COMMAND.COM C:\DOS /E:1024 /P	Specify the command interpreter and a 1 kilobyte environment.
AUTOEXEC.BAT		
	PROMPT PG	Prompt with current directory path.

SET PATH=C:\DOS;C:\WINDOWS	Search DOS and Windows directories for executable files.
SET TEMP=C:\TEMP	Directory for Windows swap files.
LOADHIGH=C:\DOS\SHARE.EXE	Permits file sharing and locking.
C:\WINDOWS\SMARTDRV.EXE	Establishes a disk cache.
SET COMSPEC=C:\DOS\COMMAND.COM	Specify the location of the command interpreter.
CD WINDOWS	Change to the Windows directory . . .
WIN	and run Windows!

Comment: with more memory space available it is possible to allocate more buffers and provide for more files.

Four-megabyte general purpose PC for DOS applications, MS-DOS 6.2

File	*Content*	*Notes*
CONFIG.SYS		
	DEVICE=C:\DOS\HIMEM.SYS	Load the DOS high memory device driver.
	DEVICE=C:\DOS\EMM386.EXE RAM NOEMS	Load the memory manager.
	DOS=HIGH,UMB	Load DOS into a UMB.
	SHELL=C:\DOS\COMMAND.COM C:\DOS /E:1024 /P	Specify the command interpreter and a 1 kilobyte environment.
	FILES=64	Allow for up to 64 files.
	BUFFERS=16	Provide 16 file buffers.
	STACKS=9,256	Set up 9 stacks each of 256 byte.
AUTOEXEC.BAT		
	PROMPT PG	Prompt with current directory path.
	SET PATH=C:\DOS;C:\TOOLS; C:\UTILITY	Search DOS, TOOLS and UTILITY directories for executable file.
	LOADHIGH=C:\DOS\SHARE.EXE	Permits file sharing and locking.
	C:\DOS\SMARTRDV.EXE	Establishes a disk cache.
	LOADHIGH C:\DOS\KEYB UK,,C:\DOS\KEYBOARD.SYS	Specify a UK keyboard layout.
	LOADHIGH C:\DOS\DOSKEY	Remember last used commands.
	SET COMSPEC=C:\DOS\COMMAND.COM	Specify the location of the command interpreter

Comment: this system is not fitted with a mouse therefore no mouse driver is specified. Within CONFIG.SYS, the command that loads DOS into high memory (DOS = HIGH,UMB) *must* follow the command that loads the high memory device driver, HIMEM.SYS.

Four-megabyte PC for network use, MS-DOS 6.2

File	Content	Notes
CONFIG.SYS		
	DEVICE=C:\WINDOWS\HIMEM.SYS	Load the Windows high memory device driver.
	DEVICE=C:\WINDOWS\EMM386.EXE RAM NOEMS	Load the memory manager.
	DOS=HIGH,UMB	Loads DOS in high memory.
	DEVICEHIGH=C:\WINDOWS\MOUSE.SYS /Y	Load the mouse driver into high memory.
	SHELL=C:\DOS\COMMAND.COM C:\DOS /E:1024 /P	Specify the command interpreter and a 1 kilobyte environment.
	FILES=80	Allow for up to 80 files.
	BUFFERS=16	Provide 16 file buffers.
	STACKS=9,256	Set up 9 stacks each of 256 byte.
AUTOEXEC.BAT		
	PROMPT PG	Prompt with directory path.
	SET PATH=C:\DOS;C:\WINDOWS	Search DOS and Windows directories for executable files.
	SET TEMP=C:\TEMP	Directory for Windows swap files.
	LOADHIGH=C:\DOS\SHARE.EXE	Permits file sharing and locking.
	C:\WINDOWS\SMARTRDV.EXE	Establishes a disk cache.
	LOADHIGH C:\DOS\KEYB UK,,C:\DOS\KEYBOARD.SYS	Specify a UK keyboard layout.
	LOADHIGH C:\DOS\DOSKEY	Remember last used commands.
	LOADHIGH C:\NETWARE\IPX.COM	Load Netware drivers into high memory (if available).
	LOADHIGH C:\NETWARE\NETX.COM	
	SET COMSPEC=C:\DOS\COMMAND.COM	Specify the location of the command interpreter

Comment: note the Netware drivers IPX.COM and NETX.COM. Both are loaded into high memory.

Four-megabyte PC dedicated to a process application running under Windows 3.1

File	*Content*	*Notes*
CONFIG.SYS		
	DEVICE=C:\WINDOWS\ HIMEM.SYS	Load the Windows high memory device driver.
	DEVICE=C:\WINDOWS\ EMM386.EXE 1024 RAM	Load the memory manager.
	DOS = HIGH,UMB	Load DOS in high memory.
	DEVICEHIGH=C:\WINDOWS\ MOUSE.SYS /Y	Load the mouse driver into high memory.
	FILES=32	Allow for 32 files.
	BUFFERS=16	Provide 16 file buffers.
	STACKS=9,256	Set up 9 stacks each of 256 byte
AUTOEXEC.BAT		
	PROMPT PG	Prompt with directory path.
	SET PATH=C:\DOS;C:\WINDOWS;C:\APPS	Search DOS, WINDOWS and APPS directories.
	WIN PROCESS	Load Windows and Start the application program.

Comment: this configuration loads Windows automatically and then runs the PROCESS application.

Two-megabyte laptop for DOS, DR-DOS 6.0

File	*Content*	*Notes*
CONFIG.SYS		
	DEVICE=C:\DRDOS\ EMM386.SYS	Load the memory manager.
	HIDEVICE=C:\DRDOS\ ANSI.SYS	Load the ANSI screen driver.
	FILES=48	Allow for 48 files.
	BUFFERS=25	Provide 25 file buffers.
	FASTOPEN=512	Reduces file access time.
	HISTORY=ON,256,OFF	Remember last used commands.
	COUNTRY=001,,C:\ DRDOS\COUNTRY.SYS	Specify a US keyboard.
AUTOEXEC.BAT		
	PROMPT PG	Prompt with directory path.
	SET PATH=C:\DOS;C:\ UTILITY;C:\APPS	Search DOS, Utility and Apps directories for executable files.

Comment: some of the DOS programs and utilities used on this system make use of the ANSI.SYS screen driver.

One-megabyte 286-based notebook controller, DR-DOS 6.0

File	*Content*	*Notes*
CONFIG.SYS		
	DEVICE=C:\DRDOS\HIDOS. SYS	Use high memory for DOS facilities.
	BREAK=ON	Enable break key checking.
	HIBUFFERS=20	Provide 20 buffers.
	FILES=60	Allow for 60 files.
	FASTOPEN=512	Reduce file access time.
	LASTDRIVE=F	Allow for drives A to F.
	HISTORY=ON,256,OFF	Remember last used commands.
	COUNTRY=044,,C:\DRDOS\ COUNTRY.SYS	Specify a UK keyboard.
	SHELL=C:\COMMAND.COM C:\ /P /E:512	Specify the command interpreter.
	HIDOS=ON	Locate DOS structures in high memory.
	HIDEVICE=C:\DRDOS\ ANSI.SYS	Load ANSI.SYS in high memory.
	HINSTALL=C:\DRDOS\ KEYB.COM UK+	Load UK keyboard driver into high memory.
	HINSTALL=C:\DRDOS\ CURSOR.EXE	Load large cursor driver
AUTOEXEC.BAT		
	@ECHO OFF	Display message
	ECHO =======	
	ECHO	Supplied by:
	ECHO	Howard Associates
	ECHO	Weybridge, England
	ECHO =======	
	PATH C:\DRDOS; C:\ MOUSE; C:\UTILITY	Search DOS, MOUSE and UTILITY directory.
	PROMPT PG	Prompt with directory path.
	MOUSE	Load mouse driver.
	MENU	Load MENU shell.

Comment: this system is used with a variety of utilities and batch files selected from a MENU shell. Break key checking has been enabled so that the user can escape if things go wrong. The system uses a hard disk with partitions A to F. The large cursor driver is useful on a machine with an LCD display.

Appendix A Glossary of terms

Access time The time taken to retrieve data from a memory/storage device, i.e. the elapsed time between the receipt of a read signal at the device and the placement of valid data on the bus. Typical access times for semiconductor memory devices are in the region 100–200 ns while average access times for magnetic disks are typically in the range 10–50 ms.

Accumulator A register within the central processing unit (CPU) in which the result of an operation is placed.

Acknowledge (ACK) A signal used in serial data communications which indicates that data has been received without error.

Active high A term used to describe a signal which is asserted in the high (logic 1) state.

Active low A term used to describe a signal which is asserted in the low (logic 0) state.

Address A reference to the location of data in memory or within I/O space. The CPU places addresses (in binary coded form) on the address bus.

Address bus The set of lines used to convey address information. The IBM-PC bus has twenty address lines (A0 to A19) and these are capable of addressing more than a million address locations. One byte of data may be stored at each address.

Address decoder A hardware device (often a single integrated circuit) which provides chip select or chip enable signals from address patterns which appear on an address bus.

Address selection The process of selecting a specific address (or range of addresses). In order to prevent conflicts, expansion cards must usually be configured (by means of DIP switches or links) to unique addresses within the I/O address map.

American Standard Code for Information Interchange (ASCII) A code which is almost universally employed for exchanging data between microcomputers. Standard ASCII is based on a seven-bit binary code and caters for alphanumeric characters (both upper and lower case), punctuation, and special control characters. Extended ASCII employs an eighth bit to provide an additional 128 characters (often used to represent graphic symbols).

Amplifier A circuit or device which increases the power of an electrical signal.

Analogue The representation of information in the form of a continuously variable quantity (e.g. voltage).

Archive A device or medium used for storage of data which need not be instantly accessible (e.g. a tape cartridge).

Assembly language A low-level programming language which is based on mnemonic instructions. Assembly language is often unique to a particular microprocessor or microprocessor family.

Asserted A term used to describe a signal when it is in its logically true state (i.e. logic 1 in the case of an active high signal or logic 0 in the case of an active low signal).

Asynchronous transmission A data transmission method in which the time between transmitted characters is arbitrary. Transmission is controlled by start and stop bits (no additional synchronizing or timing information is required).

Backplane A printed circuit board assembly on which connectors are mounted at regular intervals. Each connector is capable of accepting a printed circuit card which provides a particular function (e.g. memory, I/O, etc.).

Backup A file or disk copy made in order to avoid the accidental loss, damage, or erasure of programs and/or data.

Basic input output system (BIOS) The BIOS is the part of the operating system which handles communications between the microcomputer and peripheral devices (such as keyboard, serial port, etc.). The BIOS is supplied as firmware and is contained in a read-only memory (ROM).

Batch file A file containing a series of DOS commands which are executed when the file name is entered after the DOS prompt. Batch files are given a BAT file extension. A special type of batch file (AUTOEXEC.BAT) is executed (when present) whenever a system is initialized.

Bit A contraction of 'binary digit'; a single digit in a binary number.

Boot The name given to the process of loading and initializing an operating system (part of the operating system is held on disk and must be loaded from disk into RAM on power-up).

Boot record A single-sector record present on a disk which conveys information about the disk and instructs the computer to load the requisite operating system files into RAM (thus booting the machine).

Buffer In a *hardware* context, a buffer is a device which provides a degree of electrical isolation at an interface. The input to a buffer usually exhibits a much higher impedance than its output (*see also* **driver**). In a *software* context, a buffer is a reserved area of memory which provides temporary data storage and thus may be used to compensate for a difference in the rate of data flow or time of occurrence of events.

Bus An electrical highway for signals which have some common function. Most microprocessor systems have three distinct buses; an address bus, data bus and control bus. Some bus systems are associated with particular backplane configurations (e.g. STE and VME).

Byte A group of eight bits which are operated on as a unit.

Central processing unit (CPU) The part of a computer that decodes instructions and controls the other hardware elements of the system. The CPU comprises a control unit, arithmetic/logic unit and internal storage. In microcomputers, a microprocessor acts as the CPU (*see also* **Microprocessor**).

Channel A path along which signals or data can be sent.

Chip The term commonly used to describe an integrated circuit.

Clock A source of timing signals used for synchronizing data transfers within a microprocessor or microcomputer system.

Cluster A unit of space allocated on the surface of a disk. The number of sectors which make up a cluster varies according to the DOS version and disk type (*see also* **Sector**).

Command An instruction (entered from the keyboard or contained within a batch file) which will be recognized and executed by a system (*see also* **Batch file**).

Common A return path for a signal (often ground).

Controller A sub-system within a microcomputer which controls the flow of data between the system and an I/O or storage device (e.g. a CRT controller, hard disk controller, etc.). A controller will generally be based on one, or more, programmable VLSI devices.

Co-processor A second processor which shares the same instruction stream as the main processor. The co-processor handles specific tasks (e.g. mathematics) which would otherwise be performed less efficiently (or not at all) by the main processor.

Daisy chain A method of connection in which signals move in a chained fashion from one device to another. This form of connection is commonly used with disk drives.

Data A general term used to describe numbers, letters and symbols present with a computer system.

Data bus A highway (in the form of multiple electrical conductors) which conveys data between the different elements within a microprocessor system.

Device driver A term used to describe memory resident software (specified in the CONFIG.SYS system file) which provides a means of interfacing specialized hardware (e.g. expanded memory adapters).

Direct memory access A method of fast data transfer in which data moves between a peripheral device (e.g. a hard disk) and main memory without direct control of the CPU.

Directory A catalogue of disk files (containing such information as filename, size, attributes, and data/time of creation). The directory is stored on the disk and updated whenever a file is amended, created, or deleted. A directory entry usually comprises 32 bytes for each file.

Disk operating system (DOS) A group of programs which provides a low-level interface with the system hardware (particularly disk I/O). Routines contained within system resident portions of the operating system may be used by the programmer. Other programs provided as part of the system include those used for formatting disks, copying files, etc.

Driver In a software context, a driver is a software routine which provides a means of interfacing a specialised hardware device (*see also* **Device driver**). In a hardware context, a driver is an electrical circuit which provides an electrical interface between an output port and an output transducer. A driver invariably provides power gain (i.e. current gain and/or voltage gain); *see also* **Amplifier**.

File Information (which may comprise ASCII encoded text, binary coded data and executable programs) stored on a floppy or hard disk. Files may be redirected from one logical device to another using appropriate DOS commands.

Filter In a software context, a filter is a software routine which removes or modifies certain data items (or data items within a defined range). In a hardware context, a filter is an electrical circuit which modifies the frequency distribution of a signal. Filters are often categorized as *low-pass*, *high-pass*, *band-pass*, or *band-stop* depending upon the shape of their frequency response characteristic.

Firmware A program (software) stored in read-only memory (ROM). Firmware provides non-volatile storage of programs.

Fixed disk A disk which cannot be removed from its housing. Note that, while the terms 'hard' and 'fixed' are often used interchangeably, some forms of hard disk are exchangeable.

Format The process in which a magnetic disk is initialized so that it can accept data. The process involves writing a magnetic pattern of tracks and sectors to a blank (uninitialized) disk. A disk containing data can be reformatted, in which case all data stored on the disk will be lost. An MS-DOS utility program (FORMAT.COM) is supplied in order to carry out the formatting of floppy disks (a similar utility is usually provided for formatting the hard disk).

Graphics adapter An option card which provides a specific graphics capability (e.g. CGA, EGA, HGA, VGA). Graphics signal generation is not normally part of the functionality provided within a system mother board.

Handshake An interlocked sequence of signals between peripheral devices in which a device waits for an acknowledgement of the receipt of data before sending new data.

Hard disk A non-flexible disk used for the magnetic storage of data and programs (*see also* **Fixed disk**).

Hardware The physical components (e.g. integrated circuits) of a microcomputer system.

High state The more positive of the two voltage levels used to represent binary logic states. A high state (logic 1) is generally represented by a voltage in the range 2.0–5.0 V.

Input/output (I/O) Devices and lines used to transfer information to and from external (peripheral) devices.

Integrated circuit An electronic circuit fabricated on a single wafer (chip) and packaged as a single component.

Interface A shared boundary between two or more systems, or between two or more elements within a system. In order to facilitate interconnection of systems, various interface standards are adopted (e.g. RS-232 in the case of asynchronous data communications).

Interface system The functional elements required for unambiguous communication between two or more devices. Typical elements include: driver and receiver circuitry, signal line descriptions, timing and control conventions, communication protocols, and functional logic circuits.

Interleave A system of numbering the sectors on a disk in a non-consecutive fashion in order to optimize data access times.

Interrupt A signal generated by a peripheral device when it wishes to gain the attention of the CPU. The Intel 8086 family of microprocessors support both software and hardware interrupts. The former provide a means of invoking BIOS and DOS services while the latter are generally managed by an interrupt controller chip (e.g. 8259).

Joystick A device used for positioning a cursor, pointer, or output device using switches or potentiometers which respond to displacement of the stick in the X and Y directions.

Logical device A device which is normally associated with microcomputer I/O, such as the *console* (which comprises keyboard and display) and *printer*.

Low state The more negative of the two voltage levels used to represent the binary logic states. A low state (logic 0) is generally represented by a voltage in the range 0–0.8 V.

Memory That part of a microcomputer system into which information can be placed and later retrieved. Storage and memory are interchangeable terms. Memory can take various forms including semiconductor (RAM and ROM), magnetic (floppy and hard disks), and optical disks. Note that memory may also be categorized as read-only (in which case data cannot subsequently be written to the memory) or read/write (in which case data can both be read from and written to the memory).

Microprocessor A central processing unit fabricated on a single chip.

Motherboard A main circuit board (or system board) to which a number of expansion or adapter cards may be fitted.

Multitasking A process in which several programs are running simultaneously.

Negative acknowledge (NAK) A signal used in serial data communications which indicates that erroneous data has been received.

Network A system which allows two or more computers (or intelligent controllers) to be linked via a physical communications medium (e.g. co-axial cable) in order to exchange information and share resources.

Noise Any unwanted signal component which may appear superimposed on a wanted signal.

Operating system A control program which provides a low-level interface with the system hardware. The operating system thus frees the programmer from the need to produce hardware specific I/O routines (e.g. those associated with disk filing). *See also* **Disk operating system**.

Option card A printed circuit board (adapter card) which complies with the physical and electrical specification for a particular system and which provides the system with additional functionality (e.g. asynchronous communications facilities).

Peripheral An external hardware device whose activity is under the control of the microcomputer system.

Port A general term used to describe an interface circuit which facilitates transfer of data to and from external devices (peripherals).

Propagation delay The time taken for a signal to travel from one point to another. In the case of logic elements, propagation delay is the time interval between the appearance of a logic state transition at the input of a gate and its subsequent appearance at the output.

Protocol A set of rules and formats necessary for the effective exchange of data between intelligent devices.

Random access An access method in which each word can be retrieved in the same amount of time (i.e. the storage locations can be accessed in any desired order). This method should be compared with *sequential access* in which access times are dependent upon the position of the data within the memory.

Random access memory (RAM) A term which usually refers to semiconductor read/write memory (in which access time is independent of actual storage address). Note that semiconductor read-only memory (ROM) devices also provide random access.

Read The process of transferring data to a processor from memory or I/O.

Read-only memory (ROM) A memory device which is permanently programmed. Erasable-programmable read only memory (EPROM) devices are popular for storage of programs and data in stand-alone applications and can be erased under ultraviolet light to permit reprogramming.

Register A storage area within a CPU, controller, or other programmable device, in which data (or addresses) are placed during processing. Registers will commonly hold 8, 16 or 32-bit values.

Relay An electromechanical device which opens or closes one or more sets of switching contacts in response to an electrical input. Relays typically permit switching of currents several orders of magnitude greater than the input current. They can also provide a high degree of electrical isolation between the input and the controlled circuits. (*See also* **Solid-state relay**.)

Root directory The principal directory of a disk (either hard or floppy) which is created when the disk is first formatted. The root directory may contain the details of further sub-directories which may themselves contain yet more sub-directories, and so on.

Sector The name given to a section of the circular track placed (during formatting) on a magnetic disk. Tracks are commonly divided into ten sectors (*see also* **Format**).

Sensor A device (transducer) which provides an electrical signal from a physical input (e.g. light, sound). After appropriate conditioning, this signal is used as an input to a control or instrumentation system.

Settling time The time taken for a signal line to settle to a defined logical state when making a transition from one state to another.

Shell The name given to an item of software which aims to provide a user interface to the system (e.g. the program COMMAND.COM is the standard DOS shell).

Signal The information conveyed by an electrical quantity.

Signal level The relative magnitude of a signal when considered in relation to an arbitrary reference (usually expressed in volts, V).

Software A series of computer instructions (i.e. a program).

Solid-state relay A relay which is based on optically coupled semiconductor switching devices (*see also* **Relay**).

Sub-directory A directory which contains details of a group of files and which is itself contained within another directory (or within the root directory).

System board The system board is the mother printed circuit board which provides the basic functionality of the microcomputer system including CPU, RAM and ROM. The system board is fitted with connectors which permit the installation of one, or more, option cards (e.g. graphics adapters).

Transducer A device which converts energy from one form to another. A transducer may be used as either an input device or as an output device. Examples of input transducers are strain gauges and photo-diodes whilst examples of output transducers are actuators and lamps.

Trigger An external stimulus which is used to initiate a measurement function or control system response.

Validation A process in which input data is checked in order to identify incorrect items. Validation can take several forms including *range*, *character* and *format checks*.

Verification A process in which stored data is checked (by subsequent reading) to see whether it is correct.

Visual display unit (VDU) An output device (usually based on a cathode ray tube) on which text and/or graphics can be displayed. A VDU is normally fitted with an integral keyboard in which case it is sometimes referred to as a *console*.

Volume label A disk name (comprising up to 11 characters). Note that hard disks may be partioned into several volumes, each associated with its own logical *drive specifier* (i.e. C:, D;, E:, etc.).

Write The process of transferring data from a CPU to memory or to an I/O device.

Appendix B SI units

The International System of Units (SI) is based upon the following *fundamental units*:

Quantity	*Unit*	*Abbreviation*
Current	ampere	A
Length	metre	m
Luminous intensity	candela	cd
Mass	kilogramme	kg
Temperature	kelvin	K
Time	second	s
Matter	mol	mol

All other units are derived from the fundamental SI units. Many of these *derived units* have their own names and those commonly encountered in electronics are summarized, together with the quantities to which they relate, in the following table:

Quantity	*Derived unit*	*Abbreviation*	*Equivalent (in terms of fundamental units)*
Capacitance	farad	F	$A\,s\,V^{-1}$
Charge	coulomb	C	$A\,s$
Energy	joule	J	$N\,m$
Force	newton	N	$kg\,m\,s^{-1}$
Frequency	hertz	Hz	s^{-1}
Illuminance	lux	lx	$lm\,m^{-2}$
Inductance	henry	H	$V\,s\,A^{-1}$
Luminous flux	lumen	lm	cd sr
Magnetic flux	weber	Wb	$V\,s$
Magnetic flux density	tesla	T	$Wb\,m^{-2}$
Potential	volt	V	$W\,A^{-1}$
Power	watt	W	$J\,s^{-1}$
Resistance	ohm	Ω	$V\,A^{-1}$

Appendix C Multiples and sub-multiples

Many of the fundamental units (Appendix B) are somewhat cumbersome for everyday use, hence the following multiples and sub-multiples are commonly employed:

Prefix	*Abbreviation*	*Multiplier*
tera	T	10^{12} (=1000000000000)
giga	G	10^{9} (=1000000000)
mega	M	10^{6} (=1000000)
kilo	k	10^{3} (=1000)
(none)	(none)	10^{0} (=1)
centi	c	10^{-2} (=0.01)
milli	m	10^{-3} (=0.001)
micro	μ	10^{-6} (=0.000001)
nano	n	10^{-9} (=0.000000001)
pico	p	10^{-12} (=0.000000000001)

Examples

1 A frequency of 1430 Hz may be expressed as 1.43 kHz.
2 A distance of 0.00375 m may be expressed as 3.75 mm.
3 A light level of 0.15 mcd may be expressed as 150 μcd.

Appendix D Decimal, binary, hexadecimal and ASCII table

Decimal	*Binary*	*Hex.*	*ASCII*	*IBM extended*	*Keyboard entry*
0	00000000	00	NUL		<CTRL-@>
1	00000001	01	SOH	^A	<CTRL-A>
2	00000010	02	STX	^B	<CTRL-B>
3	00000011	03	ETX	^C	<CTRL-C>
4	00000100	04	EOT	^D	<CTRL-D>
5	00000101	05	ENQ	^E	<CTRL-E>
6	00000110	06	ACK	^F	<CTRL-F>
7	00000111	07	BEL	^G	<CTRL-G>
8	00001000	08	BS	^H	<CTRL-H>
9	00001001	09	HT	^I	<CTRL-I>
10	00001010	0A	LF	^J	<CTRL-J>
11	00001011	0B	VT	^K	<CTRL-K>
12	00001100	0C	FF	^L	<CTRL-L>
13	00001101	0D	CR	^M	<CTRL-M>
14	00001110	0E	SO	^N	<CTRL-N>
15	00001111	0F	SI	^O	<CTRL-O>
16	00010000	10	DLE	^P	<CTRL-P>
17	00010001	11	DC1	^Q	<CTRL-Q>
18	00010010	12	DC2	^R	<CTRL-R>
19	00010011	13	DC3	^S	<CTRL-S>
20	00010100	14	DC4	^T	<CTRL-T>
21	00010101	15	NAK	^U	<CTRL-U>
22	00010110	16	SYN	^V	<CTRL-V>
23	00010111	17	ETB	^W	<CTRL-W>
24	00011000	18	CAN	^X	<CTRL-X>
25	00011001	19	EM	^Y	<CTRL-Y>
26	00011010	1A	SUB	^Z	<CTRL-Z>
27	00011011	1B	ESC	^[	<CTRL-[>
28	00011100	1C	FS	^\	<CTRL-\>
29	00011101	1D	GS	^]	<CTRL-]>
30	00011110	1E	RS	^^	<CTRL-^>

Decimal	*Binary*	*Hex.*	*ASCII*	*IBM extended*	*Keyboard entry*
31	00011111	1F	US	ˆ	<CTRL-_>
32	00100000	20	SP	·	<SPACE>
33	00100001	21	!	!	
34	00100010	22	''	"	
35	00100011	23	#	#	
36	00100100	24	$	$	
37	00100101	25	%	%	
38	00100110	26	&	&	
39	00100111	27	'	'	
40	00101000	28	(	(	
41	00101001	29	)	)	
42	00101010	2A	*	*	
43	00101011	2B	+	+	
44	00101100	2C	,	,	
45	00101101	2D	-	-	
46	00101110	2E	.	.	
47	00101111	2F	/	/	
48	00110000	30	0	0	
49	00110001	31	1	1	
50	00110010	32	2	2	
51	00110011	33	3	3	
52	00110100	34	4	4	
53	00110101	35	5	5	
54	00110110	36	6	6	
55	00110111	37	7	7	
56	00111000	38	8	8	
57	00111001	39	9	9	
58	00111010	3A	:	:	
59	00111011	3B	;	;	
60	00111100	3C	<	<	
61	00111101	3D	=	=	
62	00111110	3E	>	>	
63	00111111	3F	?	?	
64	01000000	40	@	@	
65	01000001	41	A	A	
66	01000010	42	B	B	
67	01000011	43	C	C	
68	01000100	44	D	D	
69	01000101	45	E	E	
70	01000110	46	F	F	
71	01000111	47	G	G	
72	01001000	48	H	H	
73	01001001	49	I	I	
74	01001010	4A	J	J	
75	01001011	4B	K	K	
76	01001100	4C	L	L	

Decimal	*Binary*	*Hex.*	*ASCII*	*IBM extended*	*Keyboard entry*
77	01001101	4D	M	M	
78	01001110	4E	N	N	
79	01001111	4F	O	O	
80	01010000	50	P	P	
81	01010001	51	Q	Q	
82	01010010	52	R	R	
83	01010011	53	S	S	
84	01010100	54	T	T	
85	01010101	55	U	U	
86	01010110	56	V	V	
87	01010111	57	W	W	
88	01011000	58	X	X	
89	01011001	59	Y	Y	
90	01011010	5A	Z	Z	
91	01011011	5B	[	[	
92	01011100	5C	\	\	
93	01011101	5D	]	]	
94	01011110	5E	^	^	
95	01011111	5F	_	_	
96	01100000	60	`	`	
97	01100001	61	a	a	
98	01100010	62	b	b	
99	01100011	63	c	c	
100	01100100	64	d	d	
101	01100101	65	e	e	
102	01100110	66	f	f	
103	01100111	67	g	g	
104	01101000	68	h	h	
105	01101001	69	i	i	
106	01101010	6A	j	j	
107	01101011	6B	k	k	
108	01101100	6C	l	l	
109	01101101	6D	m	m	
110	01101110	6E	n	n	
111	01101111	6F	o	o	
112	01110000	70	p	p	
113	01110001	71	q	q	
114	01110010	72	r	r	
115	01110011	73	s	s	
116	01110100	74	t	t	
117	01110101	75	u	u	
118	01110110	76	v	v	
119	01110111	77	w	w	
120	01111000	78	x	x	
121	01111001	79	y	y	
122	01111010	7A	z	z	

Decimal	*Binary*	*Hex.*	*ASCII*	*IBM extended*	*Keyboard entry*
123	01111011	7B	{	{	
124	01111100	7C	¦	|	
125	01111101	7D	}	}	
126	01111110	7E	~	~	
127	01111111	7F	DEL	⌂	
128	10000000	80		Ç	
129	10000001	81		ü	
130	10000010	82		é	
131	10000011	83		â	
132	10000100	84		ä	
133	10000101	85		à	
134	10000110	86		å	
135	10000111	87		ç	
136	10001000	88		ê	
137	10001001	89		ë	
138	10001010	8A		è	
139	10001011	8B		ï	
140	10001100	8C		î	
141	10001101	8D		ì	
142	10001110	8E		Ä	
143	10001111	8F		Å	
144	10010000	90		É	
145	10010001	91		æ	
146	10010010	92		Æ	
147	10010011	93		ô	
148	10010100	94		ö	
149	10010101	95		ò	
150	10010110	96		û	
151	10010111	97		ù	
152	10011000	98		ÿ	
153	10011001	99		Ö	
154	10011010	9A		Ü	
155	10011011	9B		¢	
156	10011100	9C		£	
157	10011101	9D		¥	
158	10011110	9E		Pt	
159	10011111	9F		ƒ	
160	10100000	A0		á	
161	10100001	A1		í	
162	10100010	A2		ó	
163	10100011	A3		ú	
164	10100100	A4		ñ	
165	10100101	A5		Ñ	
166	10100110	A6		ª	
167	10100111	A7		º	
168	10101000	A8		¿	

Decimal	*Binary*	*Hex.*	*ASCII*	*IBM extended*	*Keyboard entry*
169	10101001	A9		⌐	
170	10101010	AA		¬	
171	10101011	AB		½	
172	10101100	AC		¼	
173	10101101	AD		¡	
174	10101110	AE		«	
175	10101111	AF		»	
176	10110000	B0		░	
177	10110001	B1		▒	
178	10110010	B2		▓	
179	10110011	B3		│	
180	10110100	B4		┤	
181	10110101	B5		╡	
182	10110110	B6		╢	
183	10110111	B7		╖	
184	10111000	B8		╕	
185	10111001	B9		╣	
186	10111010	BA		║	
187	10111011	BB		╗	
188	10111100	BC		╝	
189	10111101	BD		╜	
190	10111110	BE		╛	
191	10111111	BF		┐	
192	11000000	C0		└	
193	11000001	C1		┴	
194	11000010	C2		┬	
195	11000011	C3		├	
196	11000100	C4		─	
197	11000101	C5		┼	
198	11000110	C6		╞	
199	11000111	C7		╟	
200	11001000	C8		╚	
201	11001001	C9		╔	
202	11001010	CA		╩	
203	11001011	CB		╦	
204	11001100	CC		╠	
205	11001101	CD		═	
206	11001110	CE		╬	
207	11001111	CF		╧	
208	11010000	D0		╨	
209	11010001	D1		╤	
210	11010010	D2		╥	
211	11010011	D3		╙	
212	11010100	D4		╘	
213	11010101	D5		╒	
214	11010110	D6		╓	

Decimal	*Binary*	*Hex.*	*ASCII*	*IBM extended*	*Keyboard entry*
215	11010111	D7		╫	
216	11011000	D8		╪	
217	11011001	D9		┘	
218	11011010	DA		┌	
219	11011011	DB		█	
220	11011100	DC		▄	
221	11011101	DD		▌	
222	11011110	DE		▐	
223	11011111	DF		▀	
224	11100000	E0		α	
225	11100001	E1		ß	
226	11100010	E2		Γ	
227	11100011	E3		π	
228	11100100	E4		Σ	
229	11100101	E5		σ	
230	11100110	E6		μ	
231	11100111	E7		τ	
232	11101000	E8		Φ	
233	11101001	E9		θ	
234	11101010	EA		Ω	
235	11101011	EB		δ	
236	11101100	EC		∞	
237	11101101	ED		ϕ	
238	11101110	EE		ϵ	
239	11101111	EF		∩	
240	11110000	F0		≡	
241	11110001	F1		±	
242	11110010	F2		≥	
243	11110011	F3		≤	
244	11110100	F4		⌠	
245	11110101	F5		⌡	
246	11110110	F6		÷	
247	11110111	F7		≈	
248	11111000	F8		°	
249	11111001	F9		·	
250	11111010	FA		·	
251	11111011	FB		√	
252	11111100	FC		^{n}	
253	11111101	FD		^{2}	
254	11111110	FE		■	
255	11111111	FF			

Note: IBM and compatible equipment does not use standard ASCII characters below 32 decimal. These 'non-displayable' ASCII characters are referred to as 'control characters'. When output to the IBM display, these characters appear as additional graphics characters (not shown in the table).

Appendix E Bibliography

Control systems, instrumentation and measurement

Berk, A.A., *Micros in Process and Product Control*, Collins. ISBN 0 00 383296 1.

A useful introduction to microcomputer-based control systems and eminently suitable for those not possessing a formal microelectronics background.

Brindley, Keith, *Sensors and Transducers*, Heinemann Newnes. ISBN 0 434 90181 4.

A useful guide to the principles and practice of a wide range of transducers suitable for use in instrumentation and measuring systems.

Cassell, Douglas, *Microcomputers and Modern Control Engineering*, Cassell. ISBN 0 8359 4365 8.

A comprehensive guide to the principles and practice of modern control engineering.

Money, S., *Microprocessors in Instrumentation and Control*, Collins. ISBN 0 00 383041 1.

Deals in a practical manner with the application of microprocessors in the field of control and instrumentation. Assumes that the reader has a basic understanding of the principles of digital electronics.

Tompkins, Willis and Webster, John, *Interfacing Sensors to the IBM PC*, Prentice Hall. ISBN 0 13 469081 8.

Contains a considerable amount of practical data related to PC-based instrumentation.

Tooley, Michael, *Bus-based Industrial Process Control*, Heinemann Newnes. ISBN 0 434 92009 6.

Provides a general introduction to the STE bus together with representative applications.

Electronic circuits

Horowitz, Paul and Hill, Winfield, *The Art of Electronics*, Cambridge University Press. ISBN 0 521 23151 (hard cover), ISBN 0 521 29837 (paperback).

A comprehensive guide to the design of electronic circuits. The book is eminently readable and adopts a practical approach. Recommended reading for those having a limited electronic background.

Marston, *Instrumentation and Test Gear Circuits Manual*, Butterworth-Heinemann. ISBN 0 7506 0758 0.

A compendium of useful circuits covering a wide range of instrumentation and measurement applications.

Tooley, Michael, *Electronic Circuits Handbook – Design, testing and construction*, Heinemann Newnes. ISBN 0 434 91968 3.

Provides readers with a unique collection of practical circuits together with supporting information so that working circuits can be produced in the shortest possible time and without recourse to theoretical texts.

Digital techniques

Lancaster, Don, *The TTL Cookbook*, Howard Sams. ISBN 0 672 21035 5.

An invaluable collection of hints, tips, facts and figures covering all facets of TTL. A selection of the most popular TTL devices is discussed in some detail (together with pinouts for each device). The book also has a useful section on timers.

Lancaster, Don, *The CMOS Cookbook*, Howard Sams. ISBN 0 672 22459 3.

This book is similar to its TTL counterpart and makes equally good reading.

Texas Instruments, *The TTL Data Book for Design Engineers*, Texas Instruments (Europe). ISBN 3 88078 034 X.

The definitive text covering all types of TTL device. Includes electrical characteristics and pin connecting data.

Tooley, Michael, *Practical Digital Electronics Handbook*, PC Publishing. ISBN 1 870775 00 7.

A practical guide to digital circuits and microprocessor-based systems.

Towers, T.D., *The Towers' International Digital IC Selector*, Foulsham. ISBN 0 572 01179.

Provides abridged data and pin connecting information for over

13 000 digital integrated circuits. Appendices provide useful reference information on IC logic types and codings, package outlines, pinouts, manufacturers' codings, manufacturers' proprietary 'house' codings, abbreviations and a glossary.

Microprocessors and microcomputers

Berk, A.A., *The Art of Micro Design*, Heinemann Newnes. ISBN 0 408 01 403 2.

A practically orientated guide to the design of microprocessor systems with particular emphasis on control applications.

Bigelow, *PC Drives and Memory Systems*, Windcrest/McGraw-Hill, ISBN 0 8306 4551 9.

A useful guide to PC memory systems covering semiconductor RAM, floppy disk, hard disk, optical, and tape drives.

Ferguson, John, *Microprocessor Systems Engineering*, Addison-Wesley. ISBN 0 201 14657 6.

Describes the basic skills, tools and techniques required to devise, develop and implement a microprocessor-based project.

Goodman, *Memory Management for All of Us*, Sams. ISBN 0 672 30306 X.

An excellent and authoritive reference to PC memory operation, configuration and management.

Hall, Douglas, *Microprocessors and Digital Systems*, McGraw-Hill. ISBN 0 07 025552 0.

Provides an excellent introduction to microprocessor-based systems and includes chapters on the use of test equipment, digital logic gate characteristics and interfacing, flip-flops counters and shift registers, D/A and A/D converters, microprocessor structure and programming, and prototyping/troubleshooting microprocessor-based systems.

McGrindle, J.A., *Microcomputer Handbook*, Collins. ISBN 0 00 383026 8.

A comprehensive guide to all facets of microcomputers. Includes sections on Microcomputer Boards and Systems, Software, Development Systems, and Test Equipment.

Mueller, Scott, *Upgrading and repairing PCs*, Que. ISBN 0 88022 395 2.

A comprehensive guide to expanding and upgrading PCs and PC compatibles. The sections on disk storage are particularly useful.

Norton, Peter, *Programmer's Guide to the IBM PC*, Microsoft Press. ISBN 0 914845 46 2.

A useful reference which covers hardware and DOS.

Putman, Byron, *Digital and Microprocessor Electronics*, Prentice-Hall. ISBN 0 13 214354 2.

A comprehensive guide to the design and troubleshooting of modern digital and microprocessor-based systems.

Tooley, *PC Troubleshooting Pocket Book*, Newnes. ISBN 0 7506 1727 6.

A concise guide to PC troubleshooting, repair and optimization.

Whitworth, Ian, *16-Bit Microprocessors*, Collins. ISBN 0 00 383113 2.

Provides a comprehensive introduction to 16-bit microprocessors and includes details of all popular 16-bit processor families. Chapters are also devoted to interfacing, instruction sets, assembly code software and development, system software and operating systems.

Woram, *PC Configuration Handbook*, Bantam. ISBN 0 553 34947 3.

A comprehensive guide to PC configuration, troubleshooting and upgrading.

Operating systems

Microsoft Corporation, *MS-DOS Programmers Reference*. ISBN 1 55615 546 8.

Describes system functions, interrupts and structures of MS-DOS up to, and including, Version 6.

Norton, *Peter Norton's Advanced DOS 6*, Brady. ISBN 1 56686 046 6.

A useful and informative guide for DOS users.

Oets, Pim, *MS-DOS and PC-DOS – a Practical Guide*, Macmillan. ISBN 0 333 45440 5.

A concise introduction to MS-DOS and PC-DOS.

Sinclair, Ian, *Newnes MS-DOS Pocket Book*, Heinemann Newnes. ISBN 0 434 91858 X.

An invaluable source of information on using the MS-DOS operating system.

Van Wolverton, *Running MS-DOS*, Microsoft Press. ISBN 1 55615 542 5.

A comprehensive reference which covers versions of MS-DOS up to, and including, Version 6.

Programming

Coffron, James, *Programming the 8086/8088*, Sybex. ISBN 0 89588 120 9.

A useful introduction to the 8086 and 8088 for the assembly language programmer.

Craig, John, *Microsoft QuickBASIC Programmer's Toolbox*, Microsoft. ISBN 1 55615 127 6.

Comprises a powerful library of more than 250 subprograms, functions, and utilities for use within QuickBASIC programs.

Denning, Adam, *C at a Glance*, Chapman and Hall/Methuen. ISBN 0 412 27140 0.

An excellent and reasonably priced introduction to C programming. Contains numerous examples.

Feldman, Phil and Rugg, Tom, *Using QuickBASIC 4*, Que. ISBN 0 88022 378 2.

A useful hands-on guide to programming in QuickBASIC which is ideal for the beginner and which contains numerous examples.

Kerninghan, Brian and Ritchie, Dennis, *The C Programming Guide*, Prentice-Hall. ISBN 0 13 110163 3.

The definitive guide to C. An essential book for any aspiring C programmer.

Morgan, Christopher and Waite, Mitchell, *8086/8088 16-Bit Microprocessor Primer*, Byte/McGraw-Hill. ISBN 0 07 043109 4.

An excellent introduction to the characteristics, internal architecture and programming of the 8086, 8088 and associated support devices.

Norton, Aitken and Wilton, *PC Programmer's Bible*. ISBN 1 55615 555 7.

A comprehensive reference for programmers wishing to access BIOS and DOS services. Suitable for readers with little or no previous experience.

Purdum, Jack, *C Programming Guide*, Que. ISBN 0 88022 157 7.

An excellent tutorial guide eminently suitable for newcomers to the C language.

Purdum, Jack, *C Self-Study Guide*, Que. ISBN 0 88022 149 6.

A directed study guide with questions and answers. Useful as a supplement to Jack Purdum's *C Programming Guide*.

Appendix F List of suppliers

A + P Computers Ltd
35 Walnut Tree Close
Guildford
Surrey GU1 4UN
Telephone: 0483 304118
Fax: 0483 304124

Motherboards, system upgrades.

Action Computer Supplies
5–6 Abercorn Commercial Centre
Manor Farm Road
Wembley
Middlesex
HA0 1BR
Telephone: 0800 333 333
Fax: 081 903 3333

General computer supplies, media and software packages.

Amplicon Liveline Ltd
Centenary Industrial Estate
Brighton
East Sussex
BN2 4AW
Telephone: 0273 570220
0880 525 335
Fax: 0273 570215
Telex: 87563

PC expansion cards and signal conditioning products.

Arcom Control Systems Ltd
Unit 8
Clifton Road
Cambridge
CB1 4WH
Telephone: 0223 411200
Fax: 0223 410457
Telex: 94016424

STE bus products, hardware and software.

BICC-Vero Electronics
Industrial Estate
Hedge End
Southampton
SO3 3LG
Telephone: 0703 266300

BICC-Vero Electronics Inc.
1000 Sherman Ave.
Hamden
CON. 06514
USA
Telephone: 1203 228 8001

Connectors, prototyping boards, enclosures and rack systems.

Biodata Limited
10 Stocks Street
Manchester
M8 8QG
Telephone: 061 834 6688
Fax: 061 833 2190
Telex: 665608

Manufacturers of MICROLINK modules and products.

Chipboards Ltd
65 High Street
Bagshot
Surrey
GU19 5AH
Telephone: 0276 51441

PC expansion cards.

Chipboards PLC
Almac House
Church Lane
Bisley
Surrey
GU24 9DR
Telephone: 0483 797959
Fax: 0483 797702

Systems, cards and accessories.

CIL Microsystems Ltd
4 Wayside
Commerce Way
Lancing
West Sussex
BN15 8 TA
Telephone: 0903 765225
Fax: 0903 765547
Telex: 878443

Data acquisition and data logging hardware and software.

Connexions (UK) PLC
Unit 3
South Mimms Distribution Centre
Huggins Lane
Welham Green
Herts
AL9 7LE
Telephone: 0707 272091
Fax: 07072 69444
Telex: 295181

Data communications products and accessories.

Control Universal
137 Ditton Walk
Cambridge
CB5 8QF
Telephone: 0223 244447
Fax: 0223 214626
Telex: 817651

STE bus products, hardware and software.

Datel (UK)
Business Park
Wade Road
Basingstoke
Hants
RG24 0NE
Telephone: 0256 469085

Datel Inc.
11 Cabot Blvd.
Mansfield
MS 02048
Telephone: 508 339 3000

Data acquisition components and systems.

Dexion Electronics Ltd
Arkwright Road
Bedford
MK42 0LQ
Telephone: 0234 217811
Fax: 0234 213368

Industrial computer systems.

Dexion Corp.
550 Warrenville Road
Lisle
IL 60532 4387
Telephone: 708 852 8800

DSP Design Ltd
Unit 1
Apollo Studios
Charlton Kings Road
London
NW5 2SB
Telephone: 071 482 1773
Fax: 071 482 1779

STE bus products, hardware and software.

Evesham Micros Ltd
Unit 9
St Richards Road
Evesham
Worcestershire
WR11 6TD
Telephone: 0386 765180
Fax: 0386 765354

Systems, cards and accessories.

Fairchild Ltd
Eastpoint
Burgoyne Road
Thornhill
Southampton
SO2 6PB
Telephone: 042 121 6527
Fax: 042 121 6583

PC expansion cards and industrial computer systems (Distributors for Deer Mountain products).

Farnell Electronic Components
Canal Road
Leeds
West Yorkshire
LS12 2TU

Telephone: 0532 636311
Fax: 0532 633411
Telex: 55147

Electronic components, test equipment and computer products.

Keithley Instruments
1 Boulton Road
Reading
Berkshire
RG2 0NL
Telephone: 0734 861287
Telex: 847047

Keithley Inc.
28775 Aurora Road
Cleveland
Ohio 44139
Telephone: 216 248 0400

PC expansion cards and signal conditioning products (Distributors for MetraByte and IOtech products).

Maplin Electronic Supplies Ltd
PO Box 3
Rayleigh
Essex
SS6 8LR
Telephone: 0702 554155
0702 552911
0702 552961
Telex: 995695

Marconi Instruments
Longacres
St Albans
Herts
AL4 0JN
Telephone: 0727 59292

Marconi Instruments (USA)
3 Pearl Court
Allendale
NJ 07401
Telephone: 201 934 9050

Test equipment.

Memory Direct Ltd
42–44 Birchett Road
Aldershot
Hampshire
GU11 1LT
Telephone: 0252 316060

Memory upgrades, chips and SIMMs.

Microbus Manufacturing Ltd
Treadway Hill
Loudwater

High Wycombe
Bucks
HP10 9QL
Telephone: 06285 31271
Fax: 06285 31498
Telex: 846073

PC and bus products. Electronic components (including sensors and transducers).

National Instruments (UK)
Surrey House
34 Eden Street
Kingston
Surrey
KT1 1ER
Telephone: 081 549 3444

National Instruments (USA)
6504 Bridge Point Parkway
Austin
Texas 78730
Telephone: 512 794 0100

Test instruments.

Penny and Giles
6 Airfield Way
Christchurch
Dorset
BH23 3TT
Telephone: 0202 477461
Fax: 0202 484846
Telex: 41555

PC bus products.

Philips Test and Measurement
Colonial Way
Watford
Herts
WD2 4TT
Telephone: 0923 240511

Test instruments.

Portables and Upgrades Ltd
Dram House
Latham Close
Bredbury Industrial Park
Stockport
SK6 2SD
Telephone: 061 406 6486
Fax: 061 494 9125

Memory upgrades, chips and SIMMs.

Powermark PLC
Premier House
112 Station Road
Edgware
Middlesex
HA8 7AQ
Telephone: 081 951 3355
Fax: 081 905 6233

Memory upgrades, chips and SIMMs.

PPM Instrumentation Ltd
7 Riverside Business Centre
Walnut Tree Close
Guildford
Surrey
GU1 4UG
Telephone: 0483 301333
Fax: 0483 300862
Telex: 859181

Data acquisition and data logging hardware and software.

Radioplan Ltd
Unit 14
Cheltenham Trade Park
Arle Road
Cheltenham
Gloucestershire
GL50 8LZ
Telephone: 0242 224304
Fax: 0242 227154
Telex: 437244 CMINTL

Hardware and software for signal analysis.

RSC Corporate
75–77 Queens Road
Watford
Hertfordshire
WD1 2QN
Telephone: 0923 243301
Fax: 0923 237946

Systems, cards and accessories.

RS Components
PO Box 99
Corby
Northants
NN17 9RS
Telephone: 0536 201234
Fax: 0536 201501
Telex: 342512

Electronic components, test equipment and computer products.

Semaphore Systems Ltd
7 Moreland Court
Finchley Road
London
NW2 2PJ
Telephone: 071 435 6315
071 433 1255

PC expansion cards.

Siemens
Siemens House
Windmill Road
Sunbury
Middlesex
TW16 7HS
Telephone: 0932 785691

Siemens (USA)
767 Fifth Avenue
NY 10153
Telephone: 212 832 6601

Test instruments and PC-based products.

Silica Systems
1–4 The Mews
Hatherley Road
Sidcup
Kent
DA14 4DX
Telephone: 081 309 1111
Fax: 081 308 0608

Systems, cards and accessories.

SMC Computers
26 Farnham Road
Slough
Berkshire
SL1 3TA

Telephone: 0753 550333
Fax: 0753 524443

Systems, cards and accessories.

Stak Trading
Stak House
Butlers Leap
Rugby
CV21 3RQ
Telephone: 0788 577497
Fax: 0788 544584

Motherboards, system upgrades.

STC Instrument Services
Dewar House
Central Road
Harlow
Essex
CM20 2TA
Telephone: 0279 641641

Test instruments.

Visorgraph
PO Box 51
Altrincham
Cheshire
WA14 3BB
Telephone: 061 443 1846
Fax: 061 443 1597

Industrial computer systems and products.

Watford Electronics Ltd
Finway
off Dallow Road
Luton
LU1 1TR
Telephone: 0582 487777
Fax: 0582 488588

Motherboards, system upgrades.

XYCOM Europe Ltd
6 Scirocco Close
Northampton
NN3 1AP
Telephone: 0604 790 767

Industrial computer systems.

XYCOM (USA)
750 North Maple Road
Saline
Michigan 48176
Telephone: 313 429 4971

Index

146818, 41, 102
16450, 86, 102

386 enhanced mode, 146

486 chip set, 25
486DLC2, 9
486SLC, 9
486SLC2, 9

555, 271
5801A, 272, 273
590kH, 262, 263, 265

72C81, 98
74C922, 251, 252, 253
74LS04, 81
74LS138, 81, 80
74LS244, 22
74LS245, 22, 90, 233
74LS30, 81
74LS373, 28
74LS670, 28
74LS73, 260

80186, 11, 27
80286, 9, 11, 15, 16, 27
80287, 27
80386, 11, 17, 27
80386DX, 9
80386SX, 9, 15
80387, 27
8042, 33
80486, 11, 17
8080, 106
8085, 106
8086, 8, 9, 13, 10, 27, 174, 179
8086 register model, 175
8087, 25, 26, 27
8088, 8, 9, 11, 27
8089, 27
82230, 27
82231, 27
82258, 27
82284, 27
82335, 27
8234, 27
8237, 22, 27, 28
8237A, 28, 29, 33
82384, 27
8253, 27, 28, 30, 31, 32, 33, 34, 35
8254, 27, 33
8255A, 30, 32, 33, 233
8259A, 27, 31, 33, 35, 36
8284A, 27, 34, 36, 37
8288, 21, 27, 38, 39
82C100, 38, 98
82C362, 38
82C365, 38
82C461, 38
82C606, 99
8514 standard, 52, 51
85C310, 38
85C320, 38
85C330, 38

AC sensing, 266
ACK, 346
ADC, 306
ADC chip, 234

ADSTB, 28
AHOLD, 17
AIP-24, 84
ALE, 21
ALU, 12
AMI BIOS, 421, 41
ANSI.SYS, 339
APPEND, 122
ASCII, 346, 357
ASSIGN, 122
ASYST, 282
ASYSTANT, 283
AT attachment, 58
AT bus, 70
AT bus controller, 38
ATTRIB, 122
AUTOEXEC.BAT, 106, 144, 332, 340, 341
AUTOEXEC.OLD, 145
Access time, 346
Accumulator, 176, 346
Acknowledge, 346
Active high, 346
Active low, 346
Active sensor, 243
Adapter card, 6, 22, 64
Address, 346
Address bus, 4, 21, 346
Address decoder, 81, 346
Address decoding, 6, 22, 79, 80, 90
Address latch enable, 21, 28
Address selection, 346
Addressing, 13
Advanced technology, 1
Advanced diagnostic, 320
Am386DX, 9
Am386SX, 9
Am486DX, 9
Am486DX2, 9
American Standard Code for Information Interchange, 346, 357
Amplifier, 347
Analogue, 347
Analogue I/O, 233
Analogue RGB, 54
Analogue to digital converter, 234
Angular position, 238
Angular velocity, 238
Annunciator, 305, 308
Applications, 298
Archive, 347
Arguments, 211
Arithmetic logic unit, 12
Assemble command, 132
Assembler directive, 167, 168
Assembly language, 164, 166, 174, 347
Asserted, 347
Asynchronous mode, 7
Asynchronous transmission, 347
Attributes, 43
Audible warning, 270
Automatic test equipment, 217
Award BIOS, 41

BACKUP, 123
BASIC, 181
BIOS, 40, 41, 44, 104, 347
BIOS ROM, 22, 41, 47
BITBUS, 294
BIU, 8, 12, 13, 17
BREAK, 117
BS4937, 264
BUFFERS, 333
BUSY, 25, 27
Backing-up disks, 107
Backplane, 347
Backplane bus, 60, 91, 294
Backup, 347
Base 16, 5
Base memory, 46, 48
Base pointer, 177
Basic input/output system, 347
Batch command, 128
Batch file, 128, 130, 332, 347
Binary branch, 156
Binary value, 5, 357
Bit, 347
Bit-mapped graphics, 433, 43
Block transfer mode, 28
Boot, 347
Boot record, 347
Bounce, 246
Branch, 156
Buffer, 348
Bus, 348
Bus connector, 65, 66, 67
Bus controller, 21, 38
Bus expansion, 5
Bus interface unit, 8, 17
Bus system, 3
Bus transceiver, 22
Byte, 4, 348

C programming, 200
C486SRx2, 10

CALL, 187
CD, 117
CDC, 59
CGA, 51, 52, 53
CHDIR, 117
CHKDSK, 123
CLK, 35
CLOSE, 199
CLS, 118
CMOS RAM, 41, 42
CMOS battery, 41
COM1:, 109
COM2:, 109
COMMAND.COM, 4
COMP, 124
CON:, 109
CONFIG.BAK, 340
CONFIG.OLD, 144, 340
CONFIG.SYS, 144, 332, 339, 340, 341
COPY, 118, 130
CP/M, 106
CP/M-86, 106
CPU, 3, 8, 348
CPU bus, 21
CREF, 172
CRT, 52
CTRL-ALT-DEL, 112
CTRL-C, 112
CTRL-G, 112
CTRL-H, 111, 112
CTRL-I, 112
CTRL-J, 112
CTRL-L, 112
CTRL-M, 112
CTRL-P, 112, 142
CTRL-S, 112
CTRL-Z, 109, 112
Cache and DRAM controller, 59
Cache controller, 48, 38
Cache operation, 17
Capacitive promximity switch, 242
Capacitive proximity detector, 254
Capacitive proximity switch, 240
Cascade mode, 28
Cathode ray tube, 52
Cause and effect, 323
CeleSTE, 100
CeleSTE PC, 97
Central processing unit, 3, 348
Channel, 348
Character, 184
Chip, 348
Clock, 8, 21, 37, 348
Clock generator, 34
Clone, 1
Closedown routine, 194
Cluster, 348
Code page, 338, 339
Code prefetch unit, 17
Code segment, 175, 178, 179
Code segment register, 14
CodeView, 171, 174, 201
Colour, 54
Colour graphics adapter, 51
Command, 348
Command code, 224
Command groups, 222
Comment, 168
Comments, 160
Common, 348
Compare command, 132
Conditional loop, 154
Configuration, 332
Console, 109
Console I/O, 205
Control bus, 4
Control character, 112
Control panel, 144, 145
Control structure, 156
Controller, 219, 348
Conventional RAM, 40
Cooling, 25, 65
Coprocessor, 348
Copying disks, 107
Cross-reference file, 171
Cross-reference utility, 172
Crystal, 21
Crystal filter, 303, 304
Cx486DRx2, 10
Cx486S, 9, 10

DAC, 306
DACK, 7
DADisP, 285
DATE, 118
DC motor, 271
DEBUG, 27, 45, 49, 52, 131, 132, 138, 174
DEBUG.COM, 131
DEL, 119
DEVICE, 336
DIL switch, 240, 245
DIN41612, 94, 92
DIR, 21, 111, 119
DISKCOMP, 124

DISKCOPY, 107
DISPLAY.SYS, 333, 337, 338, 339
DMA, 7, 22
DMA acknowledge, 7
DMA channel, 7, 28
DMA controller, 22, 28, 33
DMA request, 7
DMAC, 28, 30
DOS, 40, 104, 349
DOS calls, 214
DPU, 59
DRAM, 43
DRIVER.SYS, 332, 336
Daisy chain, 348
Data, 348
Data bus, 4, 21, 348
Data bus buffer, 38
Data file, 198
Data patch unit, 59
Data rate, 223
Data segment, 175, 178
Data transfer, 6
Data transfer rate, 57
Debounce circuit, 247, 248
Debouncing, 245
Debugger, 174
Decimal value, 5, 357
Decoupling, 331
Default drive, 106
Define byte, 168
Define word, 168
Delay procedure, 188
Delay routine, 204, 205
Delay subroutine, 186, 187
Demand transfer mode, 28
Design, 296, 297
Desktop PC, 18
Destination index, 177
Device driver, 108, 349
Devices, 108
Differential pressure switch, 241
Diffuse scan proximity switch, 240
Digital I/O, 232
Digital RGB, 54
Direct memory access, 7, 349
Direction, 21
Direction sensing, 260
Directory, 110, 349
Directory tree, 114
Disassemble command, 136
Disck cache, 333
Disk adapter BIOS, 44
Disk backup, 107
Disk cache, 332
Disk copying, 335
Disk drive, 55
Disk files, 213
Disk operating system, 349
Display adapter configuration, 333
Display memory, 53
Display mode, 53
Display register command, 135
Distributed system, 294
Do Until...Loop, 189
Do While...Loop, 189
Do...Loop, 181, 189, 194
Do...Loop Until, 190
Do...Loop While, 190
Do...Loop Until, 159
Do...Loop While, 159
Do Until...Loop, 160
Do While...Loop, 160
Documentation, 160
Double precision, 184
Down-time, 316
Drive specifier, 111
Driver, 108, 349
Dual-422, 86
Dump command, 132, 139

ECHO, 129
ECPC, 95
EGA, 51, 52, 53
EMM386.EXE, 333, 336
EMM386.SYS, 333, 336
EMS, 41
END SUB, 187
EOP, 28
ERASE, 119
ESDI, 56
ESDI drives, 57
EU, 8, 12, 13, 17
EXE2BIN, 124
Edit bytes command, 133
Editor, 167
Electromagnetic vibration sensor, 243
Emulate processor mode, 16
Enable input, 6
Encoded keypad, 250
End of process, 28
End-of-file character, 109
Enhanced graphics adapter, 51
Enhanced mode, 145, 146
Enhanced small device interface, 57
Enter command, 133
Equipment list, 46, 47, 48

Error checking, 160
Error code, 319
Error trapping, 151
Eurocard, 91
Eurocard PC, 95
Even parity, 43
Exception, 16
Execution trace command, 135, 1340, 140
Execution unit, 8, 17
Expanded memory, 40
Expansion bus, 60, 65
Expansion bus connector, 66, 67, 70
Expansion bus signals, 73
Expansion card, 64, 73, 75, 76, 289
Expansion modules, 5
Expansion slots, 62
Extended memory, 40
Extended technology, 1
External MS-DOS commands, 122
External command, 115
Extra segment, 175, 178

FASTOPEN, 124
FDISK, 125
FIFO, 17
FIND, 125
FLUSH, 17
FM, 55
FOR, 129
FOR...NEXT, 189
FORMAT, 125, 349
FRU, 320
FWAIT, 25, 27
Fan, 25
Fault location, 328
Fault-location, 315, 321
Fault-tolerance, 316
Fetch/execute cycle, 6
Field data message, 294
Field replaceable unit, 320
File, 349
File extension, 111, 113
File manager, 144
File specification, 111
File types, 116
Filename, 111
Filetype, 111
Fill memory command, 133
Filter, 349
Firmware, 349
Fixed disk, 349
Flag register, 12, 178
Float switch, 237, 240
Floating point, 183, 184
Floating point register, 25
Floppy disk, 55
Flow, 238
Flow chart, 322
Flow sensor, 238
Flowchart, 155
Flowchart symbols, 153
Fluid sensor, 259
Format, 349
Frequency modulation, 55
Functions, 203

GEM, 278
GOSUB, 185, 186
GT200, 298, 299, 300
GUI, 144
Generic PC, 21
Glitches, 65
Go command, 133
Graphical user interface, 144
Graphics Environment Manager, 278
Graphics adapter, 51, 52, 78, 349
Graphics shell, 288

HGA, 52, 53
Handshaking, 31, 221, 350
Hard disk drive, 56, 350
Hardware, 350
Hardware debouncing, 246
Hardware design, 296
Headers, 161
Heat, 25
Hewlett Packard Instrument Bus, 217
Hexadecimal, 5, 357
Hexadecimal arithmetic command, 134
High state, 4, 350

i386, 11, 27
i386DX, 9
i386SX, 9
i486, 11
i486DX, 10
i486DX2, 10
i486DX4, 10
i486SX, 10
i486SX2, 10
I/O, 3, 4, 6, 7, 8, 22, 350
I/O card, 81, 232
I/O channel, 109
I/O functions, 205

I/O map, 101
I/O port, 30, 232
IBM, 1
IBMCACHE.SYS, 332, 333
ID byte, 49, 50
IDE drive, 56
IDE drives, 57
IEEE-1000, 91
IEEE-488, 292, 303
IEEE-488 bus, 217
IEEE-488 bus configuration, 225
IEEE-488 command code, 224
IEEE-488 command group, 222
IEEE-488 commands, 221
IEEE-488 devices, 218
IEEE-488 handshake, 222
IEEE-488 signals, 219
IEEE-488 software, 223
IF, 129
INPUT, 198
INT 21, 143, 215
INTA, 34
IOCHK, 73
IOCHRDY, 73
IOCS16, 73
IP, 14
IRQ, 7
IRR, 34
ISA, 25, 61
ISA bus, 61, 63
ISA bus extension, 57
ISR, 34
Identification byte, 49, 50
If...Else...End If, 156, 158, 171
If...End If, 156, 158
Include directive, 172
Include files, 203
Inductive proximity detector, 254
Inductive proximity switch, 242
Industrial PC, 290, 293
Industry standard architecture, 61
Input validation, 160
Input/output, 3, 350
Input/output channel, 109
Inputs, 208
Installed equipment list, 46, 279
Instruction decode unit, 17
Instruction pointer, 14, 178
Instruction queue, 8, 17
Integer variable, 183, 184
Integrated circuit, 350
Integrated support devices, 38
Interface system, 350
Interfacing, 232, 243
Interleave, 350
Intermediary bus, 59
Internal command, 115, 117
Interrupt, 179, 350
Interrupt acknowledge, 34
Interrupt controller, 33
Interrupt handling, 17
Interrupt pointer, 18
Interrupt pointer table, 18, 179, 180
Interrupt request, 7, 34
Interrupt service routine, 18

JOIN, 126
Joystick, 241, 350

KEB, 126
Key trapping, 195, 196
Keyboard, 240
Keyboard controller, 33
Keyboard entry, 191
Keypad, 240, 250
Keypad interface, 251
Keypad matrix, 250
Keystroke, 191

LABEL, 126
LCD display, 268
LDR, 264
LED, 267, 266
LIB, 173
LIM standard, 40
LINE INPUT, 198
LINK, 167, 171, 172
LPT1:, 109
LPT2:, 109
LRU, 17
LSB, 4, 5
LVDT, 239
Label, 168
Labels, 162
Lamp driver, 269
Landing zone, 56
Language extensions, 280
Last recently used, 17
Latching action switch, 249
Least significant bit, 4
Library manager, 172
Light dependent resistor, 264
Light emitting diode, 267, 266
Light level, 239
Light level detector, 265, 264
Light level threshold detector, 264

Line driver, 8
Line receiver, 8
Linear position, 239
Linear position sensor, 236, 239
Linear variable differential transducer, 239
Linear velocity, 239
Linker, 171
Liquid flow sensor, 235
Liquid level, 240
Liquid level float switch, 237
Listener, 218
Load file command, 134
Local bus standards, 58
Logic 0, 4
Logic 1, 4
Logic probe, 325, 330
Logic pulser, 326
Logical construct, 188
Logical device, 350
Long integer, 183, 184
Loop, 159, 176, 207
Loop structure, 189
Low state, 4, 350
Low-level I/O, 205
Low-level format, 58

MASM, 167, 168, 171
MASTER, 73
MBC-488, 225, 226, 229
MC146818, 41
MCA, 61
MCA bus, 63
MCGA, 51, 52, 53
MDA, 51, 52, 53
MEMCS16, 73
MFM, 56, 57
MIPS, 25
MKDIR, 120
MODE, 110, 126
MOUSE.COM, 340
MS-DOS, 2, 4, 106
MS-DOS command, 110
MS-DOS control character, 112
MS-DOS debuggers, 131
MS-DOS device, 108
MS-DOS internal command, 117
MSB, 4, 5
MSCDEX.SYS, 333
MTBF, 316
Machine identification byte, 49
Machine status word, 15
Macro, 170
Macro assembler, 170
Macro expansion, 170
Mains failure detector, 266
Maths coprocessor, 25, 27
Matrix keypad, 250
Mean time before failure, 316
Memory, 39, 43, 350
Memory search command, 135
Memory controller, 38
Memory dump, 46, 51
Memory dump command, 139
Memory management, 333
Memory manager, 336
Memory map, 44, 96, 99
Memory move command, 135
Menu, 210
Menu selection, 193
Messages, 190, 206
MetraBus, 292
Mezzanine bus, 59
Micro-channel architecture, 61
Microcomputer system, 3
Microprocessor, 3, 351
Microswitch, 241, 252
Modified frequency modulation, 56
Monitor processor mode, 16
Monochrome display adapter, 51
Most probable cause, 323
Most significant bit, 4
Motherboard, 351
Motor driver, 272
Mouse driver, 340
Move memory command, 134
Multi-colour graphics array, 51
Multiples, 356
Multi-range meter, 324
Multi-tasking, 146, 151, 351
Multimedia extension, 333
Multiplexing, 21

NAK, 351
NDP, 25
NUL:, 109
Name file command, 134
Negative acknowledge, 351
Network, 351
Networked system, 294
Nibble, 5
Noise, 154, 331, 351
Non-standard disk drives, 335
Normally closed switch, 244
Normally open switch, 244
Norton Utilities, 287

Null device, 109
Numeric variable, 183
Numerical input, 196

OPEN...FOR, 199
Object code file, 172
Odd parity, 43
Opcode, 168
Operand, 168
Operating system, 3, 104, 351
Operation code, 168
Operator push-button, 240
Operator switch, 240
Optical isolation, 261, 262
Optical proximity detector, 255
Optical proximity switch, 242
Optical sensor, 237
Option card, 6, 351
Optoisolator, 82, 261
Origin, 168
Oscillator stability, 298
Oscilloscope, 326
Oscilloscope display, 312
Output driver, 272
OverDrive, 10

PATH, 106, 107, 120
PAUSE, 129
PC, 1
PC Tools, 287
PC architecture, 18, 20
PC engine, 91
PC expansion bus, 62
PC instruments, 291, 292
PC specifications, 2
PC-35, 89
PC-AT, 1, 15, 19
PC-DOS, 106
PC-XT, 1, 19
PC-XT expansion cards, 62
PCE/2, 88
PCHK, 17
PCI bus, 59
PCSHELL, 340
PDISO-8, 81
PDP-11, 200
PIC, 31, 320
PIT, 28, 30
POSIX, 148
POST, 42, 318
PPI, 30
PRINT, 127
PRINTER.SYS, 333, 337, 339
PRN:, 109
PROMPT, 120
PS/2, 15, 19, 61
Palette, 54
Parallel I/O, 7, 232
Parallel interface, 33, 30
Parameter, 131
Parit checking, 17
Parity bit, 43
Parity error, 318
Passing parameters, 131
Passive sensor, 24, 243
Pentium, 10, 11
Performance specification, 295
Peripheral, 351
Peripheral interconnect bus, 59
Personal System/2, 19
Phoenix BIOS, 41
Photocell, 239
Photodarlington, 261
Photodiode, 239, 261
Phototransistor, 239, 261
Physical address, 14
Piezo-resistive pressure sensor, 241
Piezo-transducer, 270
Pixel, 43
Port, 30, 351
Port I/O, 205
Port address, 33
Port input command, 134
Port output command, 134
Portable operating system
interface, 148
Position transducer, 257
Possible cause, 323
Power PC, 10
Power failure detector, 266
Power supplies, 50
Power supply, 74, 77, 78, 90
Power-on self-test, 42, 318
Pressure, 241
Pressure switch, 241
Primary expansion bus connector, 66
Printer configuration, 333
Problem isolation chart, 320
Procedure, 187
Proceed command, 135, 140, 142
Processor extension not present, 16
Program code, 6
Program documentation, 160
Program headers, 161
Program Manager, 144
Program presentation, 163

Program testing, 160
Programmable interrupt controller, 31
Programmable interrupt timer, 28, 30
Programmable parallel interface, 30
Programmed I/O, 6
Programming, 149
Programming language, 150
Prompts, 190, 208
Propagation delay, 351
Protected mode, 16, 40, 151
Protocol, 351
Prototyping card, 77, 89
Proximity detector, 252
Proximity switch, 240, 242
Proximity, 241
Pseudo code, 153
Pseudo-op, 168
Push-button, 140, 244

QS0, 27
QS1, 27
Quality procedure, 315
Quartz crystal, 21
Queue status, 27
QuickBASIC, 27, 45, 46, 49, 181, 182
QuickC, 200

RAM, 3, 4, 6, 22, 45, 352
RAM diagnostic, 319
RAM drive, 332, 334
RAM fault finding, 318
RAMDRIVE.SYS, 332, 334
RAS, 30
RD, 121
README.DOC, 110
REFRESH, 22
REM, 129
RENAME, 121
RESET, 14
RESTORE, 127
RETURN, 186
RGB, 54
RLL, 56, 57
ROM, 3, 6, 22, 352
ROM BIOS, 44
ROM release date, 47, 49, 51
ROMDISK, 89
RS-232, 7, 292
RS-422, 86, 294
RS-485, 294
Random access, 351
Random access memory, 352
Read, 352
Read only memory, 3, 352
Read operation, 6
Read/write memory, 3
Real mode, 151
Real-time clock, 41
Redirection, 110
Reed switch, 242, 254
Refresh, 28, 30
Register, 176, 177, 352
Register display command, 135
Register file, 28
Register model, 175
Relay, 268, 352
Relay driver, 269
Relay output, 83
Reliability, 315, 316
Replicate memory command, 135
Resistive position transducer, 257
Resistive straing gauge, 242
Root directory, 352
Rotary position sensor, 238
Row address strobe, 30
Run-length limited encoding, 56

SA400 standard, 55
SAA1027, 275
SCPC286, 102
SCSI, 56
SELECT CASE, 195
SET PATH, 106, 107, 120, 122
SHARED, 187, 199
SIMM, 22, 43
SMARTDRV.SYS, 332, 334
SP, 14
SRQ, 221
SSR, 273, 274
ST506 standard, 56
STE bus, 60, 91, 92, 93
SUB, 187
SVGA, 52
SYS, 127
Screening, 331
Search memory command, 135
Secondary expansion bus connector, 66
Sector, 352
Segment register, 14, 175, 178
Segmentation unit, 17
Select input, 6
Select...Case...End Select, 156
Semiconductor strain gauge, 242
Semiconductor temperature sensor, 243

Sensor, 236, 243, 244, 352
Sensor interface, 243
Serial I/O, 7
Service request, 221
Settling time, 352
Shaft encoder, 238, 257, 258, 259
Shell, 352
Signal, 4, 352
Simple branch, 156
Simple logic, 297
Single in-line memory module, 22, 43
Single key input, 191
Single precision variable, 184
Single word mode, 28
Snubber network, 274
Software, 277, 352
Software debouncing, 249
Software design, 297
Software development, 151
Software development cycle, 152
Software engineering, 149
Software tools, 167, 286
Solenoid driver, 274
Solid-state relay, 273, 274, 352
Source code, 168
Source disk, 107
Source index, 177
Specifications, 2, 295
Speech annunciator, 305
Speed of response, 154
Stack pointer, 14, 177
Stack segment, 175
Standard mode, 145, 146
Start-up command, 146
Status indicator, 267
Status register, 12
Stepper motor, 275
Stock fault, 323
Strain, 242
Strain gauge, 242, 312
Strain measurement, 311
Stream I/O, 205
String input, 197
String variable, 183, 184
Structured English, 153
Sub-directory, 352
Subroutine, 185
Super VGA, 51
Supply rail, 78, 74, 90
Support devices, 25, 38
Switch, 240, 244
Switch bounce, 246
Switch debouncing, 245, 246, 249
Switch interface, 243
Symbol names, 162
Symbolic debugger, 174
Synchronous mode, 7
System address bus, 22
System board, 19, 22, 23, 24, 42, 353
System bus, 21
System configuration, 332
System controller, 38
System data bus, 22
System reset, 112
System settings, 147
System specification, 295
System unit, 18

TEST, 25
TIME, 22, 121
TOOLS.INI, 167
TPA, 44
TREE, 128
TSR, 106
TYPE, 121
Tachogenerator, 238
Tachometer, 238
Tag, 17
Talker, 218
Target disk, 108
Task switched mode, 16
Temperature, 242
Temperature sensor, 237, 262
Temperature threshold detector, 265
Terminate and stay resident, 4, 106
Test equipment, 324
Testing, 160
Text mode, 43
Thermistor, 243
Thermocouple, 242, 263
Threshold detection, 264
Timer, 33
Top-down approach, 295, 297
Touch-operated switch, 248
Tower system, 18
Trace execution command, 135, 140
Transducer, 353
Transient program area, 44
Tree structure, 114
Trigger, 353
Troubleshooting technique, 321
Turbo C, 200
Turnkey software, 280
Type K thermocouple, 264

UMA, 40

UNIX, 148, 200
Unassemble command, 136, 139
Unencoded keypad, 250
Units, 354
Up-time, 316
Upper memory area, 40
User-defined function, 188

V20, 14
V30, 14
V40, 96
VDISK.SYS, 332, 334
VDU, 353
VER, 121
VERIFY, 121
VESA bus, 25, 59
VGA, 51, 52, 53
VL bus, 59
VME bus, 60
VOL, 121
Validation, 160, 353
Variable, 183
Variable names, 183
Variable types, 183, 184
Varuable names, 162
Ventilation, 25
Verification, 353
Vibration, 243
Video RAM, 40, 43
Video display mode, 53
Video graphics array, 51
Video memory, 53
Video mode, 52
Video standard, 51
Virtual 8086 mode, 15
Visual display unit, 353
Voltage rail, 74, 90
Volume label, 126, 353

WAIT, 16, 27
WIMP, 278
WIN, 145
Warning indicator, 267
Watchdog, 316
While...Wend, 181, 189, 194
Wildcard, 115
Windows, 144, 278, 288
Windows NT, 148
Windows for Workgroups, 146, 147
Word, 4
Write, 353
Write data command, 136
Write operation, 6

XCOPY, 128
XGA, 52
XMAEM.SYS, 333

Yes/No dialogue, 191, 192

Z80, 106